Standardizing Sex

Standardizing Sex

A History of Trans Medicine

KETIL SLAGSTAD

THE UNIVERSITY OF CHICAGO PRESS   CHICAGO AND LONDON

The University of Chicago Press, Chicago 60637
The University of Chicago Press, Ltd., London

Published 2025

34 33 32 31 30 29 28 27 26 25 1 2 3 4 5

ISBN-13: 978-0-226-84322-3 (cloth)
ISBN-13: 978-0-226-84324-7 (paper)
ISBN-13: 978-0-226-84323-0 (ebook)
DOI: https://doi.org/10.7208/chicago/9780226843230.001.0001

Library of Congress Cataloging-in-Publication Data

Names: Slagstad, Ketil, author.
Title: Standardizing sex : a history of trans medicine / Ketil Slagstad.
Other titles: History of trans medicine
Description: Chicago ; London : The University of Chicago Press, 2025. | Includes bibliographical references and index.
Identifiers: LCCN 2025002981 | ISBN 9780226843223 (cloth) | ISBN 9780226843247 (paperback) | ISBN 9780226843230 (ebook)
Subjects: LCSH: Transgender people—Medical care—History—20th century. | Transgender people—Medical care—Scandinavia—History—20th century. | Gender-nonconforming people—Medical care—History—20th century. | Gender-nonconforming people—Medical care—Scandinavia—History—20th century. | Social medicine—History—20th century. | Social medicine—Scandinavia—History—20th century.
Classification: LCC RA564.9.T73 S53 2025 | DDC 362.1086/7—dc23/eng/20250317
LC record available at https://lccn.loc.gov/2025002981

Contents

INTRODUCTION

A Welfare State Story

One day in October 1921, Martha Pedersen, an office worker in her early thirties, was admitted to the Kommunehospitalet in Copenhagen.[1] Having already been arrested once for wearing women's clothes in public, Pedersen was now transferred to the psychiatric department for observation. There are no photos of Pedersen in the clinic, but one can imagine that she might have entered the hospital through the gates of the imposing neo-Byzantine-style complex with monumental arches and striped masonry that was built some sixty years earlier in response to the cholera epidemic then ravaging the Danish capital. She must have been nervous, but also excited: Assigned male at birth, Pedersen was now determined to persuade the physicians there to help her get an operation that would officially recognize her as the woman she said she was. In any case, a medical certificate would make her life easier and avoid future problems with the police.

On the surgical ward, Pedersen met Knud Sand, who was training to be a surgeon. Sand, too, must have regarded the encounter as fortuitous, since it offered him an opportunity to study and test in real life his own theories on the biology of sex and sexual development. By the time they met, Sand was already a leading medical expert in Denmark and an international figure in the booming field of endocrinology and sexual biology. In the late nineteenth century and early twentieth century, physicians and scientists in Europe began to study the mechanisms and physiology of sex and sexuality. Many of them conducted laboratory experiments on animals; as part of his own doctoral research, Sand had transplanted sex glands in rats, guinea pigs, and chickens. By examining glandular tissue under the microscope and observing what happened when they removed or transplanted sex organs in animals, researchers like Sand and Eugen

Steinach, a professor of physiology in Vienna, hypothesized that internal secretions produced by the gonads travel in the blood to different parts of the body. There, they mediate effects in the tissue leading to typical male or female sex characteristics. These theories challenged established notions of sex as defined by gonads or genitals, and sexual instinct as hardwired in the nervous system. Sexual biologists argued, to the contrary, that sex was under hormonal control, and this opened the way for a new scientific understanding of sex as plastic. If hormones regulated sex differentiation, this meant that physicians could shape sex in new ways, and the lab experiments led physicians and psychiatrists to believe that they had found a cure for what they categorized as sexual abnormality, a category in which they included homosexuals and trans people.[2] If hormonal imbalance was the cause of sexual abnormality, they reasoned that it should be possible to restore the equilibrium through surgical castration, sex gland transplants, or hormone injections.

While the animal experiments provided a scientific basis for normative ideas of sexual development, hormone theories also offered a new framework for sexual minorities to claim agency, form networks, and build advocacy. Scientific language helped to confer a sense of legitimacy on people categorized as abnormal, who could now use scientific concepts and theories to fight criminalization, promote social justice, and request medical treatment for their own purposes. This was the beginning of what today is referred to as gender-affirming treatment, and this book is a history of the emergence and formalization of medical practices to shape sex, or trans medicine, from that time until the end of the twentieth century.

Secretion Theory

The roots of these medical practices run deep. To characterize and categorize human sexuality, psychiatrists in the German-speaking world as early as the 1870s and 1880s coined new concepts and taxonomic systems based on people's self-descriptions and case stories.[3] In a book published in 1864, the lawyer, journalist, and sexual reform advocate Karl Heinrich Ulrichs, a pioneer of the nascent homosexual movement, argued that homosexuality—though he never used the term, preferring *Urning*, which referred to a female soul in a male body—was innate and congenital.[4] Carl Westphal, director of the Clinic of Psychiatry and Neurology at the Charité Hospital in Berlin, took Ulrichs's writings and combined them with case studies of his own, coining the concept "contrary sexual feeling" that encompassed cross-dressing and same-sex attraction.[5]

In psychiatry and sexology at the time, the sex of the mind and the sex of the body were indistinguishable, not separable human phenomena. When psychiatrists examined patients and created taxonomies, they looked for signs of bodily pathology that corresponded with, and therefore confirmed, new nosological categories. However, while late nineteenth-century psychiatrists often believed that sexual abnormality was caused by degeneration, secretion theory came to prominence in the early twentieth century as an explanatory framework for sexual variation.[6] One of the most ardent proponents of this theory was the German physician Magnus Hirschfeld, who invoked findings from animal experiments to advance the rights of sexual minorities.[7] He had already founded the Scientific-Humanitarian Committee in 1897 to advocate for the decriminalization of homosexuality based on advances in science and medicine; being of a minority sexual variation was neither a sin nor a pathology, he believed, but constitutionally explained on a biological basis. Broadly speaking, sex could not be reduced to one singular human characteristic but was expressed in the genitals, general physical characteristics, sex drive, and psychological characteristics.[8] A prototypical female, a *Vollweib*, for instance, produced large gametes and reflected the female phenotype in every way, and an "ideal" male, *Vollmann*, produced small gametes and exhibited all the other traits typical of men. But such sexual "absolutes" were merely abstract and idealized phenotypes, Hirschfeld argued: "It cannot be repeated often enough," he wrote in 1923, "man is not man or woman but man *and* woman."[9] The mediators of sexual differences were secretions from the sex glands, which also explained why men and women consisted of a mixture of sexual characteristics. On this basis, Hirschfeld developed a complex theory of human sex development and variation, referred to as the theory of sexual intermediates. This framework made it possible to more accurately distinguish between sexual differences in groups of people.[10] For example, by coining the term "transvestism," Hirschfeld disentangled cross-dressing from same-sex attraction, contributing to the creation of two modern identity categories: homosexuality and trans identity.[11]

Early twentieth-century sexologists, such as Sand, Steinach, and Hirschfeld, shared an intense interest in what they regarded as ambiguously sexed bodies, but they drew different conclusions from their findings. Steinach believed that separate masculinizing and feminizing gonadal cells controlled the dimorphic development of the sexes and reasoned that overactive feminizing cells caused homosexuality. The testes of homosexual and heterosexual men were different, he reasoned, and based on this logic, physicians could "restore" heterosexuality by transplanting testes from "normal," heterosexual men.[12] Some physicians, including Sand and the Viennese urologist Robert Lichtenstern, put these theories into practice by operating on homosexual men.[13] Similarly to Hirschfeld and Steinach, Sand posited that the internal secretions, and in particular those of the gonads, played a pivotal role in the development of sexual characteristics. "[T]he gonads as endocrine glands manifest hegemony and have strange power over the organism through their hormones alone," he asserted.[14]

Although Sand agreed with Hirschfeld that scientific progress would promote justice and a humanistic approach to sexual variation, they differed on what this meant in practice.[15] Even if so-called psychosexual pseudohermaphrodite characteristics such as "feminism," "excessive virility," "homosexuality," "bisexuality," and "sexual perversions" could be explained biologically, they were nonetheless pathological, reasoned Sand. As he wrote in his 1918 dissertation, it is crucial to understand the underlying biological mechanisms of "sexual perversions," since it would ease the task of assigning a "true sex" in individual cases in which the physician was in doubt.[16] In making this decision, he insisted, the physician must consider not only "generative sex" but also the "hormonal sex."[17] Sand would very soon play a key role in making decisions on the assignment of sex in the Danish medicolegal bureaucracy. In 1925, he became professor of forensic medicine at the University of Copenhagen and a member of the Medico-Legal Council of the Ministry of Justice; beginning in 1929 and for thirty consecutive years, he chaired the council that oversaw all applications for sex change and the assignment of sex in individual cases. From this position, he put his scientific theories into practice and shaped Denmark's state regulation of sex.

Once Sand had categorized a person as either female or male, he would usually recommend medical measures that would "solidify" and "strengthen" sexual characteristics of the assigned sex. For trans women, categorized by Sand as "male transvestites," this would include treatment with androgens, but not estrogens or the transplantation of ovaries.[18] By contrast, Hirschfeld and his colleagues at the Berlin Institute—founded by Hirschfeld in 1919—were much more willing to experiment with medical treatments affirming the identities of their patients. One of them was a Danish painter, Einar Wegner, later Lili Elbe, who requested medical treatment to become a woman. The exact details of their meeting are unclear, but they met in the Berlin Institute, probably in 1930. In her 1931 book, *Man into Woman*, which was translated from Danish into German (1932) and English (1933) and which drew together various autobiographical materials, Elbe described how Hirschfeld, referred to as Professor Hardenfeld, subjected her to an "hour-long questioning."[19] Hirschfeld seems to have concluded in this case that the female part was dominant and approved a castration that was conducted in Berlin.

At the time of their meeting, Hirschfeld had some experience with the topic, having already overseen sex reassignment surgery for at least three trans women, Dora Richter, Toni Ebel, and Charlotte Charlaque, and at least one trans man, Eva Katter.[20] Following the castration, Elbe was referred to the gynecologist Kurt Warnekros in Dresden.[21] Warnekros diagnosed her with pseudohermaphroditism and agreed to operate on her, amputating the "penis-like clitoris," removing the rest of the scrotum, and transplanting an ovary into the abdominal wall. When Wegner applied to the Danish Ministry of Justice to change her name to Lili Elvenes, to recognize her sex as female, and to annul her marriage, the case ended up with the Medico-Legal Council.[22] Based on a medical statement from Warnekros, Sand concluded that Elbe did not belong to the female sex, since she did not have female sex glands, but he nevertheless recommended the request be granted as Wegner belonged to the group of sexual intermediates with a strong tendency in the female direction.[23]

In the decades that followed, Elbe's book became a point of reference for many trans people, as it brought information about the possibility of these medical interventions to a wide audience. Although the historiography often presents Elbe as the "first" trans woman to undergo sex reassignment surgery, it should be noted that Elbe's operations took place almost a decade after Dora Richter underwent castration in 1923.[24] According to Warnekros's statement, it would be more accurate to categorize the procedures performed on Elbe as a case of intersex rather than trans surgery. This does not make Lili Elbe any less important in the emergence and shaping of trans medicine. On the contrary, it supports the argument put forward by historian Jules Gill-Peterson that trans people often drew on intersex discourses when requesting treatment to transition.[25]

The relatively liberal situation in Germany that allowed Hirschfeld to pursue his work would soon be subjected to a dramatic change. When the Nazis came to power in January 1933, one of the first things they did was to plunder Hirschfeld's institute. As a gay man and Jewish physician—a sexologist who promoted a theory of sexual intermediates, normalizing sexual variation—Hirschfeld represented everything the Nazis wanted to destroy. In what is often seen in the historiography as the end of the first phase of the medicalization of trans life, a public burning of books in Berlin on May 10, 1933, saw the Nazis throwing into the fire the institute library, which housed the world's largest collection of sexological literature and objects. A second phase of the medicalization of the trans experience had to wait until Harry Benjamin's work in the postwar United States in the 1940s and 1950s, culminating with the opening of the first university-based gender clinic at the Johns Hopkins Hospital in 1967. Thanks to the excellent monographs of Joanne Meyerowitz, Susan Stryker, and Rainer Herrn, we already know much about how the roots of trans medicine in Weimar Berlin informed postwar developments in the United States.[26] In a historiography dominated by the North American perspective, the story of Christine Jorgensen, an American who traveled to Copenhagen to get sex reassignment treatment and who subsequently enjoyed a high profile in the US media in the early 1950s, is often presented as a watershed moment and as the bridge that connected the German and US histories of the medicalization of the trans experience.[27] Gill-Peterson has nuanced this historiographical periodization, however, by showing how trans people in the United States had long sought medical treatment in the decades preceding the so-called second phase.[28]

Importantly, historians have overlooked the crucial role of Scandinavian countries in the development and establishment of the practices of trans medicine from the early twentieth century onward. Christian Graugaard's doctoral thesis analyzes the contribution of Knud Sand to sexual biology, forensic medicine, and the regulation of sex on the basis of requests to the Medico-Legal Council.[29] Sølve Holm used the same case materials to explore autobiographical accounts of intersex and trans life in Denmark from the early twentieth century to the 1970s.[30] As yet, however, there exists no account of the emergence of trans medicine as a distinct practice in Scandinavia. This is a very different story from the "scientific activism" and almost uninhibited hormonal and surgical experimentation in Weimar Berlin, or the prestigious, "high-risk" scientific routine of an elite medical institution in Baltimore some forty years later. In Scandinavia, trans medicine was a product of the emergence of the welfare state, rooted in eugenics, and shaped by the logic of social-medical approaches to health and disease. More than just adding a geographic or regional context to the historiography, this book asks what we learn about the practice of shaping sex in the twentieth century by looking at the Scandinavian experience with the development of trans medicine.

A Scandinavian History

With Martha Pedersen in his ward, Knud Sand was eager to test his theories. Could Pedersen's self-image and desire to become a woman have biological explanations? Were the sex glands of a mixed type contributing to an "intermediate" phenotype? For her first meeting with Sand, Pedersen had dressed in her finest clothes, wearing high heels, a hat, and a necklace; she wore perfume and carried a handbag with a handkerchief and a pocket mirror. She dressed, as Sand noted, from top to toe as a woman, with a corset, garters, and "artificial breasts." The physical examination revealed a normally developed penis and testes; Sand even made a biopsy of the testes that showed normal findings. To get a clearer understanding of Pedersen's experience, Sand wanted to look inside the body, to inspect the organs from the inside. Submitting to general anesthesia, Pedersen agreed to undergo such an examination on one condition: Sand must promise not to remove any female parts that might be discovered in the course of the procedure. "Everything feminine is in accordance with his inner nature," Sand wrote, "everything masculine is contrary to him and could be removed as quickly and brutally as possible."[31] As Sand later noted in a letter to the Copenhagen police, the examination did not reveal a definite biological explanation, however. He diagnosed Pedersen with transvestism, "one of the intermediate sexual forms," defined as "the urge to dress as what at least appears to be the opposite sex." Pedersen's "abnormality" was most likely inborn, he wrote, recommending that she be allowed to take a female name and be permitted to appear in public as a woman. Her "low sex drive and calm and peaceful character" meant that she posed minimal risk to society. In the end, there was really no way to prevent the patient from living as she did, and any prohibition would achieve very little beyond increasing the patient's feelings of unhappiness.

Pedersen's presentation of her own case, intimating that she believed she probably had female parts such as ovaries, suggests that she was familiar with the medical theory of sexual intermediates based on the works of Hirschfeld, Steinach, and Sand.[32] Pedersen wanted an operation, even if surgery in itself would not make her a woman; while she already experienced her body as that of a woman, she believed that medical treatment and legal recognition would make her life easier.[33] Sand ultimately declined to perform any further operations, however; even if he had wanted to, he assured Pedersen that he was not allowed to carry out such procedures. In the years that followed, she tried repeatedly to have an operation, but without success. When Sand sent her a written request in 1928, asking if she would be willing to act as a living case of transvestism in a lecture he was giving for the Medico-Legal Council, Pedersen took offense and broke off contact.[34]

It was around this same time that Knud Sand was laying the groundwork for a new law regulating sterilization and castration for reasons of eugenics. In the 1920s, the objective of improving the "hereditary quality" of the population found widespread support among politicians across the political spectrum, as well as among scientists and physicians, and Sand was appointed by the Danish prime minister to propose appropriate policy measures. Paradoxically, the eugenics law passed by the Danish parliament in 1929 provided the legal basis for sex reassignment, as indeed did the equivalent laws in Sweden and Norway.

Alongside the scientific theory of sex as plastic and hormonally mediated, and physicians' objective to cure sexual abnormality, the eugenics laws enabled a set of diagnostic and therapeutic practices in the 1940s and beyond that can be defined as trans medicine. Drawing on material from public and private archives, medical records, scientific articles, newspapers, activist publications, and oral history interviews with former patients, activists, physicians, and psychologists, the coming chapters attempt to weave together a thick historical analysis of the emergence of these medical practices in Scandinavia in the twentieth century.

The geographical space of Scandinavia is bound together by history, culture, trade, language, religion, and tourism—even by design—but above all by a comprehensive welfare state with universal social insurance schemes and free health care services. The central thesis of this book is that these traditions of social medicine allowed trans medicine practices to develop in a particular way in Scandinavia: as pragmatic responses to intricate medical issues and complex ethical questions within existing health care structures but always with the goal of enabling the "good society." Of course, the welfare state is not a uniquely Scandinavian phenomenon, but as Mary Hilson has noted, the Nordic governments stand apart "in the breadth and ambition of their vision for the welfare state, and their faith in their own ability to create the good society."[35] The social democratic parties that came to power in the 1920s and 1930s introduced broad social reforms with the goal of increasing prosperity, efficiency, and stability.[36] The social reforms were funded on a collectivist principle that would shape the welfare state for half a century until a more individualized rights-based and partly neoliberal principle gained momentum from the late 1970s.[37]

As historians and anthropologists have noted, a common characteristic of Scandinavian welfare states—for all of their differences—is that it makes little sense to distinguish between society and the state because the state is so deeply rooted in the idea of the "public good" and a first-person-plural identity, based on solidarity, equality, shared values, and egalitarianism.[38] According to Francis Sejersted's analysis of Sweden and Norway, the liberation of the individual in what he defines as the Social Democratic Order was "linked to the strong demand for social integration by the powerful ideal of equality."[39] But as anthropologist Marianne Gullestad has observed, there is only one word for equality and similarity in Norwegian: *likhet*. She used this distinction to characterize a particular form of "Scandinavian egalitarian individualism," a society in which people must feel similar in order to fit together.[40] In the production of similarity, the state has played a key role.[41] As noted, once again by Mary Hilson, "No other democratic societies seemed to be quite so affected by extensive state intervention into all areas of human life as the Nordic countries."[42] In writing about Sweden, Thomas Etzemüller has observed how welfare state policy has been less about standardization and disciplining than *normalization*: antitotalitarian in political practice, he argued, albeit "totalizing in its technocratic grip on the population."[43] In this sense, the production of normality becomes a distinct form of social engineering, a way for the state to enable "good lives."[44]

Traditionally, feminists have linked the objective of equality and a women-friendly policy in the establishment of the welfare state to state feminism,[45] but more recently, a new generation of scholars have criticized blind spots in welfare state feminism, in particular state responses to multiculturalism and diversity.[46] Postcolonial and queer feminist researchers have argued that the "woman-friendly state" produces otherness through heteronormative and racialized discourses on nationhood and belonging, a kind of "welfare state nationalism."[47] In alignment with this research, anthropologists have argued that it is by studying the margins of society, such as its handling of minorities, that the contours of the state become clearest.[48] This is not an entirely new argument. In sociology and criminology, for example, there is a long tradition of problematizing the shadow sides of the welfare state; Vilhelm Aubert's analysis of the "hidden society" and the welfare state's handling of its "outcasts" is just one such example.[49]

More recently, historians have turned to the role of medicine and psychiatry in regulating lives and constructing normality on the margins of the Scandinavian societies. Social medicine was key in the enactment of the notion of the good society, and it was assigned a central role in the stimulation of growth, the safeguarding of public interests, and the regulation of populations. Intellectuals like Alva and Gunnar Myrdal and socialist physicians highlighted the role of medicine in reforming society, and conversely, how socially oriented policies, in the domains, for example, of housing and nutrition, would promote a healthy population.[50] In all Scandinavian countries, science, social sciences, and medicine played a fundamental role in the creation of the twentieth-century welfare state, and proponents of social medicine rose to power in all countries after the Second World War.[51]

Looming large in the history of trans medicine are the eugenic sterilization laws introduced in the Scandinavian countries in the late 1920s and early 1930s. By contrast with Nazi Germany, however, racial biology was not the driving force for the implementation and practice of these laws; they were driven rather by the need for economic efficiency, social reform, and the goal of the "good society" enabled by social medicine.[52] Ida Ohlsson Al Fakir has argued that this same tendency was reflected in the ways in which medicine was central in categorizing and classifying Roma as "deviants" and in the construction of the "Swedish gypsies" as a scientific and medical concern.[53] Based on an analysis of letters to their doctors and the medical authorities (Medicinalstyrelsen) in 1930s and 1940s Sweden by people classified within psychiatry as psychopaths or querulants, Annika Berg traced how psychiatric knowledge and practice were crucial in shaping the "good citizen" and constructing the boundaries for normality in the emerging welfare state.[54]

Another of these margins is revealed in the ways in which the state deals with its trans people, and another way in which the state has produced normality is by regulating and standardizing sex. The history of medical expertise in administering the lives of trans people is a history of the welfare state in miniature. Because the shaping of sex characteristics involved almost all fields of medicine—from psychiatry, endocrinology, surgery, to supporting fields of expertise such as psychology and sexology—trans medicine represents a microcosm for the analysis of developments in twentieth-century medicine. It also becomes a probing rod for the increasing influence of medicine in society at large, however, and is instrumental in reshuffling the boundaries between the normal and the pathological. This is a story, in other words, of the medicalization of trans life but also of biomedicalization processes on a large scale.[55] While being paternalistic and remaining under strict control, physicians helped to create a space for sex reassignment in the face of a restrictive legal and health bureaucracy, and they mediated between bureaucracy, public institutions, the press, and the public. Therefore, the history of trans medicine in Scandinavia offers a unique insight into the *public* role of the physician in the welfare state. Physicians were not only representatives of the state or of individual patients, but acted as mediators and communicators *within* and *for* society: An important part of their role as public physicians was to promote the common good, to make good policy.[56]

A History of Practices

While historians of the welfare state and the public health care system have often taken a top-down approach, focusing on the role of grand ideas, ideology, and central public institutions, there are fewer analyses of the significance of mundane medical practices.[57] Following more recent historical scholarship inspired by science and technology studies, this book takes a bottom-up approach to the welfare state and bureaucracy by centering on medical practices: their role in the evaluation and in the distribution of welfare state benefits, and their implementation and manifold logics in regulating sex.[58] Instead of taking Scandinavia or the Nordic model as a given—based in political, economic, or historical similarities—Kristin Asdal and Christoph Gradmann have argued that one characteristic particular to the region is seen in the close intertwining of science and state, and the enacting of a *collective* through science, technology, and medicine.[59]

Medical experts traveled to neighboring countries within Scandinavia to give lectures, exchange experiences, and participate in the creation of routines and guidelines; psychiatrists and endocrinologists published case studies on sex change procedures in Scandinavian medical journals. Notably, leading American experts, such as the endocrinologist Harry Benjamin, maintained close relationships with Scandinavian colleagues: They read each other's work, corresponded about new findings, and attended the same conferences. Scandinavian physicians integrated American models in their practices but also mobilized traditions and concepts from the Scandinavian interwar period and harnessed new practices adapted to the context of the welfare state. At the heart of these developments was social medicine—a philosophy and an approach to health and illness that was rooted in society.

Social medicine has different meanings and roots in different contexts.[60] Dorothy Porter and Roy Porter argued, for example, that the main difference between social and socialist medicine is how the state is theorized. Whereas socialist medicine sees health and disease through a materialist lens and therefore the main objective is the removal of class differences, the state for social medicine is instead a technocratic and scientifically oriented enabler of health policy.[61] Yet, the historiography of social medicine is biased toward Anglo-American contexts and has primarily engaged with ideas and politics on a grand scale.

To tell the story of social medicine, trans people, and sex reassignment in Scandinavia, we need to move beyond an intellectual history of sex, medicine, and the welfare state concerned with shifting diagnostic concepts and we must turn to practice. Historians and sociologists of science and medicine have argued that a rigorous historical analysis of what has been done by scientists and doctors is impossible if we examine only what they claim to have done. We get a very limited and often mistaken understanding of practice by basing our analysis on published works, given their tendency to create a false impression of homogeneity—as Marc Berg and Annemarie Mol have argued—when what appears if you go into the field and study medicine and science in action is its diversity and plurality.[62] The scholarship of authors such as Ilana Löwy, Hans-Jörg Rheinberger, Bruno Latour, and Steve Woolgar has unpacked the black boxes of laboratory science by the ethnographic examination of mundane scientific practices.[63] In medicine, Annemarie Mol and Stefan Hirschauer used atherosclerosis and transsexuality as empirical examples to analyze how notions about the body and sex are negotiated and stabilized through material and social practices.[64] Geertje Mak's *Doubting Sex* is an excellent example of the potential utility of praxiography in medical history, analyzing how the medical practices, techniques, and technologies of "doubting sex" enacted the categories of sex, gender, sexuality, and the connections between them.[65]

By the study of the mundane practices of regulating sex in the psychiatrist's office, the hormone laboratory, the operating room, or the health bureaucracy, these authors have offered us a fresh view on the state and social medicine. As Veena Das and Deborah Poole have argued, it is in "the processes of everyday life that we see how the state is reconfigured at the margins."[66] I am inspired, moreover, by anthropologist Didier Fassin's approach to the state as a "concrete and situated reality" that is "simultaneously embodied in the individuals and inscribed in a temporality."[67] The "state," as it appears in this book, is not an all-encompassing, Weberian concentration of power but rather, as Quentin Skinner has argued, the institutional tools that enable a society to negotiate and to act.[68] The state, in other words, is neither just a top-down distribution of government power nor a faceless bureaucracy but is its people—their ethics and actions. The state representatives in this story are often physicians, but they are also lawyers and psychologists. They respond to issues within public institutions defined by laws and discourses, but they also shape decisions and discourses through their actions. A relational understanding of the state allows for an analysis of the development of the welfare state over time: By looking at its encounters with people's requests to change sex, we can see the state in action.

Materials and Ethics

Since Erwin H. Ackerknecht's plea in 1967 for a "behaviorist approach" in the history of medicine, historians have turned to medical records to analyze medical practice.[69] This book uses medical records, psychologists' notes, applications for sex reassignment, and correspondence between professionals, patients, and state offices from the 1920s to the early 2000s to interrogate the practices of trans medicine, from paper technologies, hormone analyses, and psychological testing to surgical techniques and the therapeutic practices of social medicine. Access to archive material and medical records has proved extremely difficult; the archival material also shapes the narrative with the strongest focus on Norway and Denmark and to a lesser extent on Sweden.[70] This book is not a comparative study but is instead an analysis of a certain way of practicing medicine and standardizing sex made possible by the Scandinavian context of the welfare state.

Medicine and medical understandings have played prominent and problematic roles in framing trans historiography. Some scholars have argued that transsexual identity and subjectivity were the result of social and cultural preconditions shaped by technological advances (endocrinologic and surgical).[71] Jay Prosser has criticized this approach for reducing the transsexual body to "medicine's passive effect," as "a kind of unwitting technological product" based on the notion that transsexuals were "constructed in some more literal way than nontranssexuals." Importantly, it overlooked how trans people were active in shaping their own identities and even in shaping medical practice.[72] I agree with Emmett Harsin Drager and Lucas Platero when they write that transsexuality "emerged dialectically, in conversation with medicine over the second half of the twentieth century," and that these dialectics depended on trans people sharing knowledge and shaping practice.[73] While the role of technology in the production of subjectivity and in the shaping of boundaries between what counts as normal and what counts as pathological is one area of interest for this book, this is nothing unique to trans medicine: By the end of the twentieth century, it is no exaggeration to say that all bodies and parts of society had become permeated by medical technology.

In this book, I refer to trans medicine, but—for the actors involved—the concept never existed as a category as such. To the contrary, it is an analytical concept that makes it possible to circumscribe and analyze a set of practices, logics, and thought-styles.[74] This is not a book about heroic doctors formulating their own theories and putting them into practice. The practices of trans medicine developed in a space between the clinic, the laboratory, the state bureaucracy, the press, activist organizations, and communities. It was not simply a question of physicians developing practices and technologies in the clinic or the laboratory before implementing them in a top-down manner; rather, it was the patients and communities themselves who played a role just as decisive as that of the physicians in shaping treatment standards and diagnostic categories. Patients shared some information but withheld other details for strategic reasons. They asserted their rights and access to medical care, corresponded with authorities and physicians, mobilized the press, and formed advocacy groups.

This aspect of the history is only partly visible in the archive. Medical records do not have a uniform format but consist of miscellaneous documents—autobiographical descriptions, correspondences, expert reports, lab results, X-rays, psychiatric diagnostic instruments, pre- and postoperative photographs, and drawings of surgical procedures. Physicians sometimes wrote on hospitals' letterhead paper, and this paperwork follows a standard format with stamps and signatures. Other paperwork includes notes and bullet points scribbled on random pieces of paper. Like many other medical archives, the medical archive of sex reassignment is a collection of documents on paper of variable quality and dimensions that reflects the major purpose of clinical bookkeeping as a tool for clinical practice.

Conducting oral history interviews with former patients and activists has been one way to avoid the bias that comes from basing historical analysis only on sources written by hegemonic actors.[75] Oral histories make history "more democratic," according to Paul Thompson, who added that "oral evidence breaks through the barriers between the chroniclers and their audience."[76] Excellent historical accounts of AIDS, autism, ADHD, and polio, written respectively by Richard A. McKay, Chloe Silverman, Matthew Smith, and Dóra Vargha have been an inspiration to me in this aspect of my research. Their scholarship fully demonstrates how interviews with patients, parents, activists, and medical experts can enrich traditional historical sources and provide new perspectives for investigating the boundaries between lay and professional knowledge.[77] When it came to conducting my own oral history interviews for the purposes of the present research, I have been very touched by the hospitality of people who picked me up at the train station, opened their homes and their hearts to me, and told their life stories over coffee with my recorder perched on the table between us.[78]

One of the main benefits of historical interviews of this kind is the fact that they can provide perspectives "from below" that are otherwise difficult to document, and this is particularly important when writing the history of traditionally marginalized groups of people whose stories have often not been preserved. In my experience, moreover, these interviews have also been a valuable way to trace medical practice: Details and routines were not written down but are conveyed from memory, including the importance of the waiting room for the surgeons to identify "good candidates" for surgery. Interviews with physicians and psychologists have been crucial in reconstructing how professionals met and worked—allowing me to map out professional networks across Scandinavia—how physicians examined and selected patients for treatment, and how interventions were carried out. Oral history interviews also offer a rare possibility to highlight the emotional and personal experiences of care providers, described by Nancy Tomes as "a way to demystify the large bureaucratic organizations that dominate modern medicine, returning the historian and the healthcare consumer to a more familiar and believable world of human beings."[79]

Insofar as Scandinavian trans medicine is itself characterized by ephemerality and lack of formalized structure, oral history interviews might be seen as an ideal tool for the study thereof. For many experts entering these fields, trans care was completely new, and often it was personal prejudices, experiences, and opinions that inevitably shaped how they approached their patients. Some of the professionals involved were lesbian and gay, others were feminists, and their life stories and political opinions influenced how they met with their trans patients. Many such stories would have been impossible to tell without the use of oral history interviews. There was sometimes a dissonance, however, between what care providers said they thought and did and what is documented of their actions in the medical records. This is sobering when you consider that most people who requested sex reassignment were denied treatment, with ethics boards blocking access to their medical records. There are several methodological and systemic hurdles that must be negotiated, in other words, and these increase the risk of bias in the final telling of these stories.

Another important goal of conducting the oral history interviews has been to establish an archive of trans oral history, as inspired by the groundbreaking work of the Tretter Transgender Oral History Project at the University of Minnesota.[80] The interviews of those willing to include their testimony have been deposited in the National Archives of Norway. Following the tradition of oral history writing as *community practice*, one aim of this project has been to secure and archive voices from trans history to promote community engagement, inspire political action, and to create history that can be meaningful and relevant to trans communities today. It is hoped also that such an archive can contribute to the improvement of health services for trans people today and into the future.

One limiting factor in researching this book was the lack of access to patient records, as medical records are often protected by privacy laws, even after a person has died.[81] After a year-and-a-half-long process—including several rounds of appeals—with the Regional Committees for Medical and Health Research Ethics in Norway (REK) and the Norwegian National Research Ethics Committee (NEM), I was finally granted access to medical records, but only from the 1950s and 1960s given the high likelihood that most, if not all, of the individuals are by now deceased. Records from the 1970s and later were accessed with the written consent of the former patients.[82]

The ethics committees argued that individuals who had sought medical transition were particularly vulnerable: The information in the medical records concerned "something as fundamental and private as unclear gender identity,"[83] and many had "lived with their difficulties in secret, making them very vulnerable."[84] This is true for some of the people I met in the archives, but it is not a representative description. Why are marginalized groups automatically categorized as so vulnerable that their stories cannot be told? What does it say about how minority groups are viewed today that it is unthinkable that trans people in the past could be proud of their stories and want their stories to be known? Why are these considerations not relevant to discussions of ethical historical research?

By historicizing bioethics, including the "regime of informed consent," it becomes clear how a narrow focus on informed consent overlooks issues of great ethical importance. It was not until the 1960s that informed consent became *the* central issue in medical research, preparing the ground for the self-conscious profession of bioethics. In Charles Rosenberg's understanding, bioethics has been both a recognition and a symptom of inequalities in health care systems, including specific historical abuses of medical power. While it challenged medical authority, bioethics also legitimized and strengthened medical authority by filling chairs and committees: "Bioethics has taken up residence in the belly of the medical whale."[85] Roger Cooter argued that the development of this narrowly defined practice of bioethics was premised on a "historically constructed notion of human nature in which persons/patients are defined primarily in terms of their ability to act autonomously or to make choices and take risks." The "prioritization and celebration of personhood" was made possible by a particular politico-economic context that, in Cooter's understanding, displaced other, more communal discourses.[86]

People sought help alongside assurances that information about them would remain with health care professionals; they never consented to future historians using their personal stories in research. One way to mitigate the potential violation of a patient's integrity through access to medical records—besides adhering to rigorous anonymization practices—is to give voice to that individual's personhood and perspectives by treating those historical actors as *subjects*. I have tried to treat people not as means to an end, but as ends in themselves, by trying to foreground their voices where such could be identified. To this end, I have used people's chosen names (or the pseudonymized versions thereof) and the pronouns reflecting their identity, also for the time *before* medical transition. In those cases where it is not clear which pronoun the person used, I use the pronouns "they/them." Some historians choose to highlight every time actors do not respect the identity of individuals by using the wrong pronoun, for example by adding "[*sic*]" to the relevant parts of the quotation. In my opinion, a more effective historical critique is to make it clear from the context which pronouns the person would have used, and then let the reader be the judge of the historical actors' statements. Some of these actors who have not respected the identity of trans individuals have nevertheless played a crucial role in shaping medical practices that have had profound implications for later patients and trans communities. I am not convinced that silencing these people is the most ethically sound solution.

In complex research involving living people, new questions and issues arise that were not addressed in the original research plan and that need to be addressed along the way. For example, what details of a person's story should be included to do justice to the complexity of a life? To mitigate their possible feelings of being left in a void after sharing some of their most personal stories with me, I have continued to write to research participants to keep them informed of how the research is progressing.

The correspondence with research ethics committees indicates that the research committee system and the narrow focus on informed consent as defined in health research laws (like those in Norway) are not well suited to address structural power imbalances and social justice issues in research projects, or to respond to issues of structural inequality. Even if the denial of access to medical records is intended to protect the integrity of marginalized and stigmatized individuals, in effect the ethical research system risks perpetuating exclusionary and discriminatory practices by preventing silenced stories from being told. This makes it more difficult for minority groups today to take ownership of their own histories, and it makes it nearly impossible to shed light on the misuse of power by hegemonic actors, be they physicians or authorities.

An Outline of the Book

The book begins with an examination of the foundations of trans medicine in eugenics. Chapter 1 presents the argument that the goal of social medicine in meeting with trans people in the interwar period was to control and regulate "abnormal sexuality." Hormonal and surgical treatment was intended to "restore" heterosexuality and produce a binary standard of sex, not affirm the identities of trans individuals. Paradoxically, the logic of sexual abnormality in social medicine and sexual biology supported by eugenic castration laws became a loophole for trans women to access genital surgery in the 1940s and 1950s. In other words, a regressive biopolitical legal framework was harnessed for individualized emancipatory purposes.

The future of this type of treatment was shaped not only in the clinic and in state organs but was also negotiated in court and in the medical archive. Chapter 2 charts two criminal trials in Norway in the early 1950s. Charged with the sexual abuse of minors, the defendants were diagnosed as "genuine transvestites," and forensic psychiatrists argued that medical treatment would reduce the risk of future abuse. At the same time, psychiatrists produced expert reports on the future political, social, and medical handling of trans people on behalf of the state, and the task of the psychiatrist as expert witness in court and bureaucracy in cases of sex reassignment highlights the historical significance of psychiatry's nonformalized evaluative expertise—of psychiatric practices beyond psychiatry—and, in this case, the role of forensic psychiatry in standardizing sex. The criminal cases and expert reports prescribed the medical modification of sex characteristics not primarily as a matter of caring for the individual but of protecting the public and the social fabric. Trans people and their physicians therefore made appeals to the idea of the "good of society" to defend surgical interventions in otherwise healthy bodies, not just to circumvent the legal hindrances but also to safeguard controversial interventions from professional and ethical criticism.

After Christine Jorgensen's story broke in the press, Scandinavian clinics received requests from people from all over the world seeking hormonal and surgical sex reassignment. Confronted with an urgent need for clear diagnostic criteria and lacking clinical experience, physicians turned to the medical archive to support treatment decisions. Chapter 3 analyzes how physicians worked with paper, and on paper, to create usable nosological categories. The number of cases in the scientific literature were few, so researchers included all kinds of different material in their work: letters, published cases, medical reports, and even magazine articles. With extensive national registers, national identification numbers, and state-employed physicians, the welfare state, moreover, provided optimal preconditions for researching a small patient group; in Sweden, for example, a research project drew on all these resources in one of the earliest epidemiological studies on trans people. In the 1950s and early 1960s, there were no organizations or advocacy groups for trans people, but women's magazines became a platform through which people could exchange knowledge, promote community care, and shape the medical discourse about their lived experiences. Physicians read these exchanges and included them in their research, exemplifying how knowledge production on trans issues did not happen in a closed medical space but also involved the media and the public.

Chapters 4 and 5 turn to the mundane practices respectively of endocrinology and surgery in trans health, arguing that these practices developed as pragmatic responses to complex medical and legal issues. Both endocrinologists and surgeons developed separate routines rooted in professional experience and logic for selecting candidates and shaping sex. Although the number of patients was small, the advanced and partly experimental nature of interventions helped drive the professionalization and specialization of endocrinology and plastic surgery. This would not have been possible, of course, without the willingness of patients to serve as "guinea pigs," and trans people thus managed to sometimes shape medical practice through these encounters.

Access to medical treatment was one thing, changing one's legal sex status was quite another. Chapter 6 traces how attempts were made within state apparatuses to define clear and quantifiable criteria for the assignment of sex for trans people. The deeper state bureaucrats plowed into the biology of sex, however, the more difficult it seemed to settle on universal criteria. In all Scandinavian countries, a prerequisite for any legal change of sex was castration or sterilization, even if there was no medical justification for these invasive surgical interventions. This chapter highlights an inconsistency in the logic of social medicine: that the goal of avoiding biomedical solutions to social issues only went so far, while the state was willing to violate the reproductive rights of a small minority in order to protect an ideal of a sex binary.

Inspired by lesbian and gay liberation, Nordic feminism, and other such broad social movements in the 1970s, psychologists, physicians, sexologists, psychiatrists, and social workers in Oslo and Copenhagen developed new routines and clinical approaches to sex reassignment. Chapters 7 and 8 analyze the clinical logics and routines that grew out of these initiatives. Chapter 7 highlights the reformation and adaptation of sexology and sexual health, including the adaptation and implementation of the "gender role" and "gender identity" concepts, labels originally coined by mental health professionals in the United States. Lesbian and gay activism, medicalization theory, Nordic feminism, and trans-exclusionary radical feminism created a particularly challenging situation in Norway for trans people to access treatment.[87] Chapter 8 traces the impact of psychoanalysis and projective testing on diagnostic reasoning and the processes of determining which patients were selected for treatment in the Oslo Health Council in the 1970s, a time that also saw an influx of psychologists into the health care system. According to psychodynamic reasoning, the desire to transition from one sex to another was an expression of unresolved underlying issues, and this trans-excluding framework became another hurdle for trans people to overcome when trying to convince their care providers to give them access to medical treatment.

At the same time, trans communities and activists started to organize, and in the 1970s and 1980s, they increasingly challenged medical concepts and practices pathologizing their lives. Chapter 9 analyzes how transvestite communities sought cooperation with medical experts and sometimes managed to shape medical knowledge and practice, but also how medical concepts diffused into subcultures and helped establish hierarchies and boundaries within the trans communities.

In the early 1980s, clinicians and medical researchers in Scandinavian countries and elsewhere became increasingly concerned that patients might come to regret sex reassignment, and they sought to construct prognostic factors. This step served operationable purposes, and it translated "regret" into an alleged individual psychological experience, detached from the social and political world of stigmatization and marginalization. Chapter 10 shows how the attempts to limit regret gradually came to dominate trans medicine and treatment decisions. In this context, social medicine gained importance as a clinical logic aimed at anchoring the transition process in society and the patient's environment, as an attempt to prevent regret and to secure the legitimacy of clinical practice.

It was not before 2001–2002 that the first specialized gender clinic opened in Scandinavia at the Rikshospitalet in Oslo. As will be shown in chapter 11, the opening of the clinic was not prompted by medical or scientific developments but evolved in the context of nonlinearization and in response to bureaucratic imperatives as the welfare state came under pressure to reform.[88] Similar developments in Denmark and Sweden firmly established trans medicine as a highly specialized practice in need of separate institutions and dedicated personnel. Trans medicine and the reformation of the welfare state shaped each other in co-productive ways: "Sex change" was enacted as a *totality*; sex, in other words, could only be changed completely into either female or male. Therein lies the logic behind the production of the sex standard.

CHAPTER ONE

Eugenic Beginnings

In August 1953, Christofer Lohne Knudsen, chief psychiatrist in the Office of Psychiatry in the Norwegian Directorate of Health, received an unusual request. The letter came from a physician in Ila Prison and concerned a thirty-three-year-old inmate serving a sentence for indecent exposure. Since puberty, the prisoner had felt an "irresistible urge" to put on women's underwear—doing so, the physician stated, "gave him a strong feeling of pleasure," the prisoner having "always wished he was a woman." The inmate had already escaped from prison once, and after an episode of "indecent public exposure" had agreed to undergo surgical castration. Described by the physician as a "severely moronic" person with an IQ of 60 and "obsessed with the idea of being 'transformed,'" the inmate requested permission to wear female clothing, to take a female name, and to begin hormone treatment. Was the prisoner a homosexual with abnormal tendencies? Or was this behavior merely a playing out of an "idiot's" transvestite fantasy? Unsure about the diagnosis, the physician nonetheless reasoned that hormone therapy would have a positive impact in alleviating criminal tendencies. Uncertain about the legality of such treatment, however, he now sought advice from the highest authority of the Norwegian health bureaucracy.[1]

In the early 1950s, the medical treatment for trans people was raised as an issue at the highest level of the medical and legal bureaucracy in Denmark and Norway. The letter from Ila Prison was only the first of several requests for such hormonal and surgical treatment for trans women of whom several were convicted of sexual offenses or "indecent exposure." However, such treatment represented uncharted legal territory where two legal frameworks collided: the Penal Code and the Sterilization and Castration Acts.

Castration and Sexual Abnormality

Historically, scientific and medical definitions of sex have been shaped largely through attention paid to "anomalous" embodiments and identities, such as "hermaphroditism" and "sexual inversion."[2] In interwar Scandinavia, however, the discussion shifted to a new figure: the sex offender. In the late 1920s and 1930s, all Nordic countries introduced sterilization and castration laws, similar to other European countries and several states in the United States.[3] The laws were founded on eugenics, a broad scientific and political movement that had a fair wind in its sails at the time, and they permitted the use of force when the individual was considered incompetent to consent.[4] Eugenics was not a fixed scientific-political ideology but a flexible concept and practice with variable political, social, and medical logics and justifications.[5] In Scandinavia, eugenics combined scientific reasoning and social democratic politics, and a broad coalition of politicians, scientists, and physicians promoted a eugenically informed economic and social policy as fundamental for the emerging welfare state.[6]

Although the laws had a political motive with a basis in eugenics, they were used for different purposes in practice and the rationale for sterilization or castration changed over time. The boundaries between sterilization and castration for eugenic, social, political, or economic reasons were blurry, and in clinical practice, it was often impossible to distinguish between the alleged goal of caring for the individual and that of society.[7] It would be erroneous, therefore, to reduce the logics and practices of the Scandinavian sterilization and castration laws to a eugenic goal of "optimizing" the genetic composition of the population. According to Mattias Tydén's description of the direction of travel in Sweden, for example, the laws had a stated eugenic aim and were implemented as such in the first decades, specifically targeting people classified as "asocials," but gradually shifted away from coercion and became a tool for responding to individual social and medical issues negotiated on a local level.[8] Nevertheless, the laws expressed the most authoritarian aspects of the normalizing politics of the emerging welfare state: People with intellectual disabilities—and those labeled as "abnormal"—were seen as an economic burden and an obstacle to the creation of a productive society.[9] Thus, the policy produced a clear distinction between majority and minority, and expressions of who belonged to the "good society": As Gunnar Broberg and Mattias Tydén have stated in reference to the Swedish context, "in practice the implementation of the sterilization laws came to focus on persons perceived as different."[10]

In medical and legal debates, the goals of crime prevention and eugenics were often inseparable. Genital surgery for trans women, defined by the Norwegian Ministry of Justice as measures to "transform a man into a woman," activated three interrelated logics in the interwar period: castration as a psychiatric treatment for people classified as sexually abnormal such as those identified as homosexual; castration of people registered as sex offenders as a measure of crime prevention; and castration for eugenics purposes.[11] In Sweden, the State Institute for Racial Biology opened at Uppsala University in 1922, and in Denmark, the University of Copenhagen established the Institute for Human Hereditary Biology and Eugenics in 1938. These institutes became scientific strongholds in the flourishing field of criminal biology and psychiatry in Scandinavia in the 1920s and 1930s. At the Danish institute, funded by the Rockefeller Foundation, for example, researchers established a eugenics registry of people classified as abnormal on a hereditary basis—those categorized as criminals and asocial, insane and mentally retarded, blind, and deaf—to conduct research and to provide expert testimony in individual cases.[12] In Norway and Denmark, the public debate leading to the castration laws initially focused on the threat posed by sex offenders to women and children, and the women's movement was among the most ardent advocates of harsh measures, including castration, against people registered as sex offenders to protect society.[13] "Very often," wrote a female physician in 1932—she was also a member of the Norwegian Penal Law Commission—"the purposes of crime prevention are intertwined with those of racial hygiene." The Penal Law Commission was established in 1922 to propose revisions to the law. "At present," she continued, "one of the most pertinent questions for both the learned and the unlearned is how to protect ourselves from our most dangerous criminals in the most effective and economical way."[14]

Denmark was the first Scandinavian country to enact laws regulating sterilization and castration in situations where sexual drive was perceived as a menace to the individual or society. As a member of the commission appointed by the minister of justice in 1924 to recommend restrictions on the personal freedom of people diagnosed with "degenerative dispositions," Knud Sand played a key role in shaping the legal framework in Denmark. Aimed primarily at sex offenders, it also opened the door for the castration of people classified as "mentally abnormal." A 1935 amendment shifted the law in a direction that made its purposes seem even more explicitly eugenic, and it also now included provisions for involuntary castration. The Medico-Legal Council prepared cases for the Ministry of Justice, which made the final decisions. As the council chair, Sand became the most important figure in sterilization and castration practices in Denmark.[15] The Norwegian law of 1934 mirrored the Danish law in most respects, except that it was a single law in Norway that regulated both sterilization and castration, defined as "sex operations," and it did not contain a paragraph on involuntary measures.[16] Decisions on most applications were made by the director general of health, Karl Evang—exceptions were those cases concerning minors, people classified as mentally ill, and "persons with permanently impaired mental faculties." In such cases, the Expert Council on Cases of Sex Operations, chaired by Evang, made the decision. The Swedish sterilization law was introduced in the same year, 1934, and here also, it was left to the national health authorities to make the decisions on applications.

These laws did not introduce a completely new medical routine but provided a legal framework, at least to some extent, for surgical interventions that were already being carried out. Although the hypothesis of separate masculinizing and feminizing gonadal cells controlling dimorphic sex development was disproved in the 1920s and 1930s, especially when scientists found that neither of the sex hormones were sex-specific, the idea of a hormonal explanation for homosexuality, transvestism, or other so-called sexual abnormalities did not disappear completely. On the contrary, the hormonal framework of sexual abnormality was supplemented by psychological theories.[17] On the strength of such hormonal theories promoted by physicians like Sand, for example, some men categorized as sexually abnormal were castrated in Denmark on grounds of crime prevention or for therapeutic purposes, even before the introduction of the law.[18] In his doctoral thesis, Christian Graugaard documented the fact that physicians in the 1920s had castrated three men classified as sexually abnormal. Two were convicted of the sexual abuse of young boys and underwent castration to avoid custodial sentences. The third was a priest in his mid-forties who contacted Sand in August 1923, having been dismissed from the priesthood after "exhibitionistic actions towards young boys." After an examination, Sand hoped to change the patient's bisexual tendencies in a heterosexual direction; he initially performed a ligation of the spermatic cords and proceeded the following year with a testicular transplant from a "normal man." At first, Sand refused to recommend castration, claiming that he did not have the legal basis to do so. But when the Ministry of Justice approved the request in 1929, the testes were removed.[19] The secretion theory and its hormonal explanation of sexual abnormality, in other words, had provided the scientific foundation in law for castration. The amended Danish law included a new criterion that a man could undergo castration if "sex drive" was causing "severe suffering or social deterioration," significantly expanding a "social" indication for castration. By 1954, ten homosexual men had undergone castration on this basis in Denmark.[20]

In his doctoral thesis, Sand had shown experimentally how simultaneous transplantation of "homogeneous and heterogeneous sex glands" in animals castrated in infancy had "unequivocally produced a hermaphroditic animal."[21] Age was a critical factor, however; Sand believed that sex could not be "changed" after sex characteristics were fully developed.[22] But in individuals whom he classified as hermaphrodites, or in intermediate stages thought to be amenable to change by castration, such as in bisexuality, he applied approaches adapted to each individual case. The rationale behind this praxis was the idea that in humans, "the psychosexual character," as he noted in *Die Physiologie des Hodens* [The Physiology of the Testicle], was "the most unstable of all sexual characteristics."[23] In May 1932, for example, he examined a child whose external genitals were "so malformed" at birth that the sex could not be determined with certainty. "The physical and mental development during the first five to six years was mainly in the female direction," he noted, and the parents raised the child as a girl. A thorough examination of the child including a hormonal analysis did not provide a definite answer regarding the child's sex. At the time, the child had already begun to develop in a "virile direction," and two years later, repeated hormone analysis supported this assumption. Sand now recommended a change of name and that the child should thereafter be raised as a boy.[24]

As chair of the Medico-Legal Council, Sand personally saw all people who applied for reassignment of sex status and all trans people who requested masculinizing or feminizing treatment. His influence on the regulation of this practice was therefore unmatched in Danish medicine and bureaucracy. Because Sand reasoned that sex solidified after a certain age, he rejected hormonal or surgical treatment of people he diagnosed as transvestites. Treatment would not "alleviate suffering" in these cases, he insisted, nor diminish what he defined as sexual abnormality.

On one occasion, he made an exception for a trans woman born in 1909.[25] In a letter to Sand dated December 19, 1941, she compared herself to Lili Elbe. Based on Elbe's book, *Man into Woman*, first published in Danish in 1931, and scientific literature on hormones and pharmacology, she explained her experiences and embodiment as follows: "I possessed, if not the complete female hormone-producing glandular system, at least such a large part of it that it affected the whole organism, because sometimes (most of the time), I felt more like a woman than a man, while at the same time I felt that something was fighting against it or suppressing it, preventing my own self from developing physically as a woman." Sometimes, her chest would feel tender and her breasts would swell and darken. She also sometimes felt a tenderness in her groin or back toward the rectum. To stimulate her feminine development, she had started taking estrogen pills, which she had bought on the black market, then switched to injections: "At times the daily dose was about 100,000 IE to 150,000 IE injected into the thigh muscle, partly in the groin and partly around the breast."[26]

Rejecting her explanation, Sand began treatment with androgens, which only increased her feelings of depression. When he ultimately conceded to treatment with estrogens, it made her feel happier and more at peace.[27] Yet she still wanted an operation, and several times she tried to remove her testicles by herself. Notes made later by a psychiatrist include the following line: "In one attempt, he succeeded in cutting a hole in the scrotum, which he had tied off."[28] Eventually, in 1945, Sand recommended castration, presumably as a measure to alleviate what he considered a pathological sex drive, whether a result of transvestism or homosexuality.[29] In other words, the goal was to diminish "abnormal sexuality," not to affirm his patient's identity. Some years later, when colleagues began to prescribe estrogens to trans women and androgens to trans men in order to affirm their patients' sense of self, Sand considered these to be "misguided experiments."[30]

Although psychiatrists at that time subsumed trans identity as a subcategory of homosexuality, the introduction and enforcement of the sterilization and castration laws would create new dynamics in the psychiatric discourse on sexual abnormality in general, and specifically on the distinction between homosexuality and transvestism. In 1943, the Copenhagen police brought in an eighteen-year-old for psychiatric examination. This followed the latest of several arrests, when the police had taken the tailor's apprentice to the station for being found with German soldiers while dressed in women's clothes with a scarf around the head and a bow on the forehead. With "a loose coat, ski boots, and long Plus Fours," the county physician stated, "he could easily be mistaken for a young lady dressed in sports clothes."[31] Sand was skeptical, however, that castration would change the person's "sexual disposition," and such a procedure, in any case, had to be voluntary. Hjalmar Helweg, professor of psychiatry and member of the council, doubted that this individual "was a genuine transvestite" but rather that the disguise was "a means of getting in touch with German soldiers."[32]

Among the most ardent defenders of legal castration was Louis Le Maire. He had been secretary in the Royal Danish Ministry of Justice, secretary and then advisor to the Medico-Legal Council, and from 1952 director of the Danish State Mental Hospitals. Through these roles, he was involved in the medicolegal handling of people registered as sex offenders and classified as sexually abnormal. In his view, castration was an effective regulatory tool for two reasons. First, as he stated in an article published in 1956, "de-sexing" had positive effects in terms of eugenics, removing the reproductive potential of "tainted psychopaths and otherwise inferior subjects whose descendants one dares assume would not be of any value to the community." Second, castration was an effective crime-prevention measure: "Theft, and particularly arson and other forms of delinquencies can be traced to sexual attributes, irrespective of the fact that they do not appear as actual sexual offences and in such event the question of castration would also naturally arise."[33]

In 1950s Scandinavia, this hormonal framework of sexual abnormality and castration as a eugenic and crime-prevention measure became the entry ticket for trans women to get genital surgery. One person acted as a trailblazer: Christine Jorgensen—or George William Jorgensen, which was the name she used when she signed the application for "bilateral castration" to the Danish Ministry of Justice's office on April 1, 1951.

Homosexuality as a Loophole for Trans Surgery

In May 1950, George "Chris" Jorgensen, an American photographer, set course for Denmark to get sex change surgery. "The barriers before me in my homeland were so large," she later wrote, and the Scandinavian countries were "much more progressive with such research work as I was looking for."[34] Her American doctor, Joseph Angelo, the husband of a classmate from the Manhattan Medical and Dental Assistant School, had started her on estrogen, but the path to surgery seemed blocked. Her ancestral roots in Denmark proved advantageous. Prior to her departure, she compiled the names and addresses of relatives and family friends living across Denmark in her address book. The ship to Gothenburg took nine days, and onboard M/S *Stockholm*, Chris, twenty-four, shared the cabin with two roommates. "Wonderful dinners," she wrote in a postcard to her parents, signed "Brud."[35] From Gothenburg she took the boat to Copenhagen, where she would stay with a Danish family. They immediately left for a round-trip across Europe. Jorgensen enthusiastically documented the trip, took photos, and wrote postcards to her parents from places such as Oberammergau ("all men here have beards and long hair. Almost like the reliving the Bible"), Venice ("the young Italian boys selling oranges and lemons. All very colourful"), and Geneva ("prosperous looking and gay").[36]

Soon after settling down in Copenhagen, Jorgensen reached out to physicians who might help her dream come true. She first had her ears operated on. Interviewed thirty years later, Else Sabroe, who rented a room to Jorgensen during her stay in Copenhagen, still recalled the first encounter. "The doorbell rang and up the stairs came a young man with bandaged head. I later learned that he had undergone surgery to have his ears pinned back. He was visibly nervous and smoked an enormous number of cigarettes while we talked."[37] Very early in her stay, Jorgensen wrote a letter to Knud Sand, enclosing a copy of the magazine *Medical Photography*, which featured her work. As she later explained in a subsequent letter to Sand, this was part of her plan: "This is how I hoped to meet you."[38]

Jorgensen also contacted Christian Hamburger, a hormone expert and head of the hormone department at the Statens Serum Institut, a leading institution on hormone research in Europe. Jorgensen came to their meeting on July 20 fully prepared. During a stay in California she had read up on the latest findings on the importance of the hormone glands for human physiology and psychology and was well informed on the latest hormone experiments. "I was determined in some small way to be connected with medical problems and understand some of the chemical reactions going on in the human body," she later noted.[39] Meeting with Hamburger, she knew what to ask and how to frame her request. Likely, her entry ticket was to document the cutting-edge hormone research at his lab. "I go to the Statens Serum Institut quite a bit now," she wrote to her parents on August 3, 1950. "Dr. Christian Hamburger, one of the world's best authorities on Endocrinology, thinks that I may get a good story for Eastman Kodak."[40] But for Hamburger too this encounter represented an exciting

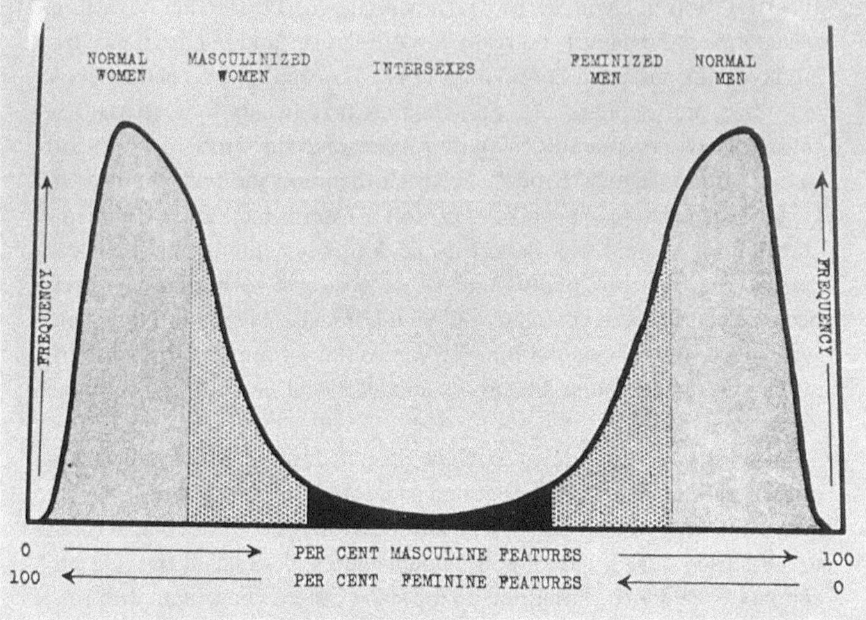

FIGURE 1.1. The spectral distribution of sex. Graph by Christian Hamburger, undated, early 1950s. Det Kgl. Bibliotek, Copenhagen.

opportunity. For a physician with a profound interest in the biology of sex and especially its "intermediate" forms, Jorgensen was the perfect case.

At the Statens Serum Institut, Hamburger followed in Sand's footsteps, and he too conducted experimental research on animals, such as rats and monkeys, as well as physiological experiments on humans. Inspired by the theories of Magnus Hirschfeld, Hamburger wrote the following in a chapter on intersexuality in a book on sexual biology published in 1954: "The notion of a 'hundred percent man' and 'hundred percent woman' is a construct that does not align with reality; both sexes possess rudimentary organs (such as nipples in men and the clitoris in women) that suggest an inherent bisexuality [*Doppelgeschlechtlichkeit*]."[41] The chapter included a U-shaped graph that depicted the transitional forms of sex between male and female. So-called "normal" men and women each made up the bulk of the U on each side with "masculinized women," "intersex," and "feminized men" interspersed in between.

The exact details of their initial meeting are unknown, but Hamburger agreed to put Jorgensen on treatment with pregnenolone and estrogen.

"He wants me to stop the tablets you sent to me and then check the return to normal," Jorgensen wrote to the Angelos back home. "In this way he can chart the hormone content of my body? Then he is going to inject strong quantities of estrogens with a hypodermic." The encounter filled her with hope; Hamburger would even look into whether it was possible to construct a vagina and who would be able to do it. "Joe, he was so nice and gave me so much hope."[42]

After a while, Hamburger petitioned the Ministry of Justice for castration. The authorities rejected the application, however, since there had been no psychiatric consultation.[43] Hamburger therefore contacted Georg K. Stürup. As a forensic psychiatrist and director of the Herstedvester Institution for Psychopaths, an internationally renowned institution for clinical criminology and sex offenders, Stürup worked with all groups of people that psychiatry categorized as sexually abnormal, but he was particularly interested in transvestism from a forensic and psychopathological perspective. Transvestism was a symptom related to manifold conditions, he argued in an article published in 1956, the same year he was appointed to the Medico-Legal Council. It ranged from the disguise of a criminal, the theft of women's underwear in the context of "partial transvestism," and voyeurism, to narcissism, "strangulation masochism," "homosexual coquetry," "pseudo-heterosexual prostitution," "baroque penis fetishism," and genuine transvestism.[44]

Among Stürup's patients was a twenty-nine-year-old trans woman. In 1938, she contacted Erling Dahl-Iversen, a professor of surgery and head of the Rigshospitalet's surgical polyclinic, framing her request for genital surgery by describing pain in the genitals for which she wanted a permanent remedy. Dahl-Iversen referred the patient to Stürup, who diagnosed her with depression and a psychosomatic condition, but did not recommend surgery.[45] Importantly, this shows that when Jorgensen arrived in Copenhagen, there was a group of physicians, other than Knud Sand, who had shared experiences and cooperated for more than a decade in the diagnosis and treatment of at least a few trans people.

Jorgensen recalled meeting Stürup three or four times. In his report to the Ministry of Justice, the psychiatrist stated that the estrogen treatment had enabled a state of mental balance and that Jorgensen's work spirit was better. Even her color photographs were "much better than they used to be." Since one could not continue to take estrogen pills for many years, he recommended castration: "This would allow him to have a quiet and peaceful life with greater chance of success in society."[46] Like

arbejdede meget omhyggeligt og langsomt, sådan som en videnskabsmand bør. Det tog sin tid, men han vidste og jeg vidste, på baggrund af hans kunnen, at vi gik i den rigtige retning. Men det var ikke hu og hej, lad-os-se-at-få-tingene-overstået. De ønskede at vide, hvor vi bevægede os hen og hvor vi ville ende. Og det synes jeg er en udmærket videnskabelig holdning.

På det tidspunkt fandtes udtrykket transsexualitet jo slet ikke. Det var først senere, da jeg kom tilbage til New York, at dr. Harry Benjamin fandt på det lille ord som forklaring.*) Så på en måde famlede vi os frem i mørke. Men det vidunderlige er, at de to herrer begge havde fremsyn til at se, at der var et eller andet, som burde undersøges nærmere. Da så hele historien kom ud, blev der jo en vældig opstand. Og jeg føler mig overbevist om, at den danske regering nok også tog på veje, som resten af verden gjorde. Men nu 34 år efter kan vi se, at transsexualitet er noget for sig selv. Vi var blot nogle af de første, der blev klar over det. Sådan er det, og jeg er meget, meget glad for at være tilbage i Danmark igen og atter møde disse to store mænd.

1984. Fra venstre: Georg Stürup, Christine Jorgensen og Christian Hamburger

Stürup: Jeg kunne godt tænke mig at føje et par ting til dine kommentarer. Det er jo ret typisk, at man ikke husker følelsesladede elementer så let. At

*) Dette er delvis ukorrekt, udtrykket transsexualitet er forinden brugt både af Hirschfeld og Cauldwell, men det er rigtigt, at det først var med Harry Benjamins bog „The Transsexual Phenomenon" 1966, at ordet fik alm. udbredelse. (PH).

115

FIGURE 1.2. Georg Stürup, Christine Jorgensen, and Christian Hamburger, 1984. Preben Hertoft and Teit Ritzau, *Paradiset er ikke til salg. Trangen til at være begge køn.* Viborg: Lindhardt og Ringhof, 1984.

Louis Le Maire, Stürup was a staunch defender of castration for those people he classified as sexually abnormal. When Stürup, Hamburger, and Jorgensen were reunited three decades later for a film and book project in 1984, Stürup said that he regretted that castration as a therapeutic tool for sexual abnormality had gone out of fashion. He still referred to Jorgensen's operation as "castration" in the context of preventing crime and treating sexual abnormality. Jorgensen disagreed with framing the operation in this way: "I don't see my surgery as 'castration,' I think it was more than that, it was a complete sex change. And the reasons were different. In the beginning, when you talked about castration, it was a punitive measure. But I wasn't punished by being castrated."[47] In a photograph of this reunion, Jorgensen and Hamburger look directly into the camera, while Stürup appears sitting in a somewhat stooped manner, eyes downcast and a little lost in thought. Looking at this photograph today, it is as though it captures in the body language of the sitters the two different logics of surgery that trans medicine pursued from the beginning: one based on sexual biology and psychiatric psychopathology, in which castration was a measure to attenuate so-called abnormal sexuality; another based on embodied experience, in which surgery was a life-affirming option.

For Hamburger, the most important outcome was the effect of treatment on overall well-being. In his statement to the Medico-Legal Council, he emphasized that estrogen therapy seemed to have a "markedly positive effect on the patient's psychological state, with more zest for life, work energy, and balance."[48] The sources don't say what Jorgensen told Hamburger and Stürup, but she probably understood that—if she wanted to get permission for surgery—she had to present herself as a homosexual man suffering from her desires and believing that castration would increase her ability to work.[49] In her statement to the Ministry of Justice in April 1951, a little less than one year after she first arrived in Copenhagen, she therefore described herself as "strongly afflicted by a homosexual inclination which is contrary to my moral convictions"; this inclination, she said, "was brought to rest during a year of hormone treatment."[50]

Only a month later, the Medico-Legal Council recommended surgical castration, on the basis that the patient must be considered a homosexual. But the minister of justice at the same time decided that only Danish citizens, in the future, could apply for sterilization and castration.[51] On September 24, 1951, approximately fourteen months after initiating hormone treatment, Jorgensen had her testes removed but the surgeon at that time inserted a pair of glass balls in the scrotum. The following year, after Jorgensen had obtained identity documents with her new name and sex from the United States, Dahl-Iversen completed the genital surgery by amputating the penis, transposing the urethra, and removing the glass balls. "Her goal," Dahl-Iversen noted, using the patient's preferred pronoun, "has always been to become feminine, which is why the glass balls are completely meaningless to her."[52] Although Jorgensen also asked

Dahl-Iversen to create a vagina in the same procedure, the surgeon decided to first observe the healing process and further developments.

While Jorgensen was being admitted to the Rigshospitalet, journalists from the United States were contacting her, offering large sums for exclusive rights to her story.[53] It is unclear why they had heard of her story in the first place; the archive does not provide an answer. Jorgensen knew that her doctors wanted to avoid publicity about her story at all costs, because they were operating in an ethical and legal gray area. However, it is not unlikely that she herself played a role in leaking the story, given how professionally she maneuvered the media in the months that followed. When the *Daily News* succeeding in getting an interview with her parents, they broke the story on December 1, 1952, with the headline "Ex-GI Becomes Blond Beauty." The news quickly became an international sensation. Soon, newspapers all over the world, including in Denmark and Norway, reported about Jorgensen's transition with headlines such as "From man to woman, from woman to man" (*Berlingske Tidene*, December 14, 1952) and "I am a normal girl with a girl's interests" (*VG*, December 2, 1952), often featuring photos of Jorgensen before and after the transition. In Norway alone, newspapers published more than 240 articles about her story within just one year.[54] As Jens Rydström and David Tjeder have noted, however, the initial enthusiastic coverage in the Swedish media soon became negative and hostile.[55]

Jorgensen was instrumental in putting not only trans medicine on the map but also a specific Scandinavian version of it. From the beginning, Jorgensen took a leading role in shaping the public image of her story, most notably in a five-part series on her life that was published in *American Weekly* in 1953. In this image, the notion of Denmark as a vanguard of scientific progress was central, as was a specific idea of Scandinavian societies as particularly progressive. Part of this strategy was to make a movie for a Danish tourist agency, Turistforeningen for Danmark, in the year before her last surgery in 1952.[56] Her strategy received a major boost when, soon after her return to the United States, the Scandinavian Societies of Greater New York unanimously named Jorgensen "The Woman of the Year" for her "outstanding contribution to the advancement of Medical Science, and because of the dignified and courageous manner in which you have deported yourself thru it all."[57]

While Jorgensen played a pivotal role in the advancement of trans medicine, she simultaneously embraced and framed her own embodiment and identity within the scientific language of the emerging medical field. "The more I talk with Dr. Hamburger the more I believe as he does—chemistry is the whole basis of life (physical—I don't mean the soul.) I think the future of medicine lies in the hands of the biochemists," she noted in a letter during the treatment. "It's wonderful to be with a man like Dr. H. He is a scientist all the way through—he's got the human machine in a test tube and yet there is nothing cold about his personality and he has the most terrific respect for life and feelings. I think it must be difficult to think of the human machine as just a bunch of chemical reactions one minute and then to think of the same machine as a human being."[58] The language of trans medicine provided scientific legitimacy and support to her decisions, also when she finally came out to her parents. In a four-page-long letter, which she wrote after the castration, she explained the role of "several small unimportant-looking glands" in maintaining physical and mental well-being and the importance of the sex hormones in sexual development. "Sometimes a child is born and to all outward appearance seems to be of a certain sex. During childhood nothing is noticed but at the time of puberty where the sex hormones come into action the chemistry of the body seems to take an opposite turn and chemically the child is not of the supposed sex but rather of the other sex." She went on to explain that she was one of the aforementioned children, that her hormonal system had been in a state of imbalance, but that hormonal treatment and surgery had restored the balance. "It was for this reason that I came to Europe and primarily to Dr. Hamburger because upon investigation I had found him to be one of the greatest gland + hormone specialists in the world." With the letter, she attached photos of how she now looked. "I have changed, changed very much as my photos will show but I want you to know that I am extremely happy and that the real me, not the physical me, has not changed. I am still the same old Bud."[59]

In February 1953, Jorgensen's medical team gave a presentation on the case at a meeting of the Danish Society for Endocrinology, the Danish Surgical Society, and the Danish Psychiatric Society. In the same year, they published two articles about the case in *Nordisk Medicin*, the Nordic medical journal, and the *Journal of the American Medical Association* in which they proposed a new routine for the medical treatment of transvestism. First, they recommended that permission be given for the wearing of women's clothing in public and for legal recognition as a woman for those who wanted to be identified as such. This should be followed by estrogen substitution, castration, and "demasculinization" surgery. The final step was the creation of a vagina.[60] "From a eugenic point of view it would do no harm if a number of sexually abnormal men were castrated and thus deprived of their sexual libido," they noted, perhaps as a strategy to fend off criticism, but certainly in line with the eugenic theories of sexual biology propounded in the preceding decades.[61]

The publications in esteemed medical journals served to reinforce the credibility of the decisions and secure the reputation of the doctors. They were aware that many colleagues questioned this treatment and that they were operating in an ethical and legal gray area. As their reputations increased, people began to seek the help of Stürup, Hamburger, and Dahl-Iversen over Denmark's leading sexologist, Knud Sand. The team gradually laid out a new strategy for the treatment of trans people, arguing that it was a palliative measure to improve conditions for the individual, but that it was also in the interests of society.[62] In 1953, Dahl-Iversen became the deputy head of the Medico-Legal Council where he was directly involved in handling the applications.[63] In April the same year, Stürup for the first time submitted a request for castration with "genuine transvestism, eonism" as an indication. The purpose of the procedure, argued Sølve Holm, was no longer to reduce homosexual tendencies, but to remove body parts that made the patient depressed and suicidal because of their classification as male.[64] Despite the skepticism of Hjalmar Helweg, the psychiatry professor who prepared the case in the Medico-Legal Council, who feared that castration would do more harm than good, the council nevertheless recommended castration:[65] "His abnormal sexual inclination (transvestism) has caused mental suffering over many years, which can only be described as severe, and which can probably only be reduced in this way."[66] Four days after the Ministry of Justice approved the request, the patient's testes were removed, and some months later, following approval by Hamburger and a psychiatric examination, a surgeon performed a vulvoplasty.

In the following years, this would be the procedure: A psychiatrist (often Stürup) would write the application for castration, stating genuine transvestism as an indication. The Medico-Legal Council would make its recommendation, which would be accepted and followed by the Ministry of Justice. After castration, surgeons were free to perform subsequent genital surgery without further approval from the authorities.[67] The Medico-Legal Council repeatedly emphasized that their role was to make the decision on castration only, based on the castration laws, and that any further treatment must be a decision between the patient and the physician, even though it understood that castration was only the first step in a series of operations.[68] With the cooperation of the patients, the Danish physicians had developed a strategy that circumvented the hurdle of castration as a prerequisite for changes of sex, and to access further genital surgery.

In the Borderlands of Medicine

In Norway too, some people tried to obtain genital surgery by first requesting castration. In the latter half of 1952, a thirty-three-year-old individual applied for castration at the Expert Council on Cases of Sex Operations. Two years after castration, she applied for genital surgery "so that I can live as a woman."[69] But compared to Denmark, the routine for genital surgery for trans people in Norway developed more slowly, as both the Ministry of Justice and the Directorate of Health remained skeptical about the procedure. The Norwegian authorities adopted a more holistic approach to the legal question of hormone treatment, castration, and genital surgery: Castration was not to be seen in isolation. As the treatment window for foreigners in Denmark closed, people started to direct their requests to the Norwegian authorities instead. One month after the request from the physician at Ila Prison about hormonal treatment for one prisoner, Christofer Lohne Knudsen in the Norwegian Directorate of Health received a second letter. This was a letter from a US citizen who had read about the Jorgensen story in the news: "Please advise whether or not treatment, such as that given to the former George Jorgensen in Denmark is permissible under the laws and customs of Norway."[70] Unsure how to respond, Knudsen consulted the Ministry of Justice.

Knudsen would come to play a key role in the administrative handling of trans medicine in Norway, but his views were not just shaped by his role in the bureaucracy. Very soon after the request from the United States, Knudsen was consulted about a similar case, but this time in his capacity as a clinician. It was not uncommon at the time, in fact, for physicians in the health bureaucracy to have side jobs as clinicians. In the autumn of 1954, Knudsen met sixteen-year-old Mattis Kvaal, who was admitted to Aker Hospital in Oslo because of bleeding from the gums. Probably because of his unique life history, the physician requested a psychiatric evaluation. Mattis's mother told Knudsen that she had given birth to a girl, but that Mattis wanted to be a boy from a very early age. During puberty, he used a tight rubber band to bind his chest and a penis prosthesis to fill his underwear. At work, Mattis was open about his past, and they in return used his chosen name and treated him as a man.[71] According to Knudsen's testimony, Mattis made a profoundly masculine impression. He wore "big tractor shoes like all boys do these days," spoke, and acted like a boy. "She repeatedly emphasizes that she feels completely like a boy and is most comfortable when she can ride her motorcycle with other boys," Knudsen noted, declining to use the patient's favored pronouns. The psychiatrist was in no doubt about the diagnosis. "This is a case of homosexuality with transvestism which, according to the available information, goes back to early childhood and therefore seems likely to be constitutionally rooted. Psychotherapy seems of no avail." He therefore recommended treatment with estrogen to change the patient's emotional life in a more "feminine direction," but this was a course of action that the patient "vehemently refused." Mattis told the psychiatrist that he had chosen to live as a boy and that under no circumstances would he be willing to begin treatment that would change this.[72] Although Knudsen now had personal experience with the group of patients and the clinical issue he was preparing in his office at the Directorate of Health, the encounter did not change how he approached the issue. To the contrary, it seemed to reinforce his opinion and deep skepticism toward hormonal and surgical treatment for trans people.

At the Ministry of Justice, the lawyers agreed on one thing, that hormonal and surgical treatments of transvestism raised different legal issues. Hormonal treatment for curative purposes was for physicians to decide, but the question of sex operations was far more difficult, as was illustrated by disagreements among the lawyers themselves.[73] One lawyer argued that surgery on the genitals, including castration and penile amputation, eliminated reproductive capacity and sex drive and therefore fell under the sterilization law.[74] Another argued that castration was not equivalent to sterilization because it also destroyed the person's "coital ability"—this amounted to mutilation and the Penal Code's definition of bodily harm.[75] In this lawyer's opinion, the ministry should not refer to existing statutes or laws, to avoid giving the impression that they approved of the treatment, but to make clear that "we distance ourselves from the project."[76] Regarding the surgery, the Ministry of Justice ended up issuing a non sequitur: "It is assumed that surgical interventions with the intention of 'transforming' a man into a woman usually have the effects specified in section 9 of the Penal Code. Therefore, the consent of the person is not sufficient to make the procedure lawful."[77]

The physician at Ila Prison had already given the indication for medical treatment. He was the clinician, after all, and just wanted confirmation that he would not fall afoul of the law. For the lawyers at the Norwegian Ministry of Justice, however, the treatment of transvestism was not simply a medical issue but was a matter of public nuisance and decency and of the prevention of sexual offenses; in their response to the Directorate of Health, the ministry included the following remark: "It seems clear that a man dressed as a woman can easily find himself in situations that could lead to—or be—criminal acts themselves."[78] Ultimately, the state lawyers saw it as a question of protecting the fabric of society, invoking a long tradition in medical and legal discourse of deploying trans femininity as a destabilizing allegory and positioning trans women as a particular threat to the social order.[79]

At the heart of these deliberations was the definition of sex. A man could not become a woman, not even through medical intervention, the state lawyers insisted: "The internal organs will remain the same."[80] It could be dangerous to encourage a man to cultivate his feminine tendencies, and what is more, the lawyers feared an "uncontrolled avalanche" of operations similar to the one to which they disrespectfully referred as "Mr. or Mrs. Jorgensen," whom they accused of having "exploited the situation in a rather unsavory way."[81] The lawyers' fears that authorizing medical treatment to change sex might set a wheel spinning out of control provides a unique insight into the "moral economy" of legal bureaucracy at the time, that is, into the values and affects shaping the state's handling of sex change.[82] On a similar note, Ann Laura Stoler argued that to understand colonial governance we need to pay attention to the state's affective dimensions.[83] By interpreting and applying the law, by determining who belongs to which sex category, and by insisting that these categories cannot be transgressed, the law acted as a moral technology that reproduced the validity of these very categories.[84] This illustrates Tal Golan's argument that much more than separate cultures, science and law are belief systems that mutually support each other with intertwined social institutions deeply invested in each other.[85] By not knowing which pronouns to use for Jorgensen, it is as if the Norwegian lawyers at the same time recognized that they were entering the landscape of medicine where the traditional legal categories of sex were no longer paramount.

First in Denmark and later, as we will see in the next chapter, in Norway, trans women succeeded in harnessing pathologizing hormonal theories and a legal framework based on eugenics for emancipatory purposes. The theory of sexual abnormality as a hormonal imbalance and the idea that hormone treatment could dampen abnormal sexual tendencies provided the medical justification for hormone treatment. The sterilization and castration laws provided the legal basis for the removal of the testes, a legal prerequisite for access to genital surgery. We are thus confronted with a paradox of mid-century social medicine, showing how the darkest, most repressive, and authoritarian policies of the welfare state provided the foundation for trans liberation.

CHAPTER TWO

Sex Change and Sex Offense

On April 24, 1954, people in Østfold County in southeastern Norway woke up to sensational news. "The defendants in the homosex case wanted to change sex," was the headline on the front page of the local newspaper. "Have always felt like women—the boy affair began after hormone pills."[1] Just over a year after Christine Jorgensen's story hit the headlines, the issue of sex change had come to Fredrikstad, a small town on the Oslo fjord. "Could the tragic outcome of the homosex case have been prevented if the accused had undergone a Christine Jorgensen operation?" the newspaper asked. Or could hormone pills unleash dangerous impulses that could lead to the sexual abuse of children?

While physicians and lawyers in the state organs were processing the first requests for hormonal and surgical treatments for trans women, lawyers, psychiatrists, and judges negotiated the future of these treatments in court. Two cases of the sexual abuse of children that were tried in court in the early 1950s concerned trans women, and it was through these cases that Norwegian psychiatrists gained their first in-depth knowledge and experience in the medical handling of trans people. They listened as the defendants spoke about their childhoods; they took notes as the defendants detailed the circumstances of the assaults. Ultimately, the psychiatrists weighed differential diagnoses and were tasked with suggesting appropriate medical responses to prevent the assaults from happening again. For decades to come, these meetings shaped the psychiatrists' approach to trans people and their requests for medical treatment. The trials established the question of sex reassignment not only as an issue concerning the proper handling of individual patients but as one of major importance to society as a whole: Correct diagnosis and treatment was about protecting the social fabric.

Transvestism in Court

The Østfold County trial involved two trans women accused of sodomy and the sexual abuse of minors, some of whom were as young as fourteen years old. The sexual abuse had been going on for a long time and followed a similar pattern: The defendants, L. Nielsen, a former ballet dancer in her early thirties, and A. Rød, a waiter in her late teens, both familiar in the gay community, took the victims home to their place where they looked at sexual images and told erotic stories. In the most severe incidents, the defendants had masturbated the victims or made them put their penises between their legs, the defendants acting as the "passive part," until orgasm. "He feels like a woman throughout," the forensic report stated about Rød.[2]

In her statement to the police and to the forensic psychiatrists, Nielsen explained that she did not see herself as a homosexual man—she identified as a transvestite and pseudohermaphrodite. In early childhood, Nielsen began to feel attraction to boys, a feeling that gradually turned into the realization that she should have been a girl. Despite devouring all the literature on sexuality she came across, she never read a description that fitted with her own experiences until she found a book by Magnus Hirschfeld, which, she said in the forensic interview, explained that a person could change sex through an operation: "I no longer lived in the darkness of uncertainty but had found a way finder."[3] In her early adulthood, she repeatedly tried to get hormones and surgery to become a woman. In the spring of 1953, having failed to get hormonal and surgical treatment in Norway, she was referred to the endocrinologist Christian Hamburger in Copenhagen, but he was unable to help due to the restrictions on treating non-Danish citizens.[4] Desperate to get hormones, she took a job on a ship to the United States. There she got hold of four boxes of estrogen pills on the black market, and to reduce facial hair growth and to stimulate her feminine development, she started taking two to three pills a week and gradually increased to one pill a day. Ultimately, she hoped it would increase her chances of getting surgery back in Norway.[5]

According to the police, it was the pills that led to the abuse. "The sexual urges were so aroused that he could not sleep at night and had to get up and go for long walks," the police wrote in their report. "The pills made him 'take anything that came his way.'"[6] If the hormone pills were causally connected to sexual crime, what did this mean for the question of legal culpability? The reformed Penal Code of 1929 stipulated that insane criminals and people diagnosed with underdeveloped and permanently weakened mental abilities should not be sentenced to prison, but to psychiatric detention.[7] For advice on this question, the court appointed two forensic psychiatric witnesses. The first, Per Anchersen, was a psychiatrist at Ullevål Hospital in Oslo. He promoted an approach to mental illness rooted in social psychiatry, based on collaboration between the patient and nonmedical experts such as psychologists and social workers. "Mental illnesses are primarily personality disorders that ruin or hinder interpersonal harmony," he wrote a few years later.[8] The other forensic witness was Jon Leikvam, a psychiatrist at Oslo Prison.

Both psychiatrists rejected a simple causal link between the intake of hormones and subsequent sexual abuse. "Transvestism very rarely leads to criminal offences, at least against minors," Leikvam told the newspaper.[9] However, hormone supplements could reinforce unconscious infantile tendencies, and therefore, the individual personality structure must be considered. In the forensic examinations the psychiatrists looked specifically for signs of immaturity: L. Nielsen, the former ballet dancer, 172 cm tall with a slim body and slender facial features, wearing cord pants and a feminine sweater, looked more like "a little girl than a grown man" with "strong emotional immaturity," "backfisch-like sexual fantasies," and "garçonne-like body type," the psychiatrists stated in their report.[10] An immature personality structure made the defendant susceptible to external stimuli, and by inducing a synthetic castration syndrome, the pills reduced control over sexual impulses. Anchersen and Leikvam concluded that Nielsen's infantile psychosexual constitution represented a contraindication to treatment, but nevertheless, based on the psychiatric evaluation and the diagnosis of genuine transvestism, the sentence was reduced on the grounds that "normal adjustment in society must have been difficult."[11]

While the psychiatrists rejected a simple connection between hormone supplements and an increased risk of sexual offenses, the potential risk secured a future role for psychiatrists in the evaluation of this treatment: Before treatment, the patient had to undergo a thorough psychiatric evaluation of personality structure and psychological resources. In other words, the courtroom confirmed the psychiatric category of transvestism and validated psychiatrists' expertise on these issues to secure social order and the common good. In this particular case, the psychiatrists turned a moral issue and the question of protecting children into a question of correct diagnosis and medical treatment. As Sheila Jasanoff noted, the courtroom is a space of reenactment and translation: The court transports something that happened in the "real world" into an impartial and trustworthy form.[12] Put differently, the use of psychiatric knowledge in court was one way for the state to negotiate and respond to complex social and ethical questions of major societal importance.

There are other examples of forensic psychiatry's increasing importance in the regulation of trans peoples' lives in the 1950s. Anchersen acted as a forensic expert in another trial that took place around the same time as the Østfold County trial. The defendant, K. Karlsen, a cook in her early thirties, was accused of performing oral sex on a fourteen-year-old boy. "No one can fully understand how horrible it was for me to undress completely with boys. I still suffer from it," Karlsen told the forensic psychiatrists.[13] During the Nazi occupation, she fell in love with a German soldier. She related to the psychiatrists how, "dressed fully as a woman, I travelled with him on holidays." When the war ended, she sought help from physicians to have hormonal and surgical treatment to become a woman. One doctor had convinced her to try treatment with testosterone, but this only made her feel miserable and increased the sex drive, and she therefore hoped that Anchersen and his cowitness would help her get the treatment to become the woman she felt herself to be.[14]

This time, the psychiatrists were sympathetic to the request. The defendant displayed considerable self-awareness and insight into the situation, and there was "every reason" to support medical interventions leading to stabilization.[15] If medical treatment could help Karlsen take on the sexual and social role of a woman, this would increase her chances of finding work and ease her path in society. In addition, the sex operation would dampen abnormal sex drives. A "criminal potential" will always be present in people like Karlsen, wrote Anchersen, but an operation would reduce the sex drive and thus "compensate" for the reduced self-control.[16] Although they reasoned as psychiatrists, basing their conclusions on psychological theories of personality formation, the justification for their recommendations reflected the original eugenic and social foundations of the sterilization law introduced two decades earlier.

The Expert Commission

In October and November 1954, the year after the first request for hormone treatment came from Ila Prison, chief psychiatrist Christofer Lohne Knudsen received three new applications for castration and genital surgery for trans women. One application was signed by Per Anchersen. The patient, Inger Martinsen, who was now in her mid-twenties, grew up in a working-class family with eleven siblings. She was confronted early with hard life experiences: Six of her siblings died at a young age. At the age of seven or eight, she first recalled feeling different from the boys around her, especially the joy she got from dressing in her sisters' clothes. Soon afterward, Martinsen began taking dance classes, and as a young adult, she studied ballet at the opera. As an adult requesting hormonal and surgical treatment, she ended up at the Rikshospitalet in Oslo in the spring of 1954. "He says he has no doubt he is a woman," the physician noted.[17] Martinsen was transferred to Anchersen for further evaluation. A physical examination revealed testicular atrophy and reduced levels of 17-ketosteroid in the urine, which could be a sign of reduced androgen synthesis. A trial of testosterone treatment resulted in sleeplessness and depression, making it impossible to continue working. On estrogen, however, the patient felt "harmonious" and was able to work, Anchersen later stated.[18]

The other two requests came from Nils Kinnerød, chief physician at Ila Prison, one concerning the same person as the previous year, and the other a person in her late twenties who was sentenced to preventive detention after repeated thefts and indecent exposure. "A few months ago, he had never told anybody about his desire to dress as a woman," Kinnerød noted. "It is very likely that this desire has led to a criminal life trajectory and made his life the tragedy it has been so far."[19] Indeed, the purpose of the thefts was to steal women's underwear; speaking with the physician, the inmate had this to say about her experience: "The thing is, when I was ten years old, I developed a deep desire to be a girl and not to be the boy that I was." At one point, she rubbed paraffin on her genitals, hoping it would burn them off. "In vain, I have always fought my feelings and desire to be a woman."[20] The inmate's approaches to women were not of a sexual nature; Kinnerød wrote that, to the contrary, "his need to approach women is that it makes him feel among equals."[21] Attached to the request was a letter written by the inmate to the director general of health, Karl Evang, in which she explained that her highest desire was to be "transformed from a man into a woman."[22] Kinnerød was willing to do everything he could to help his patient live as a woman, including hormonal and surgical treatment, but since approval to proceed on this course was a legal gray area, he sought the advice of the authorities.

With a handful of requests for sex change operations, Evang saw the need for convening leaders to discuss the issue. On April 2, 1955, he invited four of the country's most experienced psychiatrists to give their opinions on the treatment of transvestism, the chance of "social adjustment," and their recommendations for the official handling of such requests in the future.[23] As a natural authority in the field—just one month later he gave a talk about the medical treatment of transvestism at the Nordic Psychiatry Conference—Anchersen was appointed as chair. The second member was Gabriel Langfeldt, professor at the psychiatric clinic of the University of Oslo and chair of the Norwegian Board of Forensic Medicine. He was best known to the public for having diagnosed the author Knut Hamsun, Nobel Prize winner and Nazi sympathizer, with "permanently impaired mental faculties" after the war. Nils Kinnerød, who held Norway's first position as a psychiatrist in a secure detention unit, also joined the committee, having prepared two of the applications for castration. The fourth psychiatrist was Johan Bremer, director of the department for women in the Gaustad psychiatric hospital and from 1955 a member of the Expert Council on Cases of Sex Operations. The directorate also repeatedly tried to recruit an endocrinologist from the Radium Hospital, the national cancer hospital, but he was traveling and never responded to the invitation.

It must have come as a surprise to Evang when, a year later, the committee submitted not one but two reports, one of which was the dissenting opinion of Bremer. Anchersen's majority report was twenty-five pages long and invoked a long tradition of scientific writing: People familiar with the medical literature would have recognized that it followed the classic chronological tradition and structure of medical textbooks, with chapters on epidemiology, pathophysiology, etiology, diagnosis, treatment, and prognosis. Genuine transvestism usually manifested itself at an early age, the report stated, giving reason to believe that the condition had a constitutional basis. Since attempts to "bring a genuine transvestite out of his condition and back to normal heterosexuality" through psychotherapy, hormones, or surgery were unsuccessful, they suggested following the recommendations presented by Christine Jorgensen's medical team in an article in the *Journal of the American Medical Association*. Measures should be introduced in a step-by-step fashion, starting with permission to wear women's clothing, to change one's name, and to receive estrogen treatment. The goal was to improve adjustment: "To treat these patients with rejection or indifference is, in the committee's opinion, to neglect an elementary duty of the physician."[24] Selection criteria for treatment must be strict, however, and genital surgery must be reserved for a very narrow group. Selected candidates should have a stable mental state and stable living conditions, but also body type, physical habitus, and general "disposition" should be suitable for the female role. Although the committee only briefly mentioned the treatment of trans men, referred to as "female transvestites," it assumed that the principles were the same as for trans women.

At the time, Anchersen was preparing another castration request for one of his patients, Reidun Myklebust, the details of which highlight how gender norms shaped psychiatric reasoning around sex reassignment. Myklebust was twenty-seven years old, and she had been living fully as a woman for two years. The "tall, rather bony man with sharp facial features" with "depressive" and "passive traits" was "very uncritical in the way he perceives himself as womanly and charming," wrote Anchersen in a demeaning and contemptous note, "but most people would think he has a grotesque appearance as a woman." While the fact that Myklebust often became depressed when her "homosexual relationships fell apart" meant that Anchersen considered supporting the request for castration, ultimately he did not grant his support because of her unsuitable physique and disposition. Moreover, he feared that this would just be the first step and that she would continue to push for further operations.[25] The idea that only patients who fulfilled certain bodily criteria should receive treatment was in line with the reasoning of the Danish Medico-Legal Council, as well as Jorgensen's physicians, reflecting conventional gender norms, but also the logic that successful treatment depended on the individual being able to pass themselves off perfectly in society as belonging among their chosen sex.

Bremer dissented. In reviewing the literature, he found an "almost astonishing lack of follow-up studies" and concluded that demasculinizing operations were akin to "groping in the dark."[26] His skepticism toward surgical treatment was influenced by an investigation he was conducting on the effects of legal castration. Before being appointed to the commission, Bremer had worked in the Office for Psychiatry at the Directorate of Health, where he systematically interviewed and collected information about people who had been castrated for reasons to do with criminal law. Although his book *Asexualization: A Follow-Up Study of 244 Cases*, which dealt a devastating blow to the widespread practice of legal castration for "deviant sexual behavior" or for sexual offenses, was not published until 1958, he had already drawn his conclusions.[27] The indications for castration were too broad, and the procedure neither had a "sedative influence" nor a "resocializing" effect on asocial or antisocial behavior.[28] One person was only nineteen when they underwent castration. "He had been told that otherwise he would not be released from the reform school until he was 20 years old," Bremer noted, but the operation only reinforced the patient's sense of being different.[29] "It seems that he wanted to be woman, but in the real sense," Bremer noted, considering the operation a failure.[30] For Bremer, the goal of castration was to alleviate a symptom, not to validate people's sense of themselves. What is more, he reasoned that sex offenders would present themselves as transvestites and request castration in order to avoid prison.[31]

We get a better understanding of Anchersen and Bremer's arguments by looking at a debate that took place five years later in the main Norwegian medical journal, *Tidsskrift for Den Norske Lægeforening*. Here, Bremer described genital surgery for transvestism as one of the surgical "missteps" in the history of psychiatry, alongside leukotomy, lobotomy, and castration therapy. The focal infection theory had led to the "most deplorable and savage surgical orgy of all time," he wrote. As a science, psychiatry was too young: "A psychiatry that is probably at the stage of development equivalent to that of an infant should not be playing around with knives and scissors." The introduction of psychopharmaceuticals in the 1950s further contributed to his skepticism in relation to surgery. "The perverted behavior was corrected with phrenotropic drugs," wrote Bremer, referencing a case of psychopharmaceutical "conversion therapy" of a trans woman. The patient had received a multidrug cocktail consisting of 50 mg of nialamide once a day (a monoamine oxidase inhibitor), 0.40 mg of meprobamate three times a day (a tranquilizer), and 50 mg of chlorpromazine four times a day (a high-dose neuroleptic).[32] Psychopharmaceuticals have relegated surgical procedures like castration and lobotomy to the ash heap of history, he argued, hailing neuroleptics as "revolutionary therapy."[33] For Anchersen, on the other hand, the goal was to "help the transvestites, not to cure the genuine transvestism." He reasoned primarily as a clinician faced with a clinical problem with no other options available. For this group of patients, "neither human nor juridical prejudices should be allowed to prevent attempts to help these unfortunate people."[34]

Avoiding Precedent

Karl Evang now had the full support of the Ministry of Justice and of the leading expert on legal castration to proceed cautiously and enforce a restrictive policy. In November 1956, he issued his decision. No official body should have the authority to recommend or approve genital surgery.

The Sterilization Act already regulated castration and the Expert Council would evaluate these applications as usual; individual physicians would make decisions about hormone treatment.[35] Evang had long been skeptical about legal castration on a general basis. As a student of psychoanalysis and social medicine, he was an ardent defender of psychological and social interventions in health issues. The health authorities, moreover, feared setting a precedent, having received requests for surgery from abroad.

Shortly after Evang had appointed the expert commission, the Expert Council received an application for the castration of K. Karlsen signed by Jon Leikvam. Karlsen had been sentenced to security detention after the assault of the fourteen-year-old. On the application form, there were a few lines where the applicant, in their own words, explained why they requested castration: "To come to rest," Karlsen stated. "I know that nature is against me and I realize that I sooner or later will relapse."[36] For unknown reasons, as the decision was not archived, the council rejected the application. Three years later, Karlsen resubmitted the application. "My desire for boys under the legal age is so shocking to me that I do not dare to return to Society in my condition," Karlsen wrote to Leikvam. "I know that the long sentence has not helped me."[37] In a letter to the Expert Council, Karlsen wrote of being aware of the "enormous harm" it would cause to "have sexual intercourse with younger people of my own sex," but that if the application was rejected, the only alternative would be confinement in prison for many years. "A castration would help me fulfill the role in Society meant for me."[38] The Expert Council's two physicians and judge now recommended castration under doubt.[39] Bremer disagreed: "Castration is unjustifiable, since the applicant sees this as the first step in series of feminizing operations." By conceding to an operation, the "desire for new would emerge."[40] Importantly, one year later, Karlsen withdrew the original application. "He does not feel like a woman anymore," Leikvam noted. Karlsen had met a woman and now said that he had requested castration to be released from detention hoping that it would free him from his "abnormal sexual tendencies and exaggerated sexual interests."[41]

The details of this case must have contributed to the skepticism within the health bureaucracy toward hormonal and surgical treatment for trans people. The council had not yet decided on five other castration applications for people diagnosed with genuine transvestism. A year later, in the summer of 1959, Jørgen H. Vogt, director of a medical department at Aker Hospital in Oslo, sent a letter to the Office of Psychiatry concerning his patient Mattis Kvaal, the same person whom Christopher Lohne Knudsen had examined five years earlier (see chapter 1). Since Kvaal now was of legal age, Vogt wanted to know if it would be legal for him to perform a mastectomy, an oophorectomy, and a hysterectomy on his patient. In his response, Evang confirmed that his directorate was developing a general policy on the issue, but that it would take time to reach a conclusion.[42]

The vagueness of the official response about the legality of these procedures quickly proved unsatisfactory to clinicians and their patients. "Doctors will probably try to be relieved of sole responsibility to an even greater extent when performing procedures on a transvestite," observed one physician at the Directorate of Health, pointing to "uncertain prognosis" and "consequences of 'non-medical nature.' "[43] This was uncharted territory, and doctors wanted to be sure that they were not breaking the law by performing operations. With a wry turn of phrase, Vogt wrote that "one can hardly say that the processing of this is remarkably fast," two and a half years after his request and still waiting for an answer. When an answer did arrive from the psychiatrists in the health bureaucracy, he was frustrated by its vagueness: "The answer also seems to me to be somewhat unclear, since it refers to paragraphs in the Penal Code which I can hardly be expected to have at hand, and to a book by Professor Johs. Andenæs on general criminal law from 1956, page 171," he noted.[44]

The question the law professor had discussed in his textbook was the question of what qualified as a medical emergency and thus an exception to the definition of substantial bodily harm in the Penal Code. In the original edition, Andenæs used the surgery that Lili Elbe had in Dresden as an example of what counted as an illegal medical procedure. In the revised edition, however, he had changed his view and argued that it fell under the Sterilization Act.[45] The Ministry of Justice disagreed. Demasculinizing surgery was a more complex operation than castration alone—the procedures must be seen as "a whole"—and therefore consent was insufficient. This was a "borderland area" and "medical conclusions should not be the sole deciding factor," wrote the director general in the legal department.[46] The only exception was in medical emergencies, and that was a decision for physicians to make.[47] Knudsen, who disapproved of these procedures, refused to give Vogt carte blanche. The individual clinician had to make the decision—and take the responsibility—again referring to the Penal Code without explaining how this law should be interpreted in the specific case in question. "Merry Christmas!" he added in handwriting under the typed letter.[48]

In the end, the Norwegian authorities refused to make specific regulations for sex operations for trans people. The authorities did not want to entrust a public body with the responsibility, nor did they take the initiative to enforce laws to formalize and regulate the conundrum of medico-legal problems raised by this treatment. Their decision fit into a pattern of opacity and of bureaucratic inertia. In the following years, requests for surgery would go through several channels: Some were processed by the Expert Council on Cases of Sex Operations, but it is likely that many people who approached their doctors with these requests met a closed door.[49] In a letter from the Directorate of Health to Per Anchersen in November 1961, it is written that "the Director General of Health will not refrain from pointing out that he, for his part, assumes that a great balancing of interests is required in order to carry out such extensive operations as discussed in the case in question."[50]

Two years later, in 1963, plastic surgeons at the Rikshospitalet performed the first genital surgery on a trans woman in Norway. The operation included castration and penile amputation but no vaginoplasty. The file has been lost, but it is likely that the physicians assumed full legal responsibility themselves for the operation. The patient was Inger Martinsen, the ballet dancer who had requested castration in 1954. She was now thirty-one years old.

Life in Limbo

In the time from when the Directorate of Health received the first applications for hormonal and surgical treatment until they settled on a vague and ambiguous response eight years later, how was life for those people left in limbo during the intervening period? Only a few letters by those who applied for treatment have been preserved. Some people and their families wrote directly to Evang, but there was not necessarily anything unusual in people writing letters directly to the director general of health.[51] As early as 1931, he had risen to prominence in an organization of socialist physicians who advocated for sex education and abortion on demand, and as a cofounder of *Populært Tidsskrift for Seksuell Oplysning* (Popular Journal of Sexual Education), he was a defender of educating the public in sexual issues. Over the years, he answered thousands of letters from people on subjects ranging from shame and sexuality to impotence and dyspareunia.[52]

Among those who wrote to Evang was Margrete Hovd.[53] By 1959, the year she turned twenty, she had already been Anchersen's patient for two years when, in October, he recommended castration. She was already living full-time as a woman with a large circle of friends and a good job. For her, it was incomprehensible why her request for surgery was causing so many problems for the authorities. Her worst nightmare was getting ill, having to go to hospital, and there having to undress. "I take responsibility for everything because I wholeheartedly agree and consent. I long for what is just," she wrote, emphasizing that the responsibility should be hers.[54] "Over time I have learned not to take myself too seriously," Hovd wrote to the minister of justice, four years after she began consultations with Anchersen.[55] Even her mother weighed in, writing to Evang that she could not afford to send her daughter abroad to get the treatment she needed, begging that permission be given for surgery: "Her faculties would be freer, and she would be happy, for she cannot go back to what she is not." Fully supportive of her daughter, she added that "I cannot imagine her as different."[56] The gracefully handwritten letter continues with Margrete's mother suggesting that Evang meet Margrete to see for himself. One sentence has been underlined in the letter: "I am very dependent on her in many ways, since I have been alone with the children for the past 15 years."[57] In the margin, next to the underlined sentence, someone had added an exclamation mark. Was it Evang who made these marks on the letter? For an admirer of psychoanalytic theory, the mother's statement would support the idea that an absent father figure and a symbiotic relationship between the mother and her child was the root of Hovd's identity. Eventually, Hovd succeeded in getting the surgery she wanted. In 1963, she was the second person in Norway to undergo genital surgery at the Rikshospitalet's Department of Plastic Surgery.

Reading these few letters today, we see that people had no problems describing their identities and what they wanted in straightforward, colloquial ways. They wrote about their lives in a clear and natural vernacular, free of medical jargon and pathologizing terms. "The reason for my application is that I have felt like a man all my life and I want to be a man," Mattis Kvaal stated in a letter to Evang.[58] In the 1950s, those formative years when trans medicine was not yet routine and individual physicians were unsure whether they could provide treatment to their patients, writing to the health authorities was perhaps the only way people could influence their cases. When people approached the state, they must have deemed it important to invoke notions of respectability, credibility, and stability. This could mean having a secure job, a respectable position, and a large circle of friends—or even evidence of a supportive family. Kvaal's

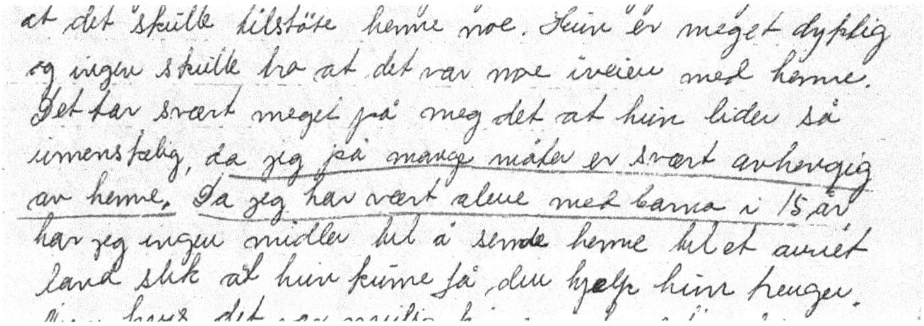

at det skulle tilstøte henne noe. Hun er meget dyktig
og ingen skulle tro at det var noe iveien med henne.
Det tar svært meget på meg det at hun lider så
umenneskelig, da jeg på mange måter er svært avhengig
av henne. Da jeg har vært alene med barna i 15 år
har jeg ingen midler til å sende henne til et annet
land slik at hun kunne få den hjelp hun trenger.

FIGURE 2.1. Letter from Margrete Hovd's mother to Karl Evang, March 16, 1960. Folder Kjønnsskifte, NAN.

letter was accompanied by a handwritten note from his mother: "I consent to the above operations on my child."[59]

Psychiatric Practices Beyond Psychiatry

In the 1950s, psychiatrists secured a key role in the evaluation of requests for hormonal and surgical sex reassignment in Norway, but it was not self-evident that psychiatrists would take on this role. Why was it not instead up to endocrinologists and surgeons to make these decisions? Part of the explanation is that the first requests came from psychiatrists in detention facilities, and they directed their requests to the Office of Psychiatry in the Directorate of Health. Another reason is that the first requests concerned people convicted of sex offenses, and psychiatrists became involved in the issue through their role as forensic experts. On a more general level, the handling of these cases—in prisons, court, and the health bureaucracy—exemplifies a much longer history of the role of psychiatrists in the welfare state. Historian of psychiatry Svein Atle Skålevåg emphasized that the legal and institutional developments of the late nineteenth and early twentieth centuries gave rise to a forensic culture consensus in which psychiatry and the law went hand in hand and acted synergistically. For example, psychiatrists worked in pairs and the court rarely heard opposing expert opinions. This culture developed in response to the goal not only of establishing accountability but of protecting society from those who were perceived as a threat.[60]

This was not an entirely new role for medicine. Physicians have held a long-standing role as society's "negotiators": Since the early modern period, European societies have depended on physicians to help govern and lead public life.[61] In the "golden age" of the welfare state around the mid-twentieth century, however, with increasing professional specialization in medicine, the psychiatrist remained in a unique position with the necessary breadth of expertise to deliberate and negotiate cases on the boundaries of morality, medicine, and social order. At the interface between society and administrative bureaucracy, between medicine and public opinion, psychiatric expertise has sought to secure public trust and bureaucratic intervention beyond the therapeutic qualifications of the psychiatrist. This expertise was about much more than the narrow role of providing an expert opinion on accountability in individual cases; it was about weighing difficult considerations, providing a reasoned response in complex issues, and providing unbiased expert opinions on sensitive issues. In short, psychiatric expertise was about making "good policy." Because the first applications for hormonal and surgical treatment for trans people in Norway in the 1950s were for people convicted of sexual offenses, psychiatrists were brought in as key witnesses, not only to make decisions on the clinical indication of the individual, but to make recommendations on behalf of the state and society. The treatment of trans people in the 1950s was precisely about protecting society from sex offenders and finding a good public solution. In other words, these court cases turned trans medicine into a subject of general interest—an issue between the public and the state, not just between the patient and the physician. However, as this chapter has shown, the "unbiased expertise" of psychiatry in meeting with trans people was predicated on transphobia and trans misogyny.[62] The psychiatric vision of "the good society" reserved the option to transition only as a stopgap measure for a few individuals.

CHAPTER THREE

Collecting Cases, Outlining Symptoms, Making Diagnoses

In the 1950s, as an increasing number of people requested hormonal and surgical treatment to modify their bodies, Scandinavian physicians sought to establish clear diagnostic criteria for clinical decisions and research purposes. This new situation in which they found little support for trans surgery in the legal, bureaucratic, and political systems provided the impetus to reformulate and modify old nosological categories. In response to this new situation, physicians proposed the category of "genuine transvestism"—later replaced by "transsexuality" or "transsexualism"—where previously there had been acknowledgment only of the general symptom of transvestism, which physicians observed in several conditions.[1] While this nosological work happened first and foremost in the clinic, at the bedside, there were too few cases, and clinicians had too little personal hands-on experience. To establish clear diagnostic criteria, physicians therefore turned to the archive. They collected cases, compared clinical descriptions, created synopses, underlined commonalities, extracted signs and symptoms, and condensed the notes into workable diagnostic tools. With operationalized diagnostic criteria, physicians were able to ask new research questions, compare patient groups, and conduct follow-up studies. This too was desk-based work: grouping findings, creating tables, and calculating significance values. Rather than stethoscopes, scalpels, or syringes, the tools of nosological classification and clinical research were pen and paper.

This is not to say that 1950s physicians were starting from scratch. As Jules Gill-Peterson has argued, mid-century discourse on transsexuality built on the concepts and framework of late nineteenth- and early twentieth-century sexual pathology, sexology, and psychoanalysis but was dressed in new terminology "without being able to extinguish their internal tensions."[2] As Magnus Hirschfeld wrote in *Die Transvestiten*, published in 1910, "the transvestic drive" must be understood as a "form of expression of the inner psychology,"[3] defined as "appearing in the clothes of the sex to which the person does not belong according to the external genitalia."[4] While Hirschfeld did not draw a sharp line between cross-dressing, identity, and embodiment, the British sexologist Havelock Ellis coined the concepts "eonism" and "sexo-aesthetic inversion," highlighting a psychological and affective experience of identifying with the opposite sex that entailed something more than the practice of cross-dressing.[5] Most likely inspired by Ellis, Hirschfeld began to distinguish a subgroup of "extreme" transvestites where he observed a "disaccord" between body and mind,[6] which he defined as "trans-sexualism."[7] "It is their bodies they find inadequate, not their *Geist*," he wrote.[8] For Hirschfeld, transvestism or trans-sexualism was not pathological but just another expression of sexual human variation with constitutional and biological explanations. By contrast, psychoanalysts such as Emil Gutheil and Wilhelm Stekel rejected Hirschfeld's biological basis of transvestism, and they sought to explain their patients' identities and behavior by traditional psychoanalytic theories such as the castration or Electra complex, repressed manifest homosexuality, and the displacement of an incestuous object onto clothing.[9]

Scandinavian clinicians and researchers read the sexological literature of the early twentieth century and referred to it in their publications. They also read the case reports of contemporaries, for example, of Swiss psychiatrists Benno Dukor and Fritz Bättig, the latter being supervised by Manfred Bleuler at the Burghölzli hospital in Zürich.[10] The Scandinavians were wary of the skepticism with which many doctors regarded genital surgery in trans people. They might have been familiar with the controversy that erupted in Germany only a few years earlier: In 1950, psychiatrist and *Dasein* analyst Medard Boss was fiercely criticized by his colleagues, including prominent psychiatrists and analysts such as Carl Jung and Alexander Mitscherlich, after he recommended genital surgery for one of his patients who identified both as a man and as a woman.[11] The controversy surrounding sex reassignment became a driving force behind attempts to create diagnostic criteria.

Working with Cases

"Will he be more content and mentally stable than before the procedure?" This was the leading question in Johan Bremer's 1956 report.[12] His choice of title for this document, "Mutilating Treatment of Transvestism," left little doubt about his conclusions. Although its objective was not nosological but to provide an expert recommendation on the medical and political handling of sex reassignment, the report enacted a clear-cut definition of genuine transvestism. As Bremer himself acknowledged, he lacked clinical experience with trans patients.[13] One option was to create an epidemiological register for research purposes. In his research on the mental health of a fishing community in northern Norway and on the long-term effects of castration of criminals and psychiatric patients, he familiarized himself with the epidemiological practices of collecting, organizing, and synthetizing large sets of clinical data. When it came to the effects of the medical treatment of transvestism, however, he lacked the evidentiary raw materials. To find the solution to a "new" clinical problem, there was no alternative therefore but to go to the archive. Bremer combed through the literature but also wrote to colleagues on the continent to share material and clinical experiences, so that the final product was a patchwork consisting of miscellaneous cases, some from medical journals, others from case series, and some cut out from correspondences. By compiling, summarizing, reformulating, numbering, and regrouping these cases, the report transformed primary literature into secondary literature, contributing to a new definition of genuine transvestism.[14]

This was nothing new. As J. Andrew Mendelsohn has argued, modern medicine has relied as much on library research as observation and empiricism in the clinic and laboratory. Since the early modern period, medicine originated in the archive because it was as much an archive of the *unknown* as the known: Physicians have shuffled, regrouped, and reclassified case stories, leading to the genre of the review of the early nineteenth century, a form of synthetized knowledge that would dominate knowledge production in medicine for centuries to come.[15]

The report was typewritten and twenty-five pages long and followed a conventional structure of introduction (pp. 1–6), main body with twenty-two cases (pp. 6–24), appraisal with recommendations (pp. 24–25), and conclusion (p. 25). For obvious reasons, Bremer could not reproduce the cases in their entirety; that would have turned the document into an unwieldy tome, jeopardizing the argumentative thrust. Since, moreover, the life stories, according to Bremer, were

> rather uniform repetitions of how they dress and act girlishly from earliest childhood, how this has remained unchanged over the years, how they gradually become aware of their oddities from the time of sexual maturity, how they

> attempt to alleviate their compulsions on their own, how they increasingly feel that they are nature's "cruel joke"—a female soul in a body with detested male attributes—and how their existence is dominated by frequent reactive depression and dysphoria, this side of their stories can easily be skipped over.[16]

Bremer next followed three steps in curating the cases. First, he selected suitable cases from the literature. Second, he extracted key information from each case story before he briefly summarized it. Finally, in the conclusion, he condensed the vignettes. The cases were organized chronologically with a dedicated number and reproduced in various lengths; the most detailed was two pages long, whereas the shortest consisted of three short sentences, for example, "No. 21. Case W, born 1905. Observation time 1¼ years. At least during this time everything seems to have been going well." Desires, doubts, delights, and disappointments were compressed into short text blocks that Bremer later used to answer questions about treatment effects.[17] By omitting, however, the lived experiences that could have provided a deeper and more complex understanding of trans life, including the challenges of growing up in a society lacking words for one's reality, a uniform concept of genuine transvestism emerged. Genuine transvestism became a homogenous disease entity, with each case confirming the generalizability of the nosological concept.

How did Bremer distill nosology in practice? Take case no. 9, which Bremer took from Fritz Bättig's dissertation *Beitrag zur Frage des Transvestitismus* from 1952.[18] Case no. 9 was born in 1900, worked in a fabric production plant, had two daughters, and lived in an unhappy marriage. In her mid-forties, she was surgically castrated, so that "the femaleness he longed for would fully evolve."[19] The surgery improved her condition, the physician noted, "masculine sides disappeared more and more," and she developed pain in the lower abdomen, "just like a woman." She now had a divorce, applied for the official change of name, and started to work in a washhouse. Nevertheless, the application for a female name was rejected, her deep voice bothered her, and people kept commenting on her in public. She also felt that she received little support from her doctors, and slowly, she started to doubt the decision. In the end, she decided to "go the way back, giving up [. . .] illusions," wrote Bremer, quoting Bättig. Thereafter, she "once more decides to become a woman"; one month later, the patient "again wants 'to become a man.' "[20] She regretted not staying at the washhouse with her female colleagues: "That was my most beautiful and happiest time."[21]

In the report, Bremer summarized the case in the following way: "No. 9, castration—observation time ca. 8 years: Reactive depressions, delusions of reference, doubts that he fits as a woman."[22] But was this really an example of the failure of a surgical procedure? In the original text, Bättig reflected that the patient's ambivalence must be related to the official refusal to recognize her female name and the difficulties in social adjustment.[23] This perspective opened the way for an understanding of the societal stigma facing trans women and an individual's search for belonging. In his conclusions, however, Bremer did not include the ambivalences, longings, and desires, the societal challenges, and the lack of medical support. To the contrary, No. 9's story was reduced to a figure for the purposes of demonstrating the futility of surgery.

Another case was originally published by the Berlin surgeon Richard Mühsam in 1926. To increase the spirit for work in a patient diagnosed with "transvestic urge" and "severe sexual neurosis," he had performed castration.[24] Following the castration, Mühsam transplanted an ovary into the abdomen, covered the penis in the perineum, and shaped the scrotum into "vagina-like structure." Two months later, the person developed "sexual feelings as a man," wrote Mühsam, and the "transvestic tendency" faded. He "gave up" living as woman, started a relationship with a woman, and in a subsequent operation, Mühsam freed the penis from the perineum.[25] Even though Mühsam defined the surgery as mutilation, he portrayed it as a success showing the benefit of surgery in the treatment of sexual neuroses. The patient perceived the penis as a "foreign body," he noted, and castration and the hiding of the penis had eased the transvestic tendency. For Bremer, on the other hand, the case demonstrated the diagnostic uncertainties and the risk of inflicting harm. What would have happened if the penis had been removed? He summarized: "Case no. 1 with reversible plastic surgical operation—observation time ca. 6 years: The transvestic drive disappeared, he continued to live as man—potent sex life after surgical reversal."[26]

Clinical vignettes compressed time: Long life stories were broken down into comprehensible and well-arranged synopses that could be used for various purposes. As Bremer collected, extracted, and condensed cases into text modules—a "work of purification," to borrow a term from Bruno Latour—he worked with paper technologies similar to those used by physicians who shaped modern nosology.[27] For the eighteenth-century physician François Boissier de Sauvages—a pioneer of modern nosology—classification was more a process than a finished product. He collected *observatio* from the literature, extracted and highlighted some findings, contributing to the creation of a nosological species.[28] Collecting cases, encircling and extracting clinical observations, condensing, summarizing, and reorganizing text on paper has been among the oldest ways of knowing in medicine, a way of knowing that cannot be classified as statistical or laboratory-based.[29]

Although there are many differences between Sauvages's and Bremer's paper technologies, they both show how important work with pen and paper has been for medical classification. By copying, extracting, and enumerating, Bremer was able to compare, contrast, and group cases into new orders. The cases became empirical facts: Cases no. 3, 8, 15, 16, 17, 18, 19, and 20 lacked sufficient observation time to draw conclusions on treatment effect, he stated. Although the immediate therapeutic result was satisfactory, the observation time was only six to fifteen months. The case numbering also made it possible to single out defining characteristics of genuine transvestism: "Five cases demonstrate an attention-seeking publicity after demasculinizing surgery (no. 4, 5, 12, 13, 14)."[30] Allegedly, cases no. 8, 9, 10, and 11 demonstrated that the request for surgery did not come from the patient but from the physician, the press, or from others who had undergone the surgery. Finally, seven cases "blatantly demonstrate how transvestites untiringly persist in their demands to be 'transformed' (nos. 1, 2, 9, 10, 11, 13, 14)."[31] The compilation of the text modules enabled Bremer to draw his conclusions, which he phrased in some particularly hostile and derogatory language: The genuine transvestite was "throughout emotionally unbalanced" and "infused by a demonstrative attention-seeking tendency."[32] Pathology and behavior were collapsed into one another: Behavior became a referent for a pathology, with the request for surgery as a defining nosological criterion.

One could object that an expert report is not an epistemological technology. As Alexa Geisthövel and Volker Hess have argued, expert reports do not generate new knowledge, but are tools for "making good politics"; they respond to concrete issues of practical societal importance.[33] In German, expert reports are referred to as *Gutachten*. The verb *achten* means to respect, regard, or have in mind, and therefore, *gutachten* is to do this in a good way. Moreover, the German word for the *Gutachten* experts is *Sachkundig* (or the Norwegian equivalent, *sakkyndig*). To be *sachkundig* means to be knowledgeable about an issue or subject; in German *Sache*, highlighting the practical/*dingliche* competence of the expert, a meaning that is lost in the English equivalent *expert* or *expert reports*.[34] Although historically the task of a *Gutachter* has not been to generate new knowledge but to provide a well-argued recommendation on a specific case, Bremer's dissent shows that the boundaries between knowledge production and the provision of expert opinions can be porous. Perhaps particularly for controversial questions in which knowledge is sparse, the line between expert recommendation and knowledge production can be indistinct. In this case, in which the expert report was produced for the state to solve a delicate issue, it became an epistemic instrument of summarizing, synthetizing, and even generating knowledge.

To be clear: The report was never published, and Bremer never became an international expert on trans medicine. Nevertheless, the report provided the scientific fundament for the state's handling of legal and medical sex change in Norway in the years to come. Bremer's rejection of surgical treatment had a profound impact on the physicians in the Directorate of Health, and the report became a reference document for the definition of genuine transvestism in the health bureaucracy. Over and over, state physicians and lawyers used the report as a reference tool for the political handling of sex reassignment on a large scale but also in individual decisions: As a member of the Expert Council on Cases of Sex Operations, which handled applications for sex operations, Bremer copied and reproduced sections from the report in the standard rejection letter for requests for genital surgery.[35] This tells us something important about the evolution of nosology between the clinic and the state bureaucracy in the mid-twentieth century: Nosological categories sometimes emerge as pragmatic responses to specific bureaucratic problems, which in turn become determinative for clinical practice.

Collecting, Condensing, Counting, Crystallizing, Tabularizing

The nosological definition was based on a paradox. The request for surgery became a crucial nosological criterion, which in turn became the main reason for rejecting the request: Patients would keep returning asking for more treatment. Both for physicians who advocated for surgery for their patients and for those who opposed it, the request for surgery became a defining characteristic of the diagnosis of transvestism. They simply mobilized the request differently: For opponents, transvestism was often progressive, and the desire for further treatment, as such, would never be satisfied. For example, in a comprehensive phenomenological treatise on "male transvestism" published in 1953, the German psychiatrists Hans Bürger-Prinz and Heinrich Albrecht and the sexologist Hans Giese reasoned that "progress is always evident, only the matter of degree varies." In their opinion, there was no clear nosological category of "real" or "genuine" transvestism: "Once a transvestitic development has been started and is in full operation, it is impossible to predict at which stage the transvestite will be satisfied."[36] Transvestism was a process, not a circumscribed pathological entity; therefore, surgeons must not accede to their patients' requests.

For advocates of surgery, by contrast, surgery was defensible precisely because genuine transvestism was distinguishable from a much broader category of transvestism. "While the male transvestite, *enacts* the role of a woman, the transsexualist wants to *be* one and to *function* as one, wishing to assume as many of her characteristics as possible, physical, mental and sexual," wrote the American endocrinologist Harry Benjamin in 1954.[37] According to Beans Velocci, the category of transsexuality emerged in the United States in the physicians' selection of patients for surgery: The desire for surgery both defined the transsexual but also defined a bad outcome. The diagnosis of transsexuality was primarily a tool for doctors to reduce the risk of regret and therefore of legal or personal revenge.[38] Scandinavian physicians corresponded with sexologists across the Atlantic, such as Harry Benjamin and Alfred Kinsey; they read each other's papers and referenced each other's work. For example, Scandinavian psychiatrists discussed nosological criteria at the eleventh Nordic Psychiatry Conference in Oslo in 1955, at which Per Anchersen presented his cases. Anchersen also used the distinction between transvestism as symptom and constitution. In 1955, Karl Evang was contacted by a man who described "tendencies of a transvestitic direction."[39] Evang referred the patient to Anchersen, who agreed to see him in his office a few days later. The patient's sexual life was "almost like a fetish," Anchersen stated, adding that there was no reason to suspect "real transvestism."[40]

Nevertheless, the number of published cases were few and physicians needed other sources for conducting research to support treatment decisions. As more people looking for surgery contacted doctors, this paved the way for new research. Those requesting treatment usually included a letter outlining their biography and emotional life, and these letters became a treasure trove for researchers within trans medicine. As the news of Christine Jorgensen broke, people from all over the world contacted the Danish team. Of the more than 1,100 letters that the endocrinologist

Christian Hamburger received, 465 were from people requesting treatment. The letters came from all over the world and were written in Danish, Dutch, English, French, German, Italian, Norwegian, Portuguese, Spanish, and Swedish. Reasoning that it would yield "highly significant information of the psychological background governing the 'desire for change of sex,'" Hamburger extracted information from the letters for research purposes and published the findings in 1953.[41] This was another way of creating a diagnostic tool on paper.

Hamburger began by categorizing and tabularizing the letters according to geography and the person's sex, age, and marital state. He then organized the findings into contingency tables, with rows for "marital status" and "sexual libido," and columns for "men," "women," "transvestic men," and "other men," depending on whether the letter referred to cross-dressing or not. All people categorized as "women" were defined as having "homosexual desires," meaning they mentioned sexual attraction to women, whereas the groups categorized as "men" and "transvestic men" were assigned into heterosexual, homosexual, bisexual, autosexual, and asexual desires. Next, Hamburger divided the letter writers into two groups—"transvestites" and "non-transvestites"—based on whether cross-dressing was a motive for surgery or not. In more than half of people classified as "transvestites," he reasoned that a diagnosis of "genuine transvestism" was justified. But even in the group of "transvestites," fewer than half had a history like Christine Jorgensen, who was the prototype for the Danish doctors' model of genuine transvestism. Among individuals who mentioned sexual attraction to men, Hamburger saw this motive "to be of major importance." In the non-homosexual group, by contrast, a common feature was that sexual life was never directly mentioned. In other words, even if the goal was to distinguish transvestism from genuine transvestism, the contingency table did not add up; the map did not fit the terrain.

As "spontaneous reactions," Hamburger reasoned, the letters provided a unique insight into the phenomenon of transvestism. They were not "filled-in questionnaires," he wrote, but evidentiary raw materials, noncurated *facts*. To illustrate his nosological arguments, Hamburger included parts of the letters in his article. "Sometimes I have masculine desires," wrote one teenager. One fifteen-year-old described how "A mi me gustan las chiquillas, me enamoro de ellas igual que un hombre" [I like little girls, I fall in love with them just like a man]. "I have found a wonderful companion who looks on me as the man I should be," wrote a third. A twenty-six-year-old person from Germany wrote of his girlfriend: "Sie ist der einzige Mensch, der mir Kraft gibt, und für den es sich lohnt, zu kämpfen, dass ich ein Mann werde; aber auch ohne dem Mädel würde ich es tun" [She is the only person who gives me strength and who it is worth fighting for to become a man; but even without the girl I would pursue it]. A twenty-year-old wrote that his "thoughts, interests and affections are those of a male."

Based on the techniques of collecting, condensing, counting, crystallizing, and tabularizing letters, Hamburger came to the conclusion that, both in men and women, the "desire for change of sex" occurred before puberty, indicating a constitutional condition. However, the distinction between cross-dressing as behavior and genuine transvestism as constitution was also a sign that the scientific understandings of sexuality and sex were changing. The mid-twentieth-century distinction between behavior and identity, reflected in the writings of Hamburger, Anchersen, and Benjamin, was a departure from the inseparability of sex and sexuality in late nineteenth-century psychiatry, such as Carl Westphal's concept of "contrary sexual feeling."[42]

The Press and the Making of Self

Bremer and Hamburger did not write for the general public. Bremer's report was an expert opinion written for the state, while Hamburger addressed fellow endocrinologists by publishing in *Acta Endocrinologica*, the journal of the main European Endocrinological Society, the Committee of the Acta Endocrinologica Countries.[43] Yet the boundaries between the medical and general public were not entirely impermeable. Information circulated between medical communities and subcultures, between experts and laypeople. In the 1950s, there were no organizations for trans people or magazines focused on trans subcultures. While the first social club and advocacy organization for trans people in Scandinavia was founded in the late 1960s (see chapter 9), Jens Rydström has argued that Swedish porn magazines in the early 1960s represented a rare free space for the negotiation of sex, sexuality, and gender in which trans people could participate.[44] Signe Bremer describes these magazines as serving as a "trans-enabling space" where individuals could share their desires, experiences and dreams at a time when there were no meeting places and outlets for trans communities.[45] There was a much more mainstream magazine in Norway, however—*Norsk Dameblad*—that had been serving a similar purpose since the 1950s.

"The wise words of the doctor in *Norsk Dameblad* have taught me to accept myself as I am," wrote a reader of the popular Norwegian women's magazine in a 1955 issue. "I no longer feel like a criminal who breaks the laws of nature, like an inferior human being."[46] In 1953, in the wake of news of celebrity patient Christine Jorgensen, the magazine initiated a series on "men in women's clothes." The series, in which people were encouraged to write and share their stories, was a great success and ran for seven years. "Did you realize that there are 'men' walking around in Oslo in women's clothes because they feel like women?" the magazine wrote in the column titled "Artemis," named after the Greek goddess of wild animals, hunting, and childbirth in the inaugural article. "At least these 'men' *know* that they are not like other men."[47] In her master's thesis, Sigrid Sandal raised the question of whether the motivation for publishing these stories was a desire to help these individuals, or to sell more magazines.[48] The subtle and empathetic coverage is nevertheless remarkable, and a sign that the editor probably viewed this topic as an important part of women's issues in general.

Norsk Dameblad was founded in 1938 and quickly became one of the most popular women's magazines in a society undergoing major demographic changes: More people were getting married, and women were engaged in unpaid domestic work, leading some historians to refer to this as a period of "housewifeization."[49] While most women's magazines at that time had a conformist profile, avoiding controversial issues and writing about female issues in indirect ways—referring, for example, to "the monthly" instead of "menstruation"—the more left/liberal-leaning female editor-in-chief of *Norsk Dameblad* did not avoid addressing pressing issues such as abortion, prostitution, alcoholism among women, or even transvestism.[50]

In its "Can We Help" column, the magazine printed several letters from trans people asking physicians for advice. For example, the director of an institute for physical culture wrote that she had received a phone call from a person with an "utterly nervous woman's voice" asking if it was possible to attend gymnastics classes for women. "You see, I am actually a man, but I am very feminine," the director recalled the person saying. Her doctor had given her the advice to dress in women's clothes and to do women's gymnastics to "activate the female spirit."[51] A gynecologist at the Rikshospitalet, to whom the magazine turned for advice, said that hormones and surgery could "activate" the feminine or masculine part of people diagnosed as transvestites. He explained that medicine was making progress in this area, that he had recently attended a lecture by Christian Hamburger in Oslo, and that he had himself traveled to Copenhagen to learn from leading experts there. Other readers found the advice of "activating the feminine" troubling. A forty-year-old person wrote of having felt, from childhood, "an irresistible desire to dress in female attire"; reading Bøe's response had led to self-questioning, whether there was something "unconsciously feminine" in "body and mind" at the root of this desire. The thought was distressing: What would happen if this person "activated" the feminine side? But the idea was liberating too: "I don't feel like myself and feel completely depressed and without energy to work when I have to dress as a man." Imagining the day when all male clothes are discarded, "only to dress as a woman from the outside in [. . .], I will be in distress, if I will have to remain a man."[52]

Born in 1898, another of its readers wrote that as a child she veiled herself in the curtains, decorated her hair with flowers, and made her cheeks rosy with moistened crepe paper. When she turned eighteen, she met another girl who was like her. They put on women's clothes, paraded the streets of Oslo, flirted with boys, and walked "arm in arm to the Palace Park like two lovers." At the time, she felt herself to be in the right environment "to be loved," she wrote. When she found out about Christine Jorgensen, she had looked back at her life with pain and sorrow: "Imagine being enchained for twenty-six years and then to be freed!"[53] In other words, medical categories produced their own feedback, having what Ian Hacking defined as "looping effects" on people, opening new paths, enabling new futures.[54] In the United States in the 1930s and 1940s, the press continued to issue news stories about medical transition, which in turn became an important source of information for people seeking this type of treatment.[55] The most significant of these stories in the Scandinavian context was that of Lili Elbe's transition, which was much discussed in Norwegian and Danish newspapers.[56] The mediatization of medicine is perhaps more the rule than the exception. As historian of medicine Jeremy Greene has emphasized, to understand the role of science and medicine in modern societies is to understand how they are mediated "all the way through," and that media offers a "capacious set of tools to reexamine matters of fact as matters of history."[57]

But the processes of sexual differentiation and subjectivization did not run only in one direction, from medical expertise to the public and subcultures; to the contrary, concepts have circulated between the clinic and communities, meaning that medicine and subcultures have shaped each other.[58] Mid-twentieth-century medical concepts have thus emerged in relational, enmeshed, and interactive ways.[59] According to Annette F. Timm, this implies that historians need to move beyond what physicians published about the issue and to ask what medical experts learned from their patients. In fact, intimate relations between individuals, communities, and medical researchers run like a thread through the history of sexology, from Magnus Hirschfeld in Germany in the early twentieth century to Harry Benjamin and Alfred Kinsey in the interwar and mid-twentieth-century United States.[60]

Norsk Dameblad offered trans women a rare chance to shape public discourse about their lives and to assert agency. In Julian Honkasalo's words, the sharing of intimate stories in letter clubs or even in public can be read as an early form of community building, care practice, and political resilience.[61] An article titled "Not a Woman and Not a Man" reported on a thirty-year-old trans woman in Sweden who described herself as a "woman trapped in a male shell." To "enhance the female element," her doctor had started hormone injections and issued a medical certificate about her condition so that she could appear as a woman in public, but the surgeon declined to perform genital surgery. The article was illustrated with two photos. In one, the subject appears wearing an oversized suit with hands clasped behind the back and appearing stiff and uncomfortable. In a more recent photo, she is dressed in a dark gown and high-heeled shoes, posing in a reclined position on the couch. On the wall above is a painting of a leopard with a menacing look. The composition of the two photos next to each other conveyed the impression of a self-confident woman asserting herself and her desires. Giving these photos to the magazine was a way of staging her story in the public eye: This was how she wanted to be seen, and the article was probably intended as a way of contacting a surgeon who could perform the desired operation.[62]

The article series became an important source of information about trans life for physicians. Per Anchersen had already examined a few trans people at Ullevål Hospital and in private practice, in addition to the people he had met in his role as a forensic expert witness. The magazine series became another crucial source for his work; he was given access to all the letters submitted by trans people to the magazine at the very time he was preparing the majority report for the director general of health. The letters provided a unique insight into the lives of people for whom he provided expert medical and political advice. He even referred to the letters in the finished report. One letter came from a woman worrying that her husband was cheating on her. As a test, she had succeeded in convincing

IKKE KVINNE OG IKKE MANN

I Norsk Dameblads forrige nr., i spalten «Kan vi hjelpe?», brakte vi et interessant brev fra en Oslo-leser, som fortalte om hvordan hennes mann var blitt helbredet for alkoholmisbruk ved så enkle midler som å kle seg i dameundertøy.

Mange av våre lesere har sikkert dradd på smilebåndet av denne betroelsen, og nektet å godta beretningen som faktum. Imidlertid har redaksjonen henvendt seg til leger som har behandlet transvestitter, og de kan fortelle at det hender slett ikke så sjelden at en slik bagatell som undertøy kan redde folk fra fysiske og psykiske lidelser.

I dag skal vi offentliggjøre en gripende beretning om hvordan en svensk mann nylig gjennomgikk en forvandling fra mann til kvinne. Men i motsetning til Christine Jørgensen, som, ved siden av den medisinske behandlingen, også ble behandlet kirurgisk, har vår svenske venn fått bare hormonbehandlinger.

Her forteller han selv — eller skal vi heretter si hun — hvordan forvandlingen er foregått.

Hun begynner sin beretning med å si:

Ja, et menneske som en grusom skjebne har plasert i grenseområdet mellom mann og kvinne, fortoner seg kanskje som en tragikomedie. Her vil jeg avsløre meg som jeg er.

Slik så jeg ut før.

Da jeg var liten, lekte jeg med dukker. Jeg var heller sammen med piker enn med gutter. På skolen var jeg den beste i klassen i håndarbeid, mens det tok meg to år å snekre sammen en fuglekasse. Jeg var ikke som andre. Jeg var veik og trakk meg unna slagsmål og hadde veldig lett for å gråte. Den gangen skjønte ingen hva som var i veien med meg, men nå vet jeg jo hva det var. Jeg hadde for mange kvinnelige kjønnshormoner. Jeg var en kvinne, innestengt i et mannlig hylster.

Militærtjenesten var et mareritt. Jeg orker ikke å tenke på den. Siden fikk jeg arbeid. Jeg flyttet til Stockholm, og begynte i restaurantbransjen, men jeg fikk sparken. Jeg hadde en alt for feminin måte å være på, sa min arbeidsgiver. Han trodde tydeligvis at jeg var homoseksuell. Men det har jeg aldri vært. Jeg forsøkte å få andre jobber. Det gikk ikke. Jeg følte meg utstøtt, jeg syntes ikke jeg hørte hjemme noen steder. Jeg stengte meg inne og så på meg selv i speilet. Jeg forsøkte å lære meg til å gå som en mann, snakke som et mannfolk. Det gikk heller ikke. Så satte jeg meg ned og spekulerte på om jeg skulle ta livet av meg. Det var en både lokkende og skremmende tanke. Men jeg fikk også en annen tanke som jeg gjerne kretset om, en tanke som virket fantastisk gal, ja, uanstendig. Men til slutt ble det den jeg stanste ved.

Jeg solgte mine mannsklær og kjøpte meg en damedrakt med skjørt og jakke. Jeg gikk hjem og tok den på meg. Men én gang følte jeg meg roligere. Så gikk jeg ut i byen. Skrittlengden passet. Jeg slappet av og ble mer selvsikker. Den kvinnelige måten å bevege seg på kom av seg selv. Det var ingen som syntes jeg så rar ut. Ingen snudde seg etter meg, unntagen en kar som kom løpende etter meg og sa:

«Unnskyld at jeg antaster Dem, men jeg liker så godt lange kvinnfolk.» Han var beruset.

Jeg visste at dette ikke var noen løsning på lengre sikt. Jeg visste at etter den første følelsen av lettelse, ville fortvilelsen komme tilbake ennå sterkere. Slik ble det også. Jeg visste ikke hva jeg skulle ta meg til. Men noe visste jeg iallfall, jeg kunne ikke gå tilbake til å bruke mannsklær. Det endte med at jeg gikk til en lege og ba om hjelp.

Han konstaterte fort hva som var i veien med meg. Meget sterkt kvinnelig innslag, hormonbalansen alvorlig forstyrret. Han ga meg innsprøytninger som skulle forsterke det kvinnelige elementet i meg. Jeg merket også at det gikk slik. Kroppen min ble stadig mer kvinnelig, og min psykiske balanse ble bedret. Jeg fikk legeerklæring for at jeg hadde lov til å bære kvinneklær «av mentalhygieniske årsaker». Jeg fikk vitenskapens velsignelse når jeg nå besluttet meg til å ta skrittet fullt ut, og leve som en kvinne, den eneste måten for meg til overhode å kunne leve.

Jeg fikk 37 sprøyter, og jeg fikk etter hvert et stort håp. Og legene som behandlet meg, understøttet dette. Jeg håpet at en dyktig kirurg ville gjøre det endelige inngrepet, det som definitivt skulle forvandle meg til kvinne, det som skulle vippe meg over fra det tåkete ingenmannsland til et kjønn som ble godkjent av alle. Den tilstand jeg var i nå, kunne jeg ikke fortsette å leve i. Jeg tenkte:

De kan vel ikke la meg i stikken nå, etter at de har gitt meg så mange sprøyter som har ført meg ennå nærmere kvinnekjønnet. De kan da vel ikke la meg bli igjen her sånn aldeles på terskelen?

Jo, det kunne de. Behandlingen sluttet med et brev fra den legen som jeg hadde betrodd meg så helt til, og som jeg hadde knyttet mine forhåpninger til. Nei, ikke fra ham selv, forresten, men fra hans sekretær.

«Dr. N. N. har bedt meg meddele at han dessverre ikke kan påta seg Deres tilfelle.»

Kaldt og følelsesløst. Han overga meg altså slik som jeg var. Et halvt menneske. En ufullstendig 30-åring. Et individ som ikke hører hjemme noe sted. Jeg er det ensomste som finns. Jeg kan ikke forstå at noen kan være mer fortvilet enn jeg er.

Hva jeg nå vil gjøre, er å be om nåde. Jeg vil be om at dere forsøker å forstå meg. At dere ikke vil vemmes. At dere ikke vil sperre opp øynene. At dere ikke vil le av meg. Jeg ber om at dere skal prøve å forstå at mitt tilfelle fra medisinsk synspunkt er like så naturlig og like så lite å skamme seg over som et benbrudd. Det er bare det, at følgene for meg er så meget mer grusomme. Her går jeg som den mest ensomme av alle, men jeg er skapt med samme krav til ømhet og samme krav til vennskap som andre. Det er bare det at jeg ikke har fått noen nøkkel som kan låse opp for meg til andre menneskers sinn.

Jeg har sagt at jeg skal komme til dere, nøyaktig som jeg er, og derfor skal jeg heller ikke skjule noen ting, og nå kommer jeg til det som er vanskeligst å fortelle om. En aften da jeg var mer fortvilet enn ellers, gikk jeg i ren desperasjon til en fornøyelsespark og stilte meg opp ved dansegulvet. Det varte ikke lenge før det kom en kar og bukket for meg. Jeg nølte et øyeblikk, jeg ble redd for meg selv og for ham og for hele situasjonen. Men så fulgte jeg med ham. Vi danset, han pratet og lo og merket ingen ting. Jeg ble grepet av stemningen, vi danset en gang til. Så gikk vi sammen mot byen.

Men så løp jeg fra ham. Han ropte navnet mitt. Det navnet som legen hadde hjulpet meg med å få, et pikenavn. Men jeg bare løp. Hvordan skulle jeg kunne fortelle ham hvordan det var fatt med meg?

Nå venter jeg ikke noe mer av livet, unntagen en smule forståelse og fred.

Det er derfor jeg ber om nåde.

I vårt neste nummer skal vi gjengi en artikkel som er sendt oss av en ung mann fra Oslo, som befinner seg i det samme tragiske grenseområdet.

Og slik er jeg blitt. I det ytre en kvinne, men blir jeg akseptert som det?

24

FIGURE 3.1. "Ikke kvinne og ikke mann," *Norsk Dameblad*, no. 14 (1955).

her husband to put on women's underwear before going to work. This way, getting undressed with other women would be impossible, she believed. But the idea had unforeseen, "fantastic consequences": Her husband, who also had a drinking problem, stopped going to the bar after work and felt more at peace.[63] The letter was a source of inspiration for other readers who also wrote to the magazine to share their experiences. "I have no doubt that this is the only reason he no longer touches alcohol," another woman wrote about her husband. "He makes no secret of the fact that this is the cause of the change."[64] Since the letters were not written by the persons themselves, we cannot know for sure what their motivations were or what they were experiencing. But for Anchersen, the letters enabled the distinction to be made between transvestism and genuine transvestism, the separation of symptom and constitution. Of the sixteen life stories, he noted, thirteen were "unmistakable fetishists" and two were "certain genuine transvestites."[65] But where did behavior end and constitution begin?

Scandinavian Epidemiology

By the end of the 1950s, the diagnosis of genuine transvestism had become well established in Scandinavian psychiatry. "Genuine transvestism," the psychiatrist Ib Ostenfeld concluded in the main Danish medical journal in 1959, was characterized by "well-defined symptoms and a quite specific abnormal personality pattern." He concluded that reports and observations from different countries "all closely match" and there was "no doubt" that the researchers referred to "the exact same condition."[66] On one level, Ostenfeld's observation was accurate: There now existed a commonly agreed phenomenological definition of the condition based on case stories, but the definition was not supported by larger patient data sets. This changed, however, with a new research program initiated by the Swedish psychiatrist Jan Wålinder a few years later, a study that would shape nosological understandings of transsexualism in Scandinavia and internationally. In 1963 at St. Jörgen's Hospital in Gothenburg, halfway between Oslo and Copenhagen, Wålinder started collecting data for his doctorate with the aim of providing a "multidimensional picture of transsexualism." One year after the publication of Benjamin's pathbreaking book *The Transsexual Phenomenon*, Wålinder finished his thesis. In the preface, Wålinder noted that he "chose to study transsexualism as it seems to be easier to delimit than transvestism, cross-dressing being a component of several abnormal conditions."[67] His aim was to give to transsexualism—a field he believed to be based on theory and speculation—an empirical basis that would reflect a more general change in medicine and psychiatry, and the quest for objectivity through large numbers.[68] The case—the oldest technology in medicine—was inadequate and seemed all the more archaic compared to the potential of large data sets. The lack of a solid epidemiological foundation for treatment decisions saw physicians making decisions based on "shaky theoretical reasoning."[69]

Wålinder started by examining the literature, that is, the published cases, much as Bremer had done some years previously. He read the publications by early twentieth-century sexologists such as Hirschfeld and Ellis, as well as the work of colleagues in Denmark and Norway, and sexological and psychoanalytic literature from the United States, notably the publications of John Money, Joan and John Hampson, and Robert J. Stoller. Wålinder also referred to Stoller's concept of "core gender identity" and to the "gender role" concept, coined by Money and the Hampsons (see chapter 7). Based on searching the literature, he identified 207 cases of transvestism or transsexualism—"185 males and 22 females"—but using Benjamin's definitions, according to which transvestism was defined by cross-dressing and transsexualism by a desire for an operation, Wålinder stated that it was impossible to separate the two.[70] Since his research, however, was driven precisely by an interest in creating diagnostic criteria for treatment selection and examining etiological factors, especially distinguishing between psychological and organic causes, he nevertheless applied the nosological concepts. Next, he extracted information from the published cases and plotted this information in cross-tables. One table, for example, contained information on psychological factors, including parental divorce, death of a parent, and so on. Another listed "physical abnormalities," such as scant facial hair, gynecomastia, and undescended testes. In yet another, he registered the frequency of "cerebral lesions" measured by EEG, demonstrating that around one-third of the patients had abnormal EEGs. This pointed to an organic cause, Wålinder reasoned.[71]

The material and the results were unsatisfactory. There was a lack of data and questions arose for which the literature review could provide no answers. While Wålinder also had to interview people himself, the advantages of conducting research in a small country with a comprehensive and well-organized public health care system and medical association now became apparent: The Royal Medical Board, the highest official medical body in the country, provided a list of all 474 psychiatrists employed by the state, and Wålinder sent a letter to all of them, asking about their experiences with patients classified as transvestites. As many as 361 responded, more than 75 percent, reporting about 91 cases. The research project served two functions: Wålinder now considered himself able to calculate the prevalence of transvestism (although in this he was mistaken, since he only had the data on those who had ever sought the advice of a psychiatrist), but he also established himself in Sweden as an expert on trans medicine. From that point onward, colleagues referred their patients to him and these cases were then included in his research project.

The welfare state context of state-employed psychiatrists, networks of communication between colleagues and state institutions, well-developed documentation systems in the bureaucracy, and the position of doctors in public life created ideal conditions for comprehensive epidemiological research projects. Wålinder managed to include forty-three patients in the study, and the participants stayed in his clinic for a week to undergo extensive medical testing, somatic and neurological examinations, body measurements such as the length of the radius and tibia, personality testing, projective tests and testing of masculinity and femininity, intelligence tests, EEGs, hormone analysis, and karyotyping of sex chromosomes. On average, the participants were interviewed two to three times, each interview lasting around one and a half hours. Wålinder also included interviews with relatives, and he collected information from parents, siblings, and other people close to the patient. In addition, he contacted other hospitals to get birth certificates and medical records from previous hospitalizations. To supplement the database, he collected data from welfare agencies, such as child welfare bureaus, social welfare bureaus, temperance boards, state insurance offices, child psychiatry institutions, and mental hospitals, including two national psychiatric departments where psychiatrists could seek advice.

Wålinder now had an enormous amount of clinical data in various formats. He began by sorting the materials, comparing findings, and weighting them before plotting data into tables and diagrams. Some diagrams were included to show the negative findings: For example, some diagrams examined correlations between body dimensions, such as the length of the tibia to radius, bi-iliac width to radial length, and bi-acromial width to tibial length, which were all within the normal range. Sex chromatin karyotyping also revealed nothing unusual. Other diagrams contained crucial clinical information. One table, for example, was supposed to demonstrate that "cross-gender behavior" began after the age of ten in more than 20 percent of cases.[72] Clinical interviews and medical records showed that 70 percent of cases experienced depressive reactions and as many as half of the sample had suicidal thoughts. Based on personality testing, Wålinder noted a high frequency of "psychoinfantilism." He observed that almost all the research subjects were "sexually aroused mentally by persons of their own sex," but noted that interviewees and their partners did not regard themselves as homosexual—"the partners regarded them as women." Wålinder noted that the concepts of homosexuality and heterosexuality acquire "another meaning in connection with transsexualism."[73] Sex, "in the usual sense of the term," generally did not play an important role in transsexualism.[74]

Wålinder did not give much thought to the confirmation bias that he built into the research design: Since only people who met the diagnostic criteria for transsexualism were included in the project—those, in other words, who requested surgery and expressed "disgust and torment" regarding their external sex characteristics—the study reproduced its own nosological presuppositions. Based on the literature and his own clinical material, Wålinder concluded that transsexualism was a "separate disease entity" with three essential characteristics. The defining element was a "conviction of belonging to the other sex," accompanied by "abhorrence of the sex attributes given by nature," leading to an "overwhelming longing for 'change in sex.'"[75] Admittedly, this definition did not deviate much from Max Marcuse's concept of *Geschlechtsumwandlungstrieb* articulated four decades earlier, or from Hans Binder's description of *Verlangen nach Geschlechtsumwandlung* ("demand for sex change").[76] However, while Marcuse's and Binder's definitions were based on phenomenological descriptions of individual cases, the nosological categories of genuine transvestism and transsexualism/transsexuality of the 1950s and 1960s were quantified concepts based on the extraction of data from large sets of published cases, autobiographies, letters, or comprehensive epidemiological studies that lent validity to the diagnosis, which could be used either for clinical or bureaucratic purposes.

Confronted in the 1950s and 1960s with an increase in the number of people requesting medical treatment to transition, physicians and state officials were looking for clear diagnostic criteria to select people for treatment and to hammer out official strategies. Clinicians and researchers collected clinical information wherever they could—in the library, patients' letters, magazines—and they initiated large epidemiological research projects. These efforts cemented a distinction between transvestism as fetishistic symptom on the one hand and genuine transvestism or transsexualism/transsexuality as a nosological entity with a constitutional basis on the other. More than anything, the nosological categorizations were attempts by physicians to find pragmatic solutions to bureaucratic problems, faced with the limits of their knowledge and experience.

CHAPTER FOUR

Hormone Architecture and Guinea Pigs

It was already late when Trygve Braatøy looked out over the crowd that had gathered in the monumental hall of the Norwegian Medical Society, a space of functionalist design. It was a cold winter's day in Oslo, early in March 1942, but many from the academic medical elite had come that evening to listen to the renowned psychiatrist and psychoanalyst at the society, just a few hundred meters from the Royal Palace. There must have been an expectant mood in the hall; Braatøy was giving a lecture on one of the most topical issues in psychiatry at the time: "In recent years," he stated, "the number of publications on sex hormones and their clinical experiments has grown like geometrical rows."[1] Synthetic sex hormones promised completely new methods of treating mental illness. Small intramuscular doses of testosterone had strong effects on the body, and stilbestrol, a synthetic nonsteroidal estrogen, could be used to fine-tune vitality in women. In some patients, it "could be experimentally turned up and down," he said. "I would say, almost at will."[2]

By the 1930s, physicians were starting to understand more of the complex architecture of the endocrine system, including the feedback loops of the pituitary gland and the adrenal cortex. Research had established that sex hormones did not exert direct antagonistic effects on sex hormone production in the gonads, but rather through feedback mechanisms in the pituitary-gonadal axis, a finding that contradicted the dogma that sex hormones were specific to women and men.[3] The chemical isolation of estrogens and androgens in urine in the late 1920s and early 1930s also marked the beginning of a new era in hormone research.[4] Writing in the early 1930s, the historian Fredrik Grøn defined endocrinology as among the most promising fields in medicine; progress was "enormous," and enduring gains had already been made.[5] When synthetic hormones came on the market in the 1930s and 1940s, physicians were able to modify bodies and regulate emotional life in entirely new ways.[6] Although clinicians had been treating patients with hormone extracts long before sex hormones were synthetized—extracts from the ovaries of young cows from Oslo slaughterhouses had been on the Norwegian market since the 1920s—synthetic hormones were cheaper and easier to manufacture.[7] Even psychoanalysts were excited. The therapeutic potential seemed limitless: Synthetic hormones offered a treatment for neuroticism, frigidity, and reactive melancholic depression. Because the drugs could be prescribed in private practices and in outpatient clinics, they offered a long-awaited alternative to invasive, expensive psychiatric inpatient treatments such as shock therapy. In a way, synthetic hormones promised a psychopharmacological revolution even before the synthetization of chlorpromazine ten years later.

By the mid-twentieth century, endocrinology had become the theater for control of the mind and body. Physicians and psychiatrists increasingly saw sex hormones as "universal drugs" for all sorts of health problems, including infantilism, infertility, anovulation, menstrual disorders, menorrhagia, frigidity, impotence, hypersexuality, hypogonadism, prostate hypertrophy—and transvestism.[8] This opened a new era of "wild experimentation" in medicine, an important backdrop to the development of trans medicine in the postwar period. The tinkering with hormone technologies involving patients and medical experts, hormone assays, and pharmaceuticals highlights a central feature of trans care in the welfare state: its *pragmatic* nature.[9] Although the nosological entity of genuine transvestism/transsexuality was established, it did not imply a single model for diagnosis and treatment. There was no preordained streamlined routine for physicians to follow; to the contrary, practices varied and shifted over time. Physicians tried different types of hormones and experimented with dosages, while patients noted the effects, kept diaries, and reported back to their care providers.

Designing Sex Hormones, Constructing Sex Boundaries

The professionalization and specialization of endocrinology took place within the hormonal infrastructure, networks, and architecture of the welfare state: public laboratories and clinics, state-controlled pharmaceutical advertising, instruments for hormone analysis, and strong state regulation of the drug market. The introduction and formalization of trans endocrinology were shaped by these general processes. Although endocrinology was born in the laboratory, more and more physicians joined the scientists and biochemists working in the field after the end of the war, and endocrinology gradually became firmly anchored in the clinic.[10] "Good cooperation between the biochemist and the physician is a prerequisite for the clinical study of hormones, and the laboratory plays crucial roles in diagnostic and therapeutic considerations," wrote Karl Fredrik Støa in 1960; Støa was director of Norway's first hormone laboratory at the Haukeland University Hospital in Bergen, which opened its doors in 1954.[11] Other hormone laboratories soon followed, such as Hormonlaboratoriet at the Aker Hospital in Oslo in 1959. Hormonlaboratoriet's director Nils Norman—unlike Støa—was a physician, having gained his PhD from McGill University, a leading hormone research unit at the time. The laboratory worked closely with Jørgen H. Vogt, the director of one of the hospital's two medical wards.

Hormonlaboratoriet began as "a small room with a few old laboratory benches" but quickly expanded, and by the mid-1960s it had well-ventilated areas and dedicated rooms for distillation and storage, refrigeration and freezing, and constant temperature rooms.[12] The staff expanded significantly during the 1960s, starting with a senior physician and a physiochemist, but soon including two extra physicians, nine technicians, three cleaners, and a secretary. From the time it opened its doors, the laboratory conducted hormone analysis in urine, including aldosterone, fractionated 17-keto-steroids.[13] Of the 7,472 hormonal analyses performed in 1965, 4,370 were for corticosteroids, 2,176 for sex hormones, and 926 for protein hormones.[14] The number of estradiol assays more than doubled between 1963 and 1968, and by 1971, clinicians could request analysis of estradiol, estrone, and aldosterone in plasma by radioimmunological techniques.

In Denmark too, hormone medicine went through major developments in the postwar period. The Danish Endocrine Society elected Christian Hamburger, Christine Jorgensen's endocrinologist, as its first president in 1947. He also directed the hormone department at the Statens Serum Institut, which, together with the Hormonlaboratoriet in Oslo, became leading institutions on the study of trans medicine and hormone transition in Scandinavia. Norwegian and Danish physicians corresponded about their trans and homosexual patients, and the cross-border circulation of tissues and biological samples connected endocrinology laboratories. Before the Hormonlaboratoriet opened, Norwegian physicians sent samples to the Statens Serum Institut for analysis, often to analyze urine for corticoids.

In 1954, for example, a psychiatrist at the Ullevål Hospital in Oslo asked Hamburger to conduct an analysis of corticoids for one of his patients: "The patient is an identical twin, and the twin brother, like our patient, is homosexual, with a marked inversion of sexual feeling, the brother even claims to have monthly bleeding."[15]

Trans endocrinology emerged through Scandinavian connections, but also in networks including researchers around the world: As the editor of *Acta Endocrinologica*, Hamburger introduced experiments on hormonal transition to the scientific public, and in 1960, he gathered together endocrinologists from all over the world in Copenhagen for the first international endocrinology congress.[16] Shared professional interests became the basis for friendships. The American psychiatrist and sexologist Richard Green (cofounder of the Harry Benjamin International Gender Dysphoria Association in 1979) visited Hamburger in Copenhagen, and Hamburger wrote a chapter on the endocrine treatment of transsexuality in the reference work on sex reassignment that Green edited with Johns Hopkins University psychologist John Money.[17] Hamburger maintained a close relationship with Harry Benjamin throughout his life, and they corresponded about professional issues as well as their personal lives. In 1969, for example, Benjamin wrote to his "dear Christian" to explain why he had not been able to embark on a planned trip to Europe that would have included a visit to Copenhagen. His wife, Gretchen, had suffered a severe flare-up of "her old tongue lesion," requiring an operation that had caused her to lose "1/3 of her tongue." Benjamin had to stay home to look after her.[18]

Technological developments made it possible for endocrinologists to conduct biochemical and physiological studies of the adrenal gland, and the determination of 17-ketosteroid levels in urine gave physicians new tools to diagnose endocrine diseases. However, new technology also reconfigured and standardized sex by drawing new hormonal boundaries between the sexes. Importantly, laboratory techniques for measuring and quantifying sex hormones raised hopes among clinicians and researchers that they could gain insights into the physiological basis of what they classified as sexual abnormality. For example, at the Burghölzli Psychiatric Clinic in Zürich, Manfred Bleuler and Fritz Bättig used analysis of ketosteroids in the examination of patients diagnosed with transvestism and to determine the "true" sex of patients with intersex conditions.[19]

Medical researchers were particularly interested in the hormonal differences between what they categorized as acquired homosexuality and congenital homosexuality.[20] Attempts to treat people classified as sexually abnormal with hormone extracts and sex gland transplants date back to the 1910s and 1920s. However, when researchers succeeded in isolating and synthetizing sex hormones and realized that none of the sex hormones were sex-specific, this sparked a new interest in experimenting with hormonal therapy in individuals whose bodies and identities did not fit with a heterosexual and binary model of sex.[21] Synthetic hormones made it possible to modify the sexed body, whether homosexed or transsexed, in more refined and goal-directed ways than glandular transplants had ever allowed.[22] Many physicians, including Karl Evang, supported experimentation with hormone treatment in homosexual patients long after the Kinsey Reports had provided an empirical basis for variation in human sexuality. In 1963, a twenty-year-old man wrote to Evang for help in making his homosexual feelings "disappear" so that he could become how "man is created sexually." In the past, Evang replied, physicians had faced the situation "helplessly," hoping that time would bring a change, but now researchers had made progress in determining the causes of the sexual abnormality. He recommended that the young man see a psychiatrist immediately to check whether the hormones were out of balance—"the younger you are, the greater the hope that it could be changed."[23]

"Looking at this tall, slender man, the shoulders appear somewhat narrow, his legs are remarkably long, and one notices that the areola mammae are slightly prominent and blue-red in color," Hans H. Bassøe, a pioneer in Norwegian endocrinology and later professor at Haukeland Hospital, noted of a patient in a 1966 letter to Karl Evang, continuing as follows: "Very little facial hair (shaves 1–2 times per week at most). External genitalia: well developed testes, relatively heavy, normal consistency. Penis well developed. Male pubic hair growth. Prostate appears to be of completely normal size, shape, and consistency."[24] Based on the physical examination, Bassøe concluded that the patient's homosexual tendency was likely caused by androgen deficiency or abnormal testicular or adrenal testosterone production. "It cannot be ruled out that this eunuchoid condition or possibly increased estrogen production may have affected his emotional life," reasoned Bassøe. To try to get to the bottom of the matter, he ordered a 24-hour urine sample to examine hormone metabolism.

The sample showed a "very high secretion of 17-ketosteroids and 17-ketogenic steroids." After a dexamethasone suppression test, in which a highly potent corticosteroid is administered to suppress the adrenal gland function, both levels decreased. However, gonadotropin levels, that is, pituitary hormones that regulate gonadal hormone production, remained within the normal range. "The secretion of estrogens at the upper limit" was "quite high for a man," the endocrinologist noted, and the high levels of androsterone was potentially caused by adrenal hyperplasia leading to suppression of testicular function.[25] "The fact that even biochemical abnormalities can be detected represents a fundamental step forward," Evang responded, thrilled by the idea that homosexuality could be explained on a hormonal basis.[26] The patient was "extremely willing" to get help for his "pathological emotional life," Bassøe stated and initiated treatment of 0.5 mg dexamethasone three times a day for three to six months to suppress adrenal secretion, hoping that the testes would regain physiological function. A year later, Bassøe reported back to Evang, enthusiastically describing the "successful" treatment of another patient with eunuchoidism and testicular atrophy who showed "no interest in women." After monthly injections of 250 mg of testosterone, the patient's homosexual tendencies had "disappeared," and he now appeared much "freer."[27] Thus, new technologies not only provided more refined hormonal diagnostics, but also enabled new therapeutic approaches and legitimized hormonal cures for what physicians categorized as sexual abnormality. The goal was to streamline sexual variation and produce a binary and heterosexual version of sex: this was one of the ways in which medicine enacted a sex standard.

Too Much of One Hormone, Too Little of Another

In the 1950s and 1960s, the interest of endocrinologists turned to the bodies of trans people, which became a laboratory for endocrinologic experimentation. The development of new technologies went hand in hand with new diagnostic and therapeutic models: As scientific fields and clinical practice, endocrinology and trans medicine were inseparable, and the framework of sexual abnormality as hormone imbalance became useful when examining trans patients. For example, one patient's "feminine disposition and appearance" and "light voice, sparse body hair, and virtually no facial hair" reinforced Bassøe's hypothesis of an underlying hormonal pathology describing the patient in the same terms and within the same conceptual understanding as the eunuchoid embodiment of homosexual patients.[28]

Beginning in the interwar period and continuing into the 1960s, psychiatrists and endocrinologists tried to cure transvestism by substituting the hormones they believed were lacking: They prescribed testosterone for those classified as "male transvestites" and estrogen for "female transvestites."[29] Many physicians who later advocated the prescription of "affirming" hormone treatment initially tried to cure their trans patients with hormone "substitution" treatment. One of them was Per Anchersen. In 1953, he became the director of the psychiatric clinic at Ullevål Hospital when Trygve Braatøy resigned. For one of his patients, a trans woman whom he diagnosed as a "male genuine transvestite," analysis of a urine sample demonstrated reduced levels of 17-ketosteroids. First, Anchersen experimented with testosterone treatment, but this only made the patient feel depressed and sleepless, and having to give up work. During the trial with estrogens, the patient recovered, felt "psychically harmonious and his working power increased."[30] Also in Copenhagen, physicians tried to cure male transvestism with testosterone. Villars Lunn, professor of psychiatry at the University of Copenhagen, for example, wrote in August 1955 about a twenty-one-year-old trans woman admitted to the psychiatric ward at the Rigshospitalet that treatment with testosterone, "with the aim of changing his sex identification," increased affect and mood lability. Estrogen tablets, on the other hand, made the patient feel "markedly relaxed and less unstable."[31]

At the Hormonlaboratoriet in Oslo, Vogt experimented with estrogen therapy in trans masculine patients, whom he diagnosed as "female transvestites." However, he quickly realized that treatment with estrogen only increased the "feeling of unhappiness."[32] In 1956, he initiated the first documented case of hormone treatment of a trans man in Norway.[33] This was Mattis Kvaal, a man who was admitted to his ward two years earlier after an episode of suspected hemoptysis. At that time, Christofer Lohne Knudsen in the Directorate of Health was asked to provide a psychiatric examination. Since then, Vogt had started treatment and noted that testosterone injections led to "a certain hirsutism, increase in muscle strength, mammary atrophy, and rarer and decreased uterine bleedings" and to general improvement of the patient's condition.[34]

For some psychiatrists, "affirming" hormone therapy became a way to keep requests for genital surgery at bay. "These patients are a major headache for us. They are pressing for surgery and seeking our help in increasing numbers," Lunn wrote to his colleagues in the Medico-Legal Council in May 1954. He had diagnosed nine people as so-called genuine transvestites, and his team had begun hormone treatment to "relax" them in order to support the "persuasive psychotherapy," that is, to talk the patients out of their request for surgery.[35] Lunn might have read Fritz Bättig's doctoral research published two years earlier, which argued that "the best solution in these cases is always to do as little as possible."[36] Christine Jorgensen's psychiatrist, Georg K. Stürup, also always sought to buy as much time as possible to make sure that the patient was "fully informed" about the consequences of any treatment options. On April 6, 1956, he wrote to his Norwegian colleague Johan Bremer that "sedative" treatment with progesterone could help delay the decision.[37]

While many psychiatrists remained skeptical about "affirming" hormone therapies for trans people, some endocrinologists were enthused by the possibilities of modifying bodies with the help of synthetic hormones. The new hormonal treatments allowed hormone experiments *in vivo*. "In man it is difficult to carry out experiments, which allow of an exact evaluation of the size and structure of the endocrine glands," wrote Hamburger and a fellow researcher in an article published in *Acta Endocrinologica* in 1951. The endocrinologist was well versed in the subject matter, having published over seventy papers on the physiology and disorders of the endocrine system since 1930. To get around this problem, the doctors proposed a study of the physiological effects of feminizing and masculinizing hormone therapy, and one case was particularly important: the hormone experiments conducted on Christine Jorgensen.

With aims that were partly therapeutic and party experimental—to examine the effects of the steroids on the testicular and adrenal function—the researchers tested various estrogens given by intramuscular injection and oral delivery, measuring 17-ketosteroid (17-KS) levels in urine as a surrogate. "All progresses at the State Serum Institute," Jorgensen wrote to Dr. Angelo and his wife in late October 1950. "Saving total urines every day is a job but well worth the coming reward. Can you realize what success will mean to literally thousands of people."[38] For more than five months, Jorgensen collected every drop of urine during the day and night. Hamburger needed to be sure that estrogen treatment was not negatively affecting hormone production in the adrenal cortex. Based on analysis of 185 24-hour urine samples following the administration of estrogen, 17-KS levels dropped significantly from normal values to "castrate values." The physiological changes were "fully reversible," and the experiments demonstrated that estrogen inhibited gonadotropin secretion, while corticotrophin secretion remained unchanged. Treatment administered orally appeared to be as effective as injections. "During periods of decreased 17-ketosteroid excretion, the testes decreased in size and became soft and flabby. Erectile potency and sexual libido were also significantly

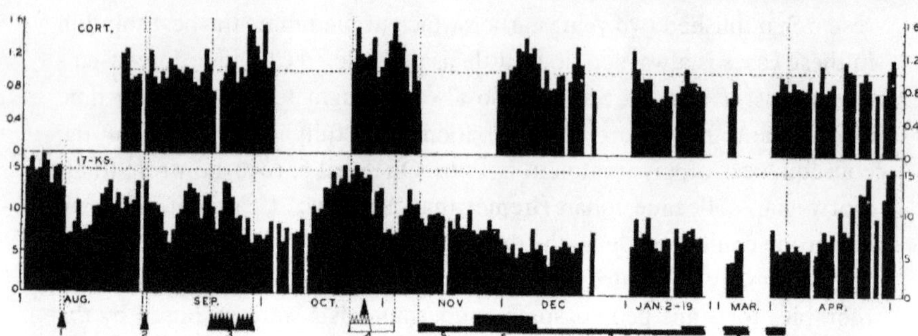

FIGURE 4.1. Figure demonstrating the experiments conducted on Christine Jorgensen and the urinary excretion results of reducing corticoids and 17-KS after the administration of estrogens in various forms. From Christian Hamburger and Mogens Sprechler, "The Influence of Steroid Hormones on the Hormonal Activity of the Adenohypophysis in Man," *Acta Endocrinologica* 7, 1–4 (1951).

impaired." At discontinuation of treatment, the testes "normalized over three to four weeks, and erections and ejaculations reappeared."[39]

While Jorgensen was aware that Hamburger was using her as a "guinea pig," she took a pragmatic attitude; she had finally found, after all, a physician who was willing to give her the treatment she wanted. In a 1984 interview, she recalled a meeting with Hamburger in his private house. For hours they talked in his garden. "I worked in laboratories and knew quite a bit about scientific research," said Jorgensen. "I had no major reservations about being used as a guinea pig and being labeled as such. I never found this designation humiliating."[40] But when Hamburger first suggested testosterone treatment, she adamantly refused. It was okay to be used as a guinea pig as long as she got the hormones she wanted. By doing so, she gave trans medicine a new direction. "He is anxious to do this for we shall keep complete records of everything," she wrote to her doctor in the United States in October 1950. "Psycyatry (spelling?) is not the answer to this problem. I think we (Dr. Hamburger and I) are fighting this thing the right way. Make the body fit the soul rather than vice-verse."[41]

There was a pragmatism in this too: If I give you something, you give me something in return. Accordingly, it "came as a shock" to Hamburger when a few years later he learned of Jorgensen's intention to publish a book based on her life. "Needless here to mention in details the enormous troubles these publications caused me," he wrote to her, referring to the enormous media coverage about her medical treatment. He himself had received two thousand letters and hundreds of visits. "For some time my reputation as an honest scientist was seriously threatened, until our medical reports had appeared." He was concerned that people "once more" would read about him "whitewashing" his decisions, particularly that he gave her "a lot of injections" in his office at the Statens Serum Institut.[42]

This was risky medicine, and one way to limit risk was to share responsibility. "I am disappointed that you have not been able to complete the treatment," Vogt wrote in August 1962 to one of his patients, Chris Dalen, a man in his early twenties, who kept missing appointments and therefore did not get the testosterone injections that had been planned for him. "Any physician can administer the injections," wrote Vogt, urging a promise from Dalen to come for control twice a year, "so that I can be convinced of the effect."[43] More than a disciplinary regime, this was a pragmatic way of navigating complex decisions in uncharted waters.

For example, three months after the press broke the story about Jorgensen's transition, the Swedish endocrinologist Rolf Luft received Max in his office. At the Karolinska Hospital in Stockholm, Luft was key in establishing a hormone lab and the specialization of endocrinology in Sweden. It is unclear if Max knew about Jorgensen's story or if it was a coincidence. In any case, Luft's decision to initiate treatment with testosterone for medical transition was not based on the Danish doctors' protocol. "I was familiar with hormones," Luft said in an interview conducted by Max's niece fifty years later. "It was male sex hormones. I had seen already how it changed women. It was used in women with breast cancer, it is muscle enhancing and vitalizing. Women who were depressed by the cancer flourished!"[44] Luft did not follow guidelines or international recommendations but "did what I thought was the best decision."[45] This decision was based on his training and practice as an endocrinologist, meeting with patients with various endocrinological illnesses, and the use of hormones in internal medicine more generally.

For all of the pragmatism that shaped these decision-making processes, they were nevertheless predicated on a binary framework of sex and sexuality. The realization that sex hormones were not sex-specific and the possibility of shaping sex characteristics with synthetic sex hormones opened the way for an understanding of sex as plastic and malleable. The plasticity of the body undermined the binarity of sex, and, as Jules Gill-Peterson argued, created a crisis in the medical understanding of sex.[46] Physicians were concerned that endocrinologic treatment of transvestism might "produce" homosexuality, which was not only classified as a psychiatric illness but was also criminalized under Norwegian law. One way to avoid this outcome, physicians believed, was to limit treatment to patients who reported little interest in sex. Physicians therefore framed the nosological category of genuine transvestism or transsexuality as characterized by a lack of desire (patients quickly understood this and told doctors what they wanted to hear, reinforcing the nosological framework). The Swedish psychiatrist Jan Wålinder, for example, authorized treatment only for those with "low libido" and warned against an "overly active approach" for patients with an active sex life, whom he considered a "high-risk group."[47]

So anxious were physicians of "producing" homosexuality that they tried to find synthetic testosterone that would not increase the sex drive in their trans patients. Vogt personally wrote to the Dutch pharmaceutical company N. V. Organon to inquire about an adrenocortical steroid for one of his patients that would induce a "more masculine appearance," especially "muscular development and hirsutism," without increasing libido as would testosterone: "We don't like the idea of testosterone treatment with or without X-ray castration, being afraid of the possibility of increasing her sexual urge towards her own sex."[48] The company, however, claimed never to have heard of such a drug: "As far as we are informed, the stimulation of the development of male sex characteristics as well as the increase of libido are typical properties of the androgenic hormones and there is no hormone in which one of these properties are present and others are absent."[49]

Synthetic Hormones and State Regulations

The experimentation within trans medicine took place within a nexus between the clinic and the laboratories on the one hand and the official goal of restricting the pharmaceutical market on the other: Pragmatism was corralled within a restrictive system of state-based medicine. Under a 1928 law, the Norwegian government had to approve all industrially produced drugs, and the Specialist Control, an agency of the state reporting to the director general of health, tested all drugs before approval. Beginning in 1941, a "need clause" required the government to approve new drugs only if they were "medically justified and necessary." Every month, the Specialist Control met to discuss requests from pharmaceutical companies for approval of their drugs.[50] State regulation of drugs included everything from dosage and methods of administration to a separate laboratory for drug testing and advertising. The Danish Parliament—the Folketing—

passed a law in 1954 that required the authorities to approve pharmaceutical brand names before they could be marketed and that the approval of drugs should be for a maximum of twenty years.[51] In addition, pharmacies and pharmacists provided another strict regulatory barrier between the industry and society more broadly. Pharmacies produced and manufactured many drugs themselves, and pharmacists saw it as their professional duty to limit the unnecessary use of drugs and to ensure that drugs were used correctly.[52]

The strict regulation of drugs in Norway and Denmark made it difficult for trans people to get the hormones they wanted outside of state institutions. Often, they had no other choice but to participate in their physicians' experiments. As Hamburger wrote in the medical record of a trans woman in her mid-thirties whom he first saw in 1952, analysis of a 24-hour urine sample showed decreased production of testicular hormones with normal production of adrenal hormones: A trial of different hormones was agreed, "without knowing which hormone would be administered. The patient was asked to keep a diary to record the somatic and psychological effects of the hormone treatment." Over a six-week period, the patient received five intramuscular injections of 500 mg testosterone isobutyrate. In the following two weeks, 200 mg progesterone was injected. Next, there was a switch to three intramuscular injections of estrogen propionate 30 mg before finally starting treatment with estrogen pills. "There is no doubt that the patient suffered from the testosterone treatment," Hamburger wrote. "He was nauseous and vomiting, restless, nervous, and sleepless." Progesterone injections did not have the same side effects; to the contrary, it caused the patient to feel better. After the estrogen injections, the patient initially had tender testicles with nausea and vomiting, then felt "remarkably well" for three weeks. On oral estrogen, the patient was calmer, slept better, and was generally in a better mood. "Of note, the desire to live as a woman did not change at all during the hormone treatment. It was just as strong during the testosterone treatment as it was before and after the treatment."[53]

In 1953, the Norwegian Specialist Control approved Etifollin for the treatment of menopausal symptoms, menstrual disorders, infantilism, and "hypoplasia of the female genitalia," and it quickly became the most widely used estrogen for the treatment of patients diagnosed with "male transsexualism."[54] The Aker team generally reserved high-dose 0.5 mg tablets for the treatment of prostate and breast cancer, and used a much lower dose (0.05 mg) two or three times a day in the treatment of transvestism.

Harry Benjamin, on the other hand, prescribed much higher doses to his patients, recommending one 0.5 mg tablet of ethinyl estradiol daily "after meals," alternatively combining low-dose tablets with 10 mg of medroxyprogesterone acetate.[55] To avoid side effects, treatment should be interrupted for a couple of weeks every two to three months, and the physician should check liver function before starting treatment. Hamburger, meanwhile, started with high doses of estrogen and gradually reduced the dose as the desired effects were achieved. Yet, as he emphasized, "it is hardly possible to generalize." Time was the most important factor: "Millions and millions of cell divisions must have taken place, before the growth of the tissue is visible." Although the effect of treatment could be monitored by analyzing hormone levels in the blood, this was usually unnecessary and clinical evaluation was sufficient. The clinician had to proceed slowly and clinically monitor the gradual demasculinization and feminization of the patient.[56]

Long-acting testosterone injections became available in the early 1950s, and in 1953 the American pharmaceutical company Upjohn applied for approval to sell long-acting testosterone cyclopentylpropionate Depo-Testosteron 50 mg/ml on the Norwegian market.[57] Fatty acids added to the solution, such as propionic acid and isobutyric acid, slowed down the absorption and elimination of testosterone, which was a "great advance in testosterone therapy," the medical director of the Specialist Control reasoned in 1953.[58] In the following years, the agency approved several long-acting testosterone preparations.[59] In 1965, it approved Triolandren, manufactured by the Swiss company CIBA, for the indication "androgen therapy."[60] Its three different testosterone esters provided a rapid effect combined with a longer duration of action, and because it could be administered intramuscularly once or twice a month, usually every third week, it quickly became a popular drug for masculinizing treatment.

New compounds and ways of administering them enabled new research questions. Research and clinical practice went hand in hand and mutually fueled further research and clinical practice. "Female transsexualism" became a clinical indication for testing different routes of testosterone administration in general. In one study, for example, Hamburger examined the effects of different shapes of testosterone suppositories on urinary 17-ketosteroid levels.[61] In another study published in 1969, Vogt and a colleague at the Hormonlaboratoriet compared plasma testosterone levels in four patients after the administration of intramuscular injection, sublingual tablets, and rectal suppositories. Due to hepatic metabolism, testosterone could not be taken orally, but intramuscular injections every three to four weeks maintained plasma concentrations at the level of other men. The authors concluded that physicians must "rely solely on its biological effects" when evaluating treatment response.[62] From initially prescribing testosterone tablets (Perandren, phenylacetate or Ultrandren, fluoxymesterone), Vogt gradually switched to intramuscular injections of Primoteston-Depot (testosterone enanthate).[63] Clinicians used to monitor testosterone indirectly by analyzing the excretion of 17-ketosteroids in urine, but this gave an unreliable picture of the concentration of active hormones. This changed in the mid-1960s when methods became available to measure testosterone directly in plasma.[64] Physicians could now study the effects of hormone treatment not only indirectly through excretion in urine but also through plasma concentrations of free and bound testosterone, and metabolic clearance rates. For example, a study published in 1969 of fourteen so-called "male transsexuals" at the Johns Hopkins Gender Identity Clinic showed that estrogen replacement worked primarily through three physiological mechanisms: first, by decreasing unbound testosterone, probably through negative feedback on the release of interstitial cell stimulating hormone—that is, follicle-stimulating hormone (FSH) and luteinizing hormone (LH); second, by reducing the amount of unbound testosterone in the blood—that is, the physiologically active fraction; and lastly, by reversing the ratio of testosterone to androstenedione.[65]

Tinkering as Care

Mid-century trans medicine was a conglomerate of things, practices, and people: hormonal and surgical infrastructures, networks, architectures, hospitals, laboratories, operating rooms, technologies for hormone analysis in urine and plasma, techniques for tissue transplantation, pharmaceuticals, different methods of administering drugs, pharmaceutical advertising, and technologies for the anastomosis of vessels and nerves. Material, spatial, and technological infrastructures—from scalpels and pills to waiting rooms—shaped endocrinology and plastic surgery both as forms of knowledge and clinical practice.[66] Medical technologies enabled the reconstruction of bodies and identities, but patients also shaped medical knowledge and practice.[67] Although endocrinology and surgery had much deeper roots, their professionalization and specialization progressed rapidly in the postwar decades. The hormone treatment of trans people played a key role in these processes, not because the number of patients was large—it was not—but because the "new" clinical problem of medical transition stimulated tinkering with old ways of knowing and doing, while incorporating these thought-styles and practices into new routines and adapting them to novel technologies.

A restless experimentation with ways of administering drugs, with low-dose regimens, combining drugs, gradually reducing hormone doses, adjusting doses according to biological and psychological effects, and switching between treatment plans—all of these reflect a cautious and pragmatic form of care based in physiological and experimental reasoning.[68] Physicians agreed on a step-by-step approach to treatment starting with social transition, followed by hormone therapy, and ending with surgery. Hormone therapy, reasoned Anchersen, would give patients time to adjust to social and physical changes before surgery, and the pharmacologically induced "hormonal castration syndrome" would reduce the risk of adverse effects before a definite decision to proceed with surgical castration.[69] A six-month period of estrogen therapy before surgery was a "sine qua non," according to Hamburger, Stürup, and Dahl-Iversen.[70]

Pragmatism also dictated that problems be solved as they arose: "A patient of mine with transvestism is unable to work due to his condition and is therefore unable to pay for his Primoteston medication. Would you be interested in sending me 12 ampoules of 250 mg?" wrote Vogt to the pharmaceutical company Tollef Bredal A/S in February 1962 with reference to his patient Chris Dalen.[71] When Dalen was sentenced to twenty-one days in jail for drunk driving, Vogt advocated there should be no custodial sentence because of the difficulties that would inevitably be encountered in either women's or men's prisons, and which would clearly "disrupt the precarious social and psychological adjustment."[72] Vogt even took out a loan for Dalen as help to buy an apartment: "I would rather give him a personal loan than have to guarantee a normal bank loan," he said in a letter to the bank. "Of course, I cannot have full security for the loan, perhaps it may have to be disbursed on the basis of my confidence in his reliability."[73] Certainly, Vogt was exceptionally kind and supportive, but his actions also reflected something more fundamental: an empathic, pragmatic form of medicine.

As physicians increasingly used synthetic sex hormones in medical treatment from the 1940s onward and set up specialized laboratories and clinics for hormone diagnostics and treatment in the 1950s, feminizing and masculinizing treatments also developed into a specialized form of medical care. Diagnostic tests (e.g., serum and urine hormone assays), pharmaceuticals (e.g., estrogen pills and testosterone injections), clinical research (e.g., diagrams and charts of hormone levels), and the architecture and technology of endocrinology (e.g., radioimmunological techniques, immunoassays, and gas chromatography) shaped clinical knowledge and practice. Because patients were willing to act as guinea pigs to achieve the physical changes they desired, they also provided the necessary clinical credibility and economic support for the emerging field of endocrinology and pharmaceutical therapy. Without ever being defined as "trans patients," the tinkering with new ways of measuring hormone levels in blood and urine grouped these patients as a collective.[74] This collective allowed physicians and researchers to develop technologies and techniques in clinical practice while caring for the individual, "each person in the collective being simultaneously an object and a subject of care," as Myriam Winance wrote in a text about care and disability.[75] This was the basis for the shift in thinking from the model of "curing" a sexual abnormality to "enabling" or "affirming" an identity.

CHAPTER FIVE

The Hospital Home and Surgical Pragmatism

Sophia still remembers the hours spent sunbathing on the balcony of the plastic surgery department at the Rikshospitalet in the late 1960s. A long journey had come to an end as she finally began sex reassignment surgery, and soon a new journey would begin. She was among the first women to undergo this surgery in Norway since the first genital reconstructive operations in 1963. When she looked at the Royal Palace from the balcony across the park, she felt complete in a way she had never experienced. Thinking back to the time she grew up in Oslo in the 1950s, she could remember few concepts or role models to help her understand her own experiences of being different. In her own words, Sophia recalls: "I remember a boy stomping on the ground in frustration, saying, 'But you're really a girl!'" She remembers classmates fondling and caressing her. Once, when she was at the hairdresser's, he stuck the clippers in her neck. "It felt completely wrong. I almost felt like it was some kind of assault that they were removing my beautiful hair." The experience of being a girl preceded the cognitive understanding of one's identity: "I felt it was completely wrong before I understood that it was wrong."[1]

The development of endocrinology into a clinical discipline coincided with the emergence of plastic surgery into a clinical specialty in its own right. In the 1950s and 1960s, plastic surgery gained recognition and esteem among the public and the professional hierarchy of medicine. The press published laudatory articles about plastic surgery and how surgeons were able to optimize bodies, lives, and identities in ever new ways. In both endocrinology and plastic surgery, the possibilities of medicine seemed limitless. It was no coincidence that these two professions emerged at the same time as the assessment and treatment of trans patients became routine: The diagnosis and treatment of patients seeking hormonal and surgical treatment for transition was a marginal but crucial task for both professions. Undoubtedly, trans surgery was only a small part of plastic surgeons' overall workload, but the specialized techniques—whether vaginoplasty, breast enlargement, phalloplasty, or facial feminization surgery—accelerated the specialization and professionalization of plastic surgery. Plastic surgeons did not simply perform technical tasks on behalf of other medical professionals but had their own ideas about sex reassignment and about how best to proceed with transitions. Surgeons were actively involved in deciding who should get treatment. They selected patients based on body phenotype, appearance, gait, and behavior, but also on what was technically possible.

Historians often cite World Wars One and Two as the primary impetus for the professionalization of plastic surgery: Severed limbs, burns, and disfiguring injuries to soldiers and civilians presented surgeons with challenges unprecedented in the modern era.[2] This is usually how plastic surgeons have told their own histories. In the words of American surgeon Maxwell Maltz, "The Great War opened up an entirely new era in restorative surgery."[3] Preliminary debridement and improved methods of wound dressing were among the main achievements coming out of World War One. Other historians, however, argue that plastic surgery was in the making long before the First World War as a feature of the American culture of self-representation.[4] Similarly, Sander L. Gilman argued that aesthetic surgery emerged in Europe and the United States in the late nineteenth century and spread throughout the world along the "paths of colonial and economic expansion and the domination of Western medical theory and practice."[5]

In Scandinavia, plastic surgery became a separate clinical specialty only after the Second World War.[6] The Serafimerlasarettet, the oldest hospital in Sweden, opened a special department for plastic surgery in 1944, and four years later the Swedish Medical Association officially listed plastic surgery as a separate specialty.[7] In Norway, plastic surgery became a clinical subspecialty in 1948, and a year later the Rikshospitalet created a position for a specialist in plastic surgery.[8] In Denmark, the institutionalization of plastic surgery was slower, with the discipline becoming recognized as a proper specialty only in 1962. Plastic surgery remained a niche within Scandinavian medicine, and it was an uphill struggle for surgeons to establish restorative and aesthetic surgery as a field to be taken seriously among other surgical disciplines.[9] Although trans surgery contributed to the legitimization and formalization of the emerging profession of plastic and reconstructive surgery and drove specialization processes, a central feature of this surgery in the welfare state was its mundanity. Through the second half of the century, none of the Scandinavian countries had a gender identity clinic like those in the United States, and generalists performed all procedures: Plastic surgeons handled these tasks as part of their general practice, which included cleft lip and palate operations, and surgery to treat congenital deformities and burn injuries.

As for sex reassignment surgery, the plastic surgeon was not simply a technician following the recommendations of the psychiatrist. To the contrary, he—it was always a man—took part in diagnostic considerations and the selection of patients, shaping knowledge and practice in this type of care. At the plastic surgery department in the Rikshospitalet in Norway, the psychiatrist Per Anchersen cooperated with Henrik Borchgrevink, the man who performed all genital surgeries at this hospital. From 1963 to 1985, fifty-one patients with an average age of twenty-eight years old underwent sex reassignment surgery at the Rikshospitalet: Twenty-one were female-to-male operations and thirty were male-to-female.[10] At the Aker Hospital, the endocrinologist Jørgen H. Vogt worked with the hospital's surgeons, and his mostly trans masculine patients underwent chest surgery there. In Denmark, genital operations were performed on trans women in different locations. Christine Jorgensen was castrated at the Amtssygehuset (the Copenhagen County Hospital under the direction of Hans Wulff) and a year later, Dahl-Iversen performed genital reconstructive surgery with vulvoplasty at the Surgical Department of the Rigshospitalet. In the following years, Poul Fogh-Andersen performed most of his operations either at the Rigshospitalet or the Diakonissestiftelsen. Most patients in Sweden had surgery at either the Karolinska Hospital in Stockholm or at the St. Jörgen Hospital in Gothenburg. As relatively small countries with comprehensive public health care systems, physicians who enjoyed high status in society and state bureaucracy, and a legal situation in which surgeons could operate without fear of or the risk of litigation, the Scandinavian countries provided conditions where the role of the surgeon in the welfare state could easily be distinguished from the role of counterparts in the United States.[11]

The Waiting Room

The plastic surgery department at the Rikshospitalet in Oslo was opened in 1953. Where previously fifteen beds had been distributed among the general surgical wards, the new department had twenty-four beds and a separate operating room. Due to increasing demand, the department subsequently moved in 1966, the same year as the Gender Identity Clinic opened at Johns Hopkins Hospital.[12] This relocation to the old Military Hospital, a majestic three-story patrician villa from the mid-nineteenth century, doubled the number of beds, provided a new operating theater, and, with twenty-eight new positions, tripled the capacity.[13] The old villa, however, did not provide optimal conditions for a modern surgical department; just one year after its opening, Chief Surgeon Gunnar Eskeland therefore demanded that the department move to the main hospital.[14] Such demands did not produce the desired outcome; in May 1980, a newspaper referred to conditions at the old villa as "prehistoric."[15] It provided, nevertheless, a unique context for the development of specialized diagnostic and therapeutic practices. As John Law argued: "Care depends not so much on a formula as a repertoire that allows for situated action."[16] The villa's architecture, its interior design and floor plan, would shape the surgical practice of trans medicine for years to come.

The diagnostic and therapeutic decision-making process began in the waiting room when the surgeon called the patient. Describing his first meeting with Borchgrevink in the villa in the late 1970s, Isak says: "I was in this office for the first time before it was decided whether they would allow this or not, and then he told me that the immediate first impression had a lot to say. He had to be personally convinced, he wasn't just a surgeon." According to Isak, Borchgrevink told him, "When I call you and see who stands up, I can immediately see if I think this is the right decision or not." The first impression was crucial: "He observed the gait and the way I walked, and the first thing he told me was that I looked very convincing. I remember that, because it was a very nice thing to hear."[17] At the time, Isak had yet to start taking testosterone, but he had transitioned socially and was dressing as a man.

The surgical assessment included the patient's motivation, self-representation, and assertiveness. Astrid remembered waiting for Borchgrevink in the villa sometime in the mid-1980s. Three or four other women were sitting in the same room. Since she still had not been able to change her name, her old name was printed on her identity documents. "I waited," says Astrid, "hoping that he would call me by my last name, but then he came and used my first and last name. The other women around me looked around and they didn't see a man in the room. I started looking around as well. It was a horrible experience, nobody gets up. Then he calls out again, and I had to go in there, so I stood up. The other women looked at me with astonished looks on their faces. It was extremely embarrassing."[18] In his office, she told Borchgrevink that she would have preferred that he use only her last name: "He got angry. I was the first ungrateful person in the patient group, he said. I thought to myself that I don't want the person who was going to operate on me to become an enemy, so I apologized."[19] Besides demonstrating the power hierarchy in these decisions, the two episodes illustrate how the surgeon's assessment of the patient's phenotype, body type, attitude, and appearance began as soon as the patient rose from the chair in the waiting room.

Isak's and Astrid's experiences say something important about the surgical logic of sex reassignment: Paradoxically perhaps, for the plastic surgeon, it was not the surgery in itself that defined the moment of the change of sex. Much more, the change spanned the time leading up to the operation and the surgical evaluation by the surgeon. "She keeps the Foley catheter," was the final sentence of the surgical description of vaginoplasty, coded as "feminisatio," conducted at the Rikshospitalet's Department of Plastic Surgery in 1981. But "feminisatio" had begun long before the patient reached the operating table. Before the operation, the surgeons usually used the patients' preferred pronouns. "The patient is admitted for sex converting operative therapy. She has never been seriously ill or admitted to hospital," the physician noted in the admission note.[20] Hence, "sex change" took place *before* genital surgery but *after* the first meeting in the waiting room. In other words, sex reassignment was a process that did not depend on the use of scalpels or surgery in itself, but on the surgeon's selection of patients and decisions to operate.

The Surgical Villa

The villa had three floors. The first floor housed the offices of the administrative staff and the head nurse, and there were separate floors for the female and male wards. Patients undergoing sex reassignment surgery were admitted to the ward that corresponded to their identity. For many of the patients, this was the first time they had met other people in the same situation as their own. Isak recalled an episode when a young man, who must have been around eighteen, returned from the women's floor. "He had been upstairs, and that's where they took people who went the other way, from men to women. He said he had seen these strange women walking

FIGURE 5.1. Oslo Militære Sykehus, photo from the period between 1935 and 1940, when the villa still housed a military hospital. Photo: Karl Harstad. Oslo Museum, http://oslobilder.no/OMU/OB.F12033a, CC BY-SA 3.0 NO, https://creativecommons.org/licenses/by-sa/3.0/no/deed.en.

around. I realized then that it was not so easy for everyone, and I told him to shut up and stop talking like that."[21]

The old villa architecture provided a friendly atmosphere that sometimes felt more like a home than a hospital. "It was very nice there. It was old and strange, which created a special atmosphere," says Isak. "There were not so many rooms to be in, so we sat in the corridor. I almost miss it, you got close to each other in a special way, and those who worked there were very friendly."[22] Due to the complexity of reconstructive surgery, patients would spend weeks or even months in the hospital, returning for successive operations. Spending so much time together in a home-like environment led to the formation of bonds between patients, and between patients and staff. Sophia recalls how the department would throw small parties for the patients. "The men would come down in the evening. There was a piano, and we were a group of young people sitting together," she recalls. "I remember a very handsome guy who showed up, some weeks must have passed, and he managed to lure me into the operating room for a little smooch. And the nurses followed the romance with this handsome man. They found it very amusing." More than fifty years after her genital reconstructive surgery in the late 1960s, there are other patients whom Sophia still remembers: a soldier with a hand injury, a person with burns, and a young girl who needed repeated surgeries. "It felt like family."[23]

The family feeling was not a coincidence but an integral component of surgical care. In the 1960s and 1970s, sex reassignment surgery was rare, even for the staff. "Surgical sex change is very special and requires a particular routine, not only from a surgical-technical point of view, but also from a nursing point of view, with human demands on the entire staff at the plastic surgery department," Borchgrevink stated in 1990, looking back on three decades of experience since he performed the first genital operations.[24] To prevent the spread of rumors in the workplace, every member of staff, from nurses to cleaners, was informed of planned operations, about which the patients themselves were kept in the dark. This facilitated teamwork, a professional family feeling. "None of the patients knew," says Sophia. "The staff had taken a vow of silence, even the cleaning lady knew. She was a corpulent woman from the west coast of Norway, and she kept telling me how beautiful I was."[25] On the day of the operation, the staff introduced themselves to the patient. "They paraded into my room, where I was alone in my bed. It was like a line-up. Dr. Eskeland and Dr. Borchgrevink stood at the head of the bed."[26] The procedure of involving the entire staff in the treatment reflected an integrated and comprehensive, yet matter-of-fact approach to plastic surgery: The optimal surgical outcome required the entire staff to support the procedure, and the involvement of the entire hospital staff was a pragmatic way for the surgeons to ensure the integrity of a controversial intervention.

Trans Feminine Surgery

In the US gender identity clinics that opened their doors in the 1960s and 1970s, surgical sex reassignment focused on genital reconstruction: The genitals became the essential site through which sex was changed. As Eric Plemons and Chris Straayer have argued, the medical discourse of transsexuality reduced trans life to the process of surgical transformation by foregrounding the "fulcrum of surgery."[27] For example, in 1974, Milton T. Edgerton, one of the founders of the Johns Hopkins Gender Identity Clinic, wrote: "[The] removal of the male genitalia and construction of a vagina is the most paramount request of the male transsexual."[28] Beginning in the mid-1980s, however, facial feminization surgery became increasingly popular among surgeons and patients, a development that Plemons sees as a shift from a dominant narrative of being "born in the wrong body" to the self-realization of an "invisible me." The therapeutic goal now shifted from a gender norm centered on genitalia to a model of gender that emphasized interaction and recognition in society.[29]

Harry Benjamin had already argued in the early 1950s that genitals were "sources of disgust" among transsexual people. In Scandinavia too, an aversion to genitals became a defining feature of the nosological entity of transsexualism.[30] Anchersen stated that "genuine transsexuals usually have an aversion to their own genitals,"[31] and Wålinder emphasized "abhorrence of the sex characteristics given by nature" as one of three pathognomic criteria.[32] Sophia recalls that when he examined her in his office in the late 1960s, Anchersen asked if she "hated" her genitals. "But I did not feel that way," she says. "I was not disgusted with my body. It just didn't feel right that it was there."[33] Nevertheless, disgust with genitals became a ticket to get genital surgery. "I know of one person who told that she was happy enough with her equipment," says Sara, a woman who transitioned in the 1980s. "She wanted surgery, but she had no problems using what she already had. She did not get surgery."[34]

By the late 1950s, vaginoplasty had become a standard part of genital reconstructive surgery, but there seemed to be a range of different views among surgeons when it came to the importance of sexual relations for their patients. When Sandra Mesics requested sex reassignment surgery at Pennsylvania Hospital in 1974, she encountered doctors who were "cold and dismissive," and they brushed away her questions about postoperative sensation and orgasmic function. "They were very noncommittal and almost insulted that I dared ask about that," she later wrote.[35] For Borchgrevink, on the other hand, making provisions for penetrative sex was an integral part of surgical care. "It was very important to Borchgrevink that I would enjoy sex," Sophia explains. "We talked a lot about sex, and he had thought about what should be preserved."[36] The dilator he made was just a preliminary measure. "He explained that when you have a regular sex partner, you don't need to use it."[37] But Borchgrevink also learned this from his patients and implemented it into surgical practice. In an abstract presented at an international congress for plastic surgeons in 1987, Borchgrevink argued that "leaving some spongious tissue anteriorly in the vulva seems to give both the patient and her partner the sensation of

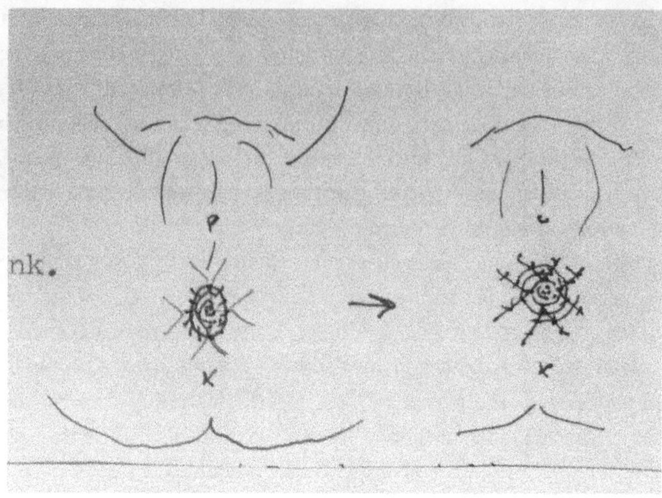

FIGURE 5.2. Surgical drawing of V-Y plasty for postoperative vaginal stenosis. Case 1007, OUHA. © Oslo University Hospital.

a clitoris."[38] If the vaginal introitus was too narrow after surgery to allow for penetrative sex, the patient was readmitted for V-Y plasty with four incisions at 3, 6, 9, and 12 o'clock, usually under sacral spinal anesthesia.

In fact, the depth of the vagina became a crucial measurement for surgeons in arguing for change of legal sex and national identity number (see chapter 6). In a letter to the Ministry of Justice in 1970, Borchgrevink stated that his patient had undergone a "complete conversion from male to female as far as is surgically possible." He explained which parts had been removed, how the scrotal skin had been used to create "naturally looking female genitals," how the urethra was in the "natural position of a woman," and that he had constructed a "vagina of natural width and depth in its natural place." He added that preoperative hormone treatment had given her "beautiful and satisfactory breasts" and that her demeanor was "sweet and feminine."[39]

At the same time, there was no strict boundary—at least not in Norway—between a genital and facial model of sex reassignment surgery. From the beginning, both patients and surgeons recognized the importance of societal recognition—or "passing"—as a treatment goal. Broadly speaking, male-to-female (MtF) surgery included facial feminization surgery, breast augmentation, and genital reconstructive surgery (vulvo- and vaginoplasty). Of the 219 sex reassignment operations Borchgrevink performed from 1963 to 1985, only 81 were on the genitals. For example, most male-to-female patients were offered breast implants some months after the genital surgery.[40] "Incision 5 cm below the papilla bilaterally, dissection to the pectoral fascia, blunt undermining and implantation of 180 ml Surgitec Snyder, closure with Dexon and Prolene. Dry padded elastic plaster dressing," the surgeon noted in 1982.[41] There was nothing spectacular in itself in the technicalities of chest surgery for trans women: The procedure was easy to perform, and the surgeons used the same techniques as for breast augmentation for other indications.

Surgical change of sex was inseparable from the goal of societal adjustment and integration, and facial feminization surgery, such as rhinoplasty and thyroid chondrolaryngoplasty, was a natural part of the treatment package to achieve this goal. One patient returned to the surgeon several years after the initial genital surgery to have her nose corrected. "She also has a somewhat large, drooping nasal tip that she would like to have modified," the surgeon noted. In the operating room, he began by resecting the cartilage: "The nose was reduced in size by resection of the cartilage in the nasal septum and lateral cartilage and by chiseling off the anterior nasal bridge. The bridge of the nose was chiseled and ground down to make it slightly deeper, but no infracture of the nasal bone was performed."[42] For the plastic surgeon, this was business as usual. As the accompanying drawing in the medical record underscores, this was a straightforward matter of straightening a nose, a mundane form of care. The fact that patients could simply return to the hospital, even years after the initial operation, is another facet of surgical pragmatism: If a new problem arose, the surgeon took care of it. By contrast with Hopkins surgeons like Howard W. Jones, who encouraged colleagues to "combat the tendency of these patients to desire polysurgery," Borchgrevink approached this surgery like any other type of surgical care, at least if he decided to operate on them in the first place.[43]

The earliest documented cases of vaginoplasties on trans women were performed at the Institute für Sexualwissenschaft in Berlin in 1930 and 1931.[44] In an article published in 1931, Felix Abraham, a sexologist at the institute, described how, after removing the testicles and penis, a cavity was dissected in the perineum. The surgeon then inserted a rubber sponge wrapped in so-called Thiersch's skin grafts into the cavity. Two to three weeks later, the sponge was removed.[45] Trans women benefited from experiments conducted on patients with vaginal agenesis, a rare condition in which the vaginal cavity and internal reproductive organs don't fully

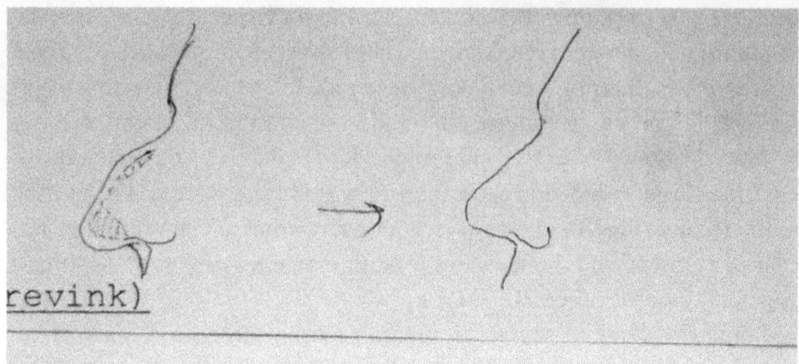

FIGURE 5.3. Surgical drawing of rhinoplasty. Case 1007, OUHA. © Oslo University Hospital.

develop. For example, to avoid graft shrinkage and contraction, plastic surgery pioneer Archibald Hector McIndoe and his colleague J. Bright Banister inverted a split-thickness skin graft onto a form.[46] The main challenge was the tendency of the vagina to shrink due to spontaneous healing of the surgical wound, so McIndoe recommended leaving a form in the vagina for six months.[47]

In Denmark, the first genital reconstructive surgeries for trans women in the early 1950s were limited to castration, penis amputation, and vulvoplasty. Jorgensen underwent castration at København Amts Sygehus in Gentofte on September 24, 1951. One year later, in November 1952, the genital reconstruction was completed. So, under pentothal anesthesia, chief surgeon Erling Dahl-Iversen amputated the penis; this procedure, that raised so many ethical and legal questions for doctors and lawyers, was described by him in a few concise sentences. He began by dissecting the urethra down to the scrotum. He then divided the scrotum to create the labia and sutured the urethra, so that the opening pointed downward instead of forward. Next, he sutured the two corpora cavernosa to appear just below and attached to the symphysis, "as a kind of clitoris but placed subcutaneously."[48] Dahl-Iversen described Jorgensen's desire to have a vagina as having to be "put on hold"; he first wanted to monitor the effects of the initial surgery for a few years.

In Scandinavia, the first vaginoplasty on a trans woman was likely performed in Copenhagen in 1954. A non-Danish national in her forties had traveled to Denmark to get surgery similar to Jorgensen's; when she learned that this was impossible, she studied a textbook on veterinary surgery and conducted the castration herself. When castrations had already taken place, Danish doctors would—according to Alex Bakker—make some exceptions to the rule against offering surgery to non-Danish nationals and would operate on foreign patients, even after Jorgensen's surgery.[49] Three months after the self-surgery, following estrogen treatment under the direction of Christian Hamburger and Georg K. Stürup, Poul Fogh-Andersen performed a vaginoplasty on this patient by inverting the penile skin and using it as a full-thickness skin graft on a form, according to the McIndoe principles.[50] This was long before the Hopkins teams described this technique in a 1970 article.[51] At the Rikshospitalet in Oslo, the first two male-to-female genital operations were performed in 1963, and Borchgrevink used the same technique with skin grafts.[52] Thereafter, surgeons experimented with various techniques for constructing a vagina, from simple reconstruction techniques to intestinal transplants, pedicle flaps, and free grafts.

The medical records of the first women undergoing genital surgery in Norway have been lost. The earliest file I have been able to locate is from 1981. Here, Borchgrevink noted how he began by amputating the penis at the base and removing the testes and epididymis. Next, he sutured the corpora cavernosa around the end of the urethra and used the scrotal skin to form the labia majora. He then dissected a cavity in the perineum between the anus and the urethra, using the same approach as for perineal prostatectomy. The surgeons next began to create the vagina and vulva. First, they resected the scrotal skin, "1½ cm on each side of the midline," before pulling it down to the symphysis in the midline. There they sutured it, creating a sag, and sutured the urethra with a thinner suture on the posterior side. They then prepared a ten-centimeter-deep cavity with room for two fingers in the perineum. They were careful not to injure the rectum during the preparation and used a thick Hegar dilator in the rectum and a catheter in the urethra to separate surrounding structures. "The pelvic floor muscles were very well developed and were cut sharply, making it easier to dissect from the top."

In constructing a vagina, surgeons relied on what they already knew and the techniques they already mastered. They used their fingers to measure the depth and width of the cavity, and they used scalpels to resect, dissect, and shape the labia. Sometimes they used techniques and instruments developed for other purposes: a Hegar dilator, for example, normally used to dilate the cervix, was used in the rectum; a catheter might be placed in the urethra to protect the surrounding tissue. "After most of the loose subcutaneous tissue was removed, the penile skin tube was

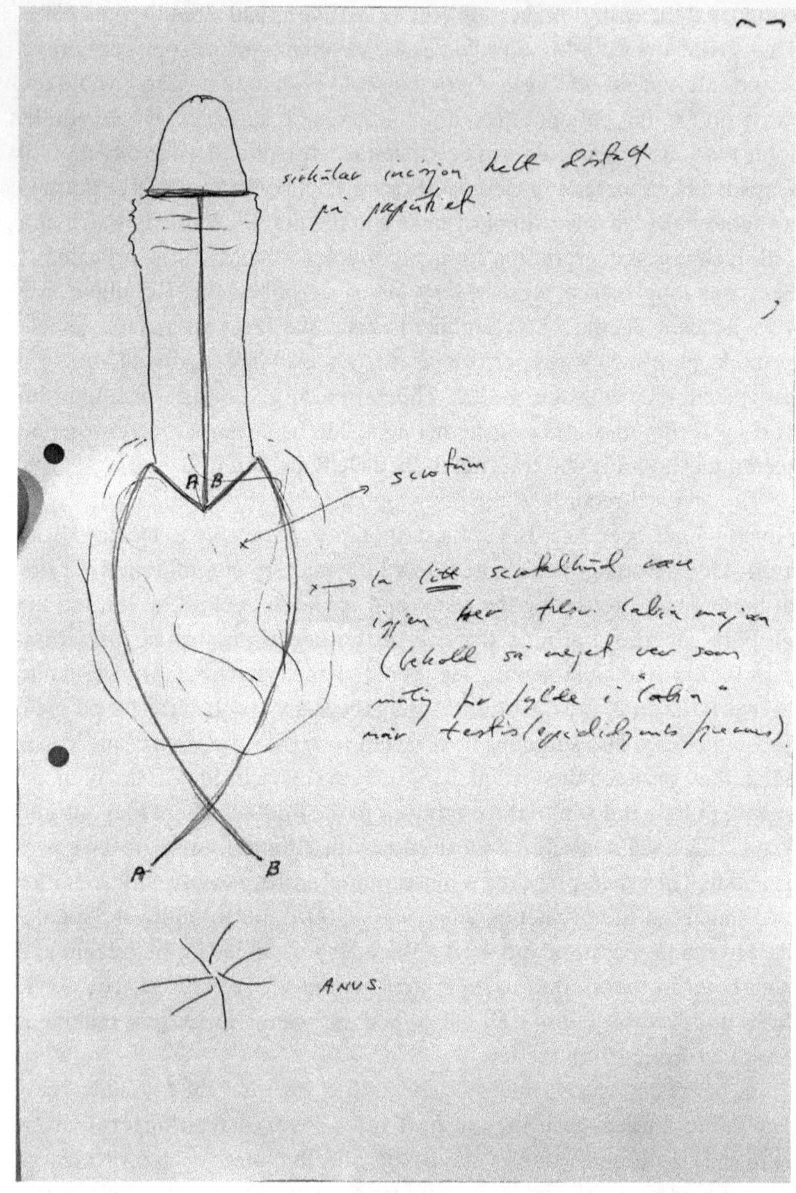

FIGURE 5.4. Surgical drawing of MtF genital surgery. Henrik Borchgrevink, IHPA.

inserted into the cavity as a full-thickness graft. The skin was threaded over a mold made of acrylic cotton wadding and Vaseline gauze, and the skin was pierced with numerous holes for drainage," the surgeons noted. They then sutured the outer edge of the skin to the perineal skin with degradable suture material, and the mold was held in vertical position by suturing from side to side.[53]

This was not surgery-as-spectacle, or at least this is not an aspect visible in the writing. "Writing catches something and simultaneously loses almost everything," reflected John Law in an essay on care and killing in veterinary practices.[54] We cannot know what the surgeons or surgical nurses felt as they amputated the penis or dissected a new vagina, but the concise, poignant, and dry style of writing suggests a mundane form of surgical care. The penis was "skinned" before being placed on a mold and inserted into the new vagina. Next, the skin was pierced to prevent fluid retention, which could lead to infection, and to reduce the risk of graft rejection. As in other surgical procedures, the goal of vaginoplasty was to do what was necessary without doing too much, without causing harm, and to limit the risk of complications.

After the surgery, the patient had to stay in bed with the mold in the vagina and with a special metal cage placed over the lower body in bed to protect the graft and prevent the patient from accidentally removing the mold. Sophia recalled that the nurses gave her morphine and paraffin to prevent constipation. A week later, the surgeons removed the mold, again under anesthesia. "Later, the cavity must be kept open and wide and deep enough by continued dilation during the shrinkage phase of the graft, and regular blocking is recommended even after three months," noted Borchgrevink.[55] Patients were instructed to use a special dilator to keep the vagina open. "This was groundbreaking work," Sophia says. "Borchgrevink made the mold in a plastic material, but it didn't look like a penis. It was pink and looked like a tongue from the front, with a depression in the middle and rounded ends."[56] The indentation at the top was probably there to prevent the dilator from putting pressure on the urethra in the vaginal roof. Since the original dilator kept falling out, Borchgrevink created this to fit better to Sophia's body. When a problem arose, the surgeons tried to find a solution.

Postoperative care was an integral part of the surgery. In the seminal textbook *Transsexualism and Sex Reassignment* (1969), a separate chapter written by Patricia Hadley Thompson, the head nurse of the gynecological operating room at Johns Hopkins Hospital, was devoted to various molds and vaginal forms. The mold could be made of a variety of materials, she noted, from silicone rubber and polymethyl methacrylate to balsa wood with a "foam rubber sheathing, stockinette and plastic covering." Thompson recommended a special bracket and harness made of "straps of amber surgical tubing" and noted that a "muslin binder reinforced with adhesive tape is used as the medium to which the bands area attached with safety pins" to keep the form in place and maintain proper depth. "Upon discharge, however, this arrangement is far from satisfactory," wrote Thompson. It was bulky, difficult to manage, and the necessary materials were not readily available outside of the hospital. Therefore, the clinic had developed a special "apparatus" consisting of an elastic girdle, elastic bands, brackets, and hooks, which combined a "girdle, stocking support, and form-holding device, all in one."[57] This included a discrete two-inch band of elastic for nighttime use, and patients could purchase a separate, thinner panty.

Both in the United States and in Norway, these steps were at the heart of the pragmatic nature of trans surgery: the midline incision, the dissection of the cavity, the resection of the scrotal skin, the creation of the labia from the scrotum, the selection of different sutures, the dilatation of the vagina with a mold, the adjustment of the mold to the individual anatomy, the construction of a special cage to prevent the patient from touching the graft, and giving the patient instructions on how to use the mold with lubricant. This is not the typical "tender love" and "warm hands" kind of care, but a matter-of-fact, mundane type of care, rooted in technology.[58] Yet, as John Law has argued and as the example of plastic surgery shows, the first requirement for care is a great deal of *effort*: "What may sometimes appear simple from the outside is never that in practice."[59] When Borchgrevink, Eskeland, Dahl-Iversen, and Fogh-Andersen performed these surgeries, they had no previous experience and only a couple of published descriptions of techniques to follow. They approached a new clinical problem in their role as clinicians: They were familiar with the elasticity of different skin types; they knew how to make various incisions and dissect bluntly to avoid damage to nerves and vessels; and they used what they already knew from performing prostatectomies and transplants, and the surgical treatment of intersex conditions, burns injuries, cleft palates, and congenital deformities. Along the way, they adapted techniques to new clinical problems, such as developing a mold that would not fall out but would properly sit in the vagina. This was also a kind of tinkering in plastic surgery.

Trans Masculine Surgery

"The others think I am a boy, and so far, everything has been fine," Chris Dalen told Jørgen H. Vogt in one of their first meetings. They first met in 1958 on the day Dalen turned eighteen. In the cafeteria where Dalen worked, his female colleagues wore aprons and men wore overalls. "I hunch my back and make silly arm movements to hide my breasts. You must be flat there! It makes me tired and nervous, thinking all the time about covering it up, and it makes my back hurt too."[60] Three years later, Vogt prescribed treatment with testosterone, and in the spring of 1963, Dalen underwent chest surgery. After the surgery, he returned to work and felt "considerably 'freer' than before."[61]

Bodies were different and required different approaches and solutions; surgeons accordingly tinkered with different techniques. In chest surgery, the main crux was the placement of the nipples and areolas. "The common tendency is to place the nipples in too high and too medial a position," noted Hoopes, the Hopkins surgeon. "A location overlying the fifth intercostal space not less than eleven centimeters from the midsternal line has proven satisfactory."[62] At the Rikshospitalet in Oslo, surgeons used the areolae as full-thickness free grafts. "We started by dissecting 50 øre large bits of skin on both sides," noted Borchgrevink in relation to a twenty-three-year-old admitted in 1968. After removing the breasts with a "downward convex arcuate line from the anterior axillary fold to the sternum," he transplanted the grafts into corresponding deepithelized areas on the chest.[63]

In Sweden, at least one trans man was operated on as early as the mid-1940s. Our only source for this case is an article published in 1961 by surgeons and gynecologists at Karolinska Hospital. The authors reported on the results of several hormonal and surgical treatments for transvestism, three FtM and two MtF. While four of these operations were performed in or after 1954, one year later than the news of Jorgensen's procedures had made international headlines, the fact that a trans man had surgery as early as in the mid-1940s further decenters the role of Jorgensen in the history of trans surgery and shows that these operations originated in the interwar period and were performed as part of regular treatment—not on a grand scale—but also without much fuss or attention. At the time of publication of the 1961 article, the patient had been followed for sixteen years, and the surgeons concluded that the results had been satisfactory. Since early childhood, he had played with boys and preferred to be dressed as a boy. At the first meeting, he said that his appearance had become more masculine during puberty; his voice had become deeper, his muscles more developed, and he had begun to shave. He had gradually begun to live as a man in public. At the age of twenty-five, he had undergone an endocrinologic and gynecological examination with normal results. Psychiatric treatment was considered superfluous and his request for surgery was accepted. The surgeons removed the uterus, the fallopian tubes and ovaries, and performed a masculinizing operation on his chest. Subsequent histologic examination of the ovaries revealed no pathology, and he began treatment with testosterone, which was discontinued eight years later, because it did not cause further masculinization or beard growth.[64]

In the beginning, phalloplasty was not routinely offered as part of treatment for trans men. Of the next thirty patients referred to the Rigshospitalet in Copenhagen subsequent to the first operation in 1956, the Danish team accepted only eight for surgery. All underwent chest surgery, but only two had genital surgery. According to the doctors, most patients refrained from this surgery because of the risk of complications. However, a close reading of their justifications provides another explanation. "All eight imagined their clitoris to be an underdeveloped penis and none of them perceived the clitoris as a feminine element that should be removed by surgery," noted the psychiatrist Thorkil Sørensen in a follow-up study published in 1981.[65] In other words, people perceived their bodies as already sexed according to their identity, an embodied form of sex not dependent on surgical interventions, which explicitly contradicted a key diagnostic criterion of transsexuality: hatred of or disgust toward one's own genitals.

Like Borchgrevink, Vogt seldom recommended phalloplasty to his patients. The surgery was too complicated, the outcomes too unsatisfactory, and the complications too many. Yet Borchgrevink was not entirely opposed to it. Of one twenty-three-year-old trans man whom he examined in 1968, he wrote to the psychiatrist that he was not immediately convinced that such an operation should be offered, but that it was possible after two or three preliminary operations. The problem of urethral strictures could be avoided by leaving the urethral orifice intact rather than constructing a urethra in the penis. Importantly, he emphasized that the indication should be made by the psychiatrist.[66] The "riskier" the procedure, the more solid the indication had to be; ideally, this meant sharing the responsibility with other experts. In this case Borchgrevink turned to the psychiatrist, whom he asked to decide on the indication. One year after chest surgery, he conceded to perform phalloplasty. This was done in two sessions, first by forming a skin tube on the left side of the abdomen from the symphysis to the iliac crest. Three weeks later, he detached the superior end from the iliac crest and inverted it over its medial basis.[67]

One of the first described documented attempts to construct a penis was published in *Zentralblatt für Chirurgie* in 1936. During the war, thousands of soldiers and civilians in the Soviet Union suffered the loss of their penis through injury or mutilation, noted the surgeon Nikolaj A. Bogoras: "The severe psychological trauma, the inability to perform sexually, the impossibility of continuing family life, and the inability to reproduce often have a corrupting effect on the patient, and a significant number end their life in suicide."[68] In the article, Bogoras described how he constructed a penis around a cartilage pin with an abdominal skin tube inserted into the hole of the former penis. This, however, was not the method used for trans men at the Johns Hopkins Hospital or at the Rikshospitalet in Oslo. Surgeons at these institutions followed the abdominal pedicle flap technique developed by Sir Harold Gillies and D. Ralph Millard and later published in the 1957 reference work *The Principles and Art of Plastic Surgery*.[69]

The 1970s and 1980s witnessed huge forward leaps for plastic surgery. New transplantation techniques, such as using the muscle and musculocutaneous flaps, and microvascular technology allowed surgeons to cover defects and construct body parts in new ways.[70] Gunnar Eskeland, the chief surgeon at the Rikshospitalet, described the development as "explosive."[71] Many of these methods were applied in his department, from clitoroplasty and the extension of the urethra to the clitoris (so-called Dennis Brown II) to the creation of a scrotum by fusing the labia majora.[72] Beginning in the early 1980s, surgeons implemented double flap surgery for phalloplasty, which involved the fusion of two skin tubes, which were shaped on the lower abdomen. "Two skin tubes were constructed according to plan on the abdomen, with a common base at the mons pubis and laterally upwards to just above the ventral iliac spine on each side," noted the surgeon about a procedure conducted on a patient in 1982.[73]

The patient was then able to leave the hospital and return some weeks later for the tubes to be fused. "Both skin tubes were cut off laterally upward, opened along the suture line, where on the left side there was a moderate necrotic wound in the suture, this was excised." As in other surgeries, the surgeons solved problems as they arose: If some of the tissue was necrotic, it was resected to promote growth and healing. "The unfolded skin tubes were then sutured together with Vicryl subcutaneously and

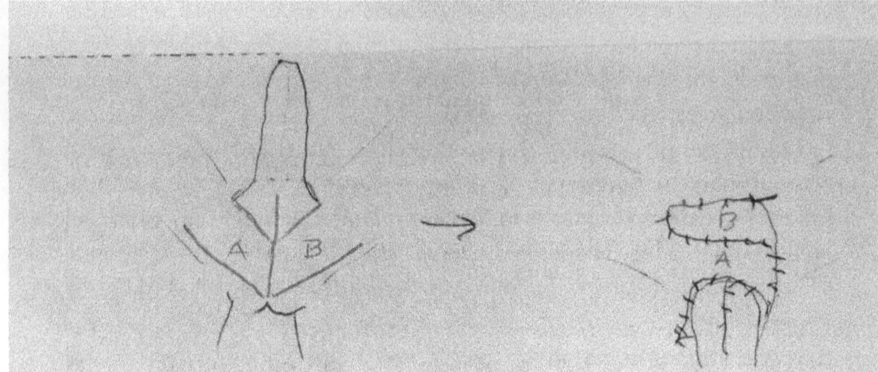

FIGURE 5.5. Surgical drawing of the transposition of the neopenis. Case 1005, OUHA. © Oslo University Hospital.

Miralene in skin; the end of the merged tube was not sutured again to allow drainage and a Baker's drain was placed."[74] These were common surgical techniques applied on a "new" medical problem: Draining the surgical wounds prevented the accumulation of debris, liquid, and pus that could cause inflammation and infection and thus jeopardize the healing of the new organ.

Three weeks later, the fused tube was transplanted to the pubic area. "The neopenis was circumcised at the base so that it was attached to the symphysis by a subcutaneous stalk only. The skin down to the symphysis towards the anterior parts of the labia majora was prepared as lateral skin flaps." The surgeons then inverted the penis downward and moved the flaps upward, aligned them side by side, and filled the defect from the former base of the skin tube. "Foley catheter in urethra, mastichs, gauze, and a light and porous dressing holding the neopenis in a down-and-forth position. Circulation is good. The residual wound at the tip of the neopenis from the previous surgery was excised and sutured directly."[75]

In surgical prose, a groundbreaking procedure became something mundane, even dull: Surgeons were literally creating a new organ but they went about their business in an entirely unsentimental manner. Detaching the fused tube, flipping it over, filling wound defects with new skin, choosing different types of sutures, and making sure that the bandage kept the penis in place after surgery—all of these were ways of tinkering with surgical techniques in order to get to the goal of creating a penis. This was the final step performed under anesthesia. In a subsequent procedure using local anesthesia on an outpatient basis, a glans was created by shaping the tip of the penis into a corona. Two diamond-shaped incisions were made on the front and two rectangular incisions on the back of the penis, and the surgeon sutured the tip of the skin tube and tied it together to form the head of the penis.

Borchgrevink left the vagina and clitoris intact, thus preserving sensitivity and orgasmic function. The penis, however, did not have erectile function, and the urethra was not usually transplanted into the penis. "To put it bluntly, if sex is very important to you, then you should not have this surgery," says Jonas in an interview today. He had the surgery in the early 1990s. For him, getting naked with his partner has always been complicated. "What gets easier," he says, referring to the treatment, "is that you can be a man among men. You don't have to constantly pretend to be someone who you are not. In the end, that becomes incredibly exhausting, and you end up not being around people, you end up taking detours. I used to take naps during my lunch breaks. I was so exhausted."[76] Isak, who had surgery in the early 1980s, had a different experience. "I miss being able to have erections and sexual intercourse. I always thought I would have to forget about it, but I wish I could pee standing upright, for example when you're on a boat." Before the surgery, he was very shy about his body. "I did not want to show myself to anyone, it was a bit lonely. But I enjoyed skin-to-skin contact a lot." After the surgery, he found ways to enjoy sex with his girlfriend, whom he later married. "I did not know then how important sex would become," he says. "It was something I had completely shut out."[77]

The Baltimore surgeons used to construct a urethra with a skin tube inside the penis, but this technique often led to the formation of urethral

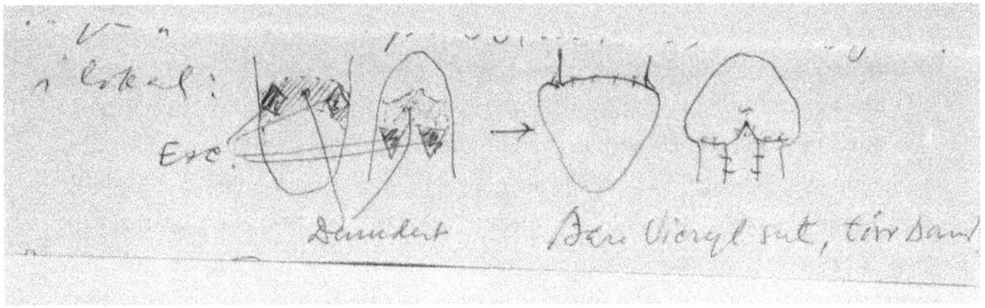

FIGURE 5.6. Surgical drawing of the construction of the corona of glans penis. Case 1005, OUHA. © Oslo University Hospital.

strictures. In some patients, Hoopes used the same technique as was used at the Rikshospitalet, but this "leaves the surgeon with a sense of totally unacceptable compromise," he noted in 1974.[78] So why did this become the most important technique in Norway, performed on fifteen patients by 1985? One possible explanation lies in the contextual differences between the elite institution of the Johns Hopkins Hospital with its high-profile gender identity clinic and a public hospital in the welfare state. The "Norwegian penis" was a purely cosmetic penis. The "Hopkins penis," by contrast, had to be functional. Anything else was considered an "unacceptable compromise."

Creating a functional penis was extremely labor intensive and very expensive, requiring the most advanced surgical techniques. It may be that surgeons in the hospital villa did not have the skills for this type of experimental surgery. However, Norwegian surgeons were no less innovative or creative than their North American colleagues, but their practice was rooted in a pragmatic form of care. Perhaps working in a public health care system, moreover, where the patient did not have to pay and where there was less risk of litigation, allowed surgeons to pursue surgical procedures that would have been considered suboptimal and unprofitable in a market-based health care system? Although the Norwegian penis did not have all of the functionality of other penises, and perhaps not everything a patient wanted, it did fulfill some needs and desires; some patients found it satisfactory. It filled the underwear and fulfilled the "social" goal of being recognized as a man on the beach or in the locker room. Why shouldn't this be a pragmatic solution to a technical problem that medicine had yet to solve?

The Aesthetics of Surgery

Another facet of welfare state plastic surgery was its refusal to be categorized as either cosmetic or curative medicine; it was both.[79] Reconstructive surgery for trans patients developed alongside general processes of professionalization and specialization in plastic surgery. For plastic surgeons in the public health care system, it was of paramount importance to them that their interventions were embedded in medicine proper. "If you are under the misapprehension that plastic surgery is only about human vanity, it is high time to revise your opinion," declared a Norwegian left-liberal tabloid in 1974. The article was illustrated with a photograph of a patient who had been severely burned in early childhood, resulting in limited mobility due to constrictions. The title, "Changes the Lives of Many," underscored Eskeland's message: There was no sharp line between cosmetic and reconstructive surgery. For Eskeland, it was a question of capacity: "The long waiting lists prevent plastic surgery departments from accepting patients for purely cosmetic surgery." Plastic surgeons argued that their operations improved lives, and that was at the heart of good medical practice. This included operations to "change sex," which for some people was an "essential intervention" to give some selected patients a "life worth living."[80]

But for the surgeons too, these operations were compelling. "I remember how incredibly exciting it was for Borchgrevink when he removed the tampon," says Sophia.[81] Borchgrevink identified strongly with his work. He lived not far from the hospital, and he would come to see a patient immediately if there was a complication. When patients came in for checkups, he routinely photographed their bodies, and he kept a collection of the photographs in his office. Sophia recalls that he asked her to model for an international team of surgeons visiting the department. She was living proof of surgical success—and of his surgical skills. When she had completed surgery, her psychiatrist Per Anchersen told her that she had to "preserve what was beautiful, the gracility of her body." Sophia took note of the word, which she had never heard before. "You have to take care of your body," he told her. "As if one could lose one's body," she says.[82]

The statement pointed to a central goal of trans surgery in the 1960s and early 1970s: The creation of something beautiful was inseparable from the objective of selecting patients for treatment. Only those who could meet the physical requirements for a "successful" woman or man were selected for surgery. Certainly, the treatment protocol was based on—and reproduced—narrow norms of gender expression, but these norms were not strictly aesthetic; they also included functionality. This is evident from the criteria in the first Norwegian follow-up study on the cohort of patients from 1963 to 1985. In female-to-male patients, Borchgrevink noted the appearance of the pectoral region, the shape of the areola, and looked for scarring. They measured the length and circumference of the penis, the size and appearance of the scrotum, the shape and consistency of the testicular prostheses, and the size and function of the clitoris. In male-to-female patients, they noted the shape of the nose and Adam's apple, and the size, consistency, and look of the breasts. They examined the vulva and the shape of the labia, the appearance of the new clitoris, and the position and function of the urethral orifice. And they measured the width of the

Lørdag 11. mai 1974

Det plastisk kirurgiske behandlingstilbud:

FORANDRER LIVET FOR MANGE

Fleksnes ønsket seg en ny nese – og fikk det: Husker vi ikke feil så var det i den aller første episoden i den første Fleksnes-serien i TV. Plages De av rynker og overveier en ansiktsløftning? Flere enn Dem har overveid det og gjort det. For stor eller liten byste? Plastikk-kirurgene vet råd for det også.

Tilfellet Liv

Liv i dag, 20 år gammel. Etter en lang rekke plastisk-kirurgiske operasjoner har hun nå normal førlighet.

Liv fotografert på Rikshospitalet, ca. ett år gammel. Arrene på ryggen hindrer henne i å rette seg opp.

FIGURE 5.7. "Forandrer livet for mange," *VG*, May 11, 1974. © Dagbladet.

Lørdag 11. mai 1974

17

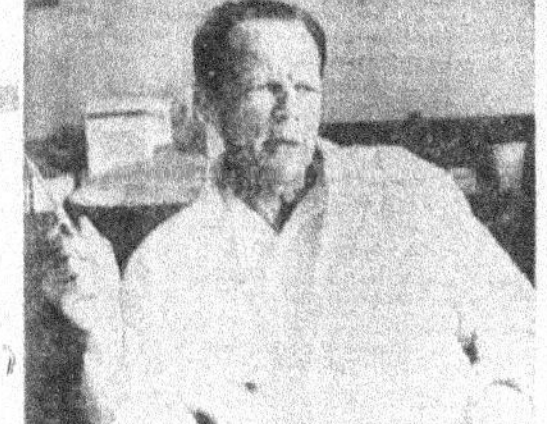

OVER-LEGE

dr. med. Gunnar Eskeland — behov for flere plastisk-kirurgiske spesialavdelinger. I Norge har vi i dag bare fire selvstendige plastisk-kirurgiske avdelinger — foruten Rikshospitalets avdeling — med ca. 50 senger.

To verdenskriger

fram i lyset er imidlertid lagt og fremst å gi et eksempel på hva man i dag kan rette på av fysiske skader og skavanker ved hjelp av plastisk-kirurgiske metoder og vise at plastikk-kirurgien er mer enn en avansert form for skjønnhetspleie.

Dr. med. Gunnar Eskeland, overlege ved Rikshospitalets plastisk-kirurgisk avdeling:

— Plastisk kirurgi grenser til mange andre områder innen kirurgien og er følgelig vanskelig å definere. Men i korthet er det en rekonstruktiv kirurgi som tar sikte på å korrigere defekter og deformiteter på kroppens overflate, uansett årsak. Det kan dreie seg om medfødte tilstander, men det kan også være resultatet av skader.

Den plastiske kirurgien er et relativt nytt spesialfelt innen medisinen. For så vidt kan man takke de to verdenskriger for mye av den utvikling som har skjedd innen faget. Pionerarbeidet ble gjort i England under og etter første verdenskrig med krigsskadde pasienter. Den annen verdenskrig satte ny fart i arbeidet for å finne fram til nye behandlingsmåter og -teknikker.

Lange ventelister

Her hjemme hører imidlertid den plastiske kirurgien ennå til hva ekspertene kaller underutviklede områder innen medisinen.

I Norge har vi i dag bare fire selvstendige plastisk-kirurgiske sykehusavdelinger. Foruten Rikshospitalet som ble opprettet så sent som i 1953 og som pr. i dag har ca. 50 senger, er det bare ved Haukeland sykehus i Bergen, ved Sentralsykehuset for Rogaland i Stavanger og ved Porsgrunn Lutherske sykehus man finner egne avdelinger for plastisk kirurgi.

Nesa til Fleksnes

til å ta imot pasienter til rent kosmetiske operasjoner, sier han. det nok — for å gå tilbake til Fleksnes — hender man på medisinsk grunnlag opererer neser med deformiteter. På medisinsk psykologisk grunnlag hender det også man tar imot kvinnelige pasienter med over- eller underutviklet byste.

Tekst: KARIN HAUGEN
Foto: KNUT SNARE og Rikshospitalets arkiv.

Er det overutvikling som er problemet går behandlingen ut på å bringe de kvinnelige attributter ned på et rimelig nivå ved å fjerne overflødig fett og kjertelvev. Motsatt kan det være aktuelt å hjelpe kvinner som fra naturens side er sparsomt utstyrt i bysteveien ved å operere inn en protese bestående av en pose fylt med væske eller med en gelé-aktig substans.

I mange tilfelle vil imidlertid slike operasjoner måtte henvises til privatklinikker som har tilknyttet plastisk kirurgisk ekspertise. Når det gjelder såkalte ansiktsløftninger for å fjerne rynker og gjenvinne ungdommelig utseende, vil det alltid være tilfellet. Spesialavdelingene har ikke kapasitet til å beskjeftige seg med rent kosmetiske operasjoner.

Hvilke pasientkategorier er det så de plastisk kirurgiske sykehusavdelingene fortrinnsvis beskjeftiger seg med?

Tre kategorier

Overlege Eskeland: — Pasientene kan stort sett deles i tre kategorier. For det første har vi pasienter med medfødte misdannelser. Av denne kategorien utgjør pasienter med leppe/ganespalte eller det som ofte betegnes som hareskår det store flertall ved vår avdeling. Leppe kjeve/ganespalte er nemlig en av de hyppigste av alle medfødte misdannelser: Hvert år fødes det her i landet 120—140 barn med leppe/ganespalte.

Andre medfødte misdannelser kan ha rammet ører, neser, øyelokk — men også hender og føtter. De mange variasjoner av misdannelser har etterhånden tvunget fram en form for spesialisering innen den plastiske kirurgien. Håndkirurgien f.eks. anses i dag som et eget spesialfelt, og ved Rikshospitalets avdeling har man i flere år forsøkt å få opprettet en egen stilling for en håndkirurg — men uten å lykkes: Man har bare til disposisjon en håndkirurg på deltid.

faktisk ikke tenke meg noen investering innen helsesektoren som med så små utgifter ville gi en slik samfunnsøkonomisk gevinst. Hvis De spør om dette betyr at håndskader i Norge ikke får så god behandling som de bør og kan få, er svaret ja.

Nytt ansikt

Den andre pasientkategorien er pasienter med skader og følgetilstander av skader, det være seg etter trafikkulykker og bedriftsulykker eller etter brann- og skuddskader. I noen tilfelle kan pasienten ha fått ødelagt hele ansiktet f.eks. ved en trafikkulykke, og plastikk-kirurgenes oppgave blir så å bygge opp et nytt ansikt av restene av det gamle.

I andre tilfelle kan pasienten ha fått ødelagt huden på store deler av kroppen ved brannskade. I slike tilfelle kreves et møysommelig arbeid med transplantasjon av hud fra andre steder på kroppen for å hjelpe pasienten.

Den tredje pasientkategorien omfatter en lang rekke tilstander så som såkalt elefantsyke, kroniske leggsår, godartede og ondartede svulster i huden, strålskader, etc. Til «diversegruppen» hører også de relativt sjeldne kjønnsskifte-operasjonene.

Kjønnsskifte

Overlege Eskeland: — Jo da — kjønnsskifte-operasjoner er faktisk også foretatt her i landet, ved Rikshospitalets avdeling en 6—8 ganger i løpet av de siste 10—15 år. Pasientene har både vært menn som ønsket å bli kvinner og kvinner som ønsket å bli menn. Operasjonene er blitt foretatt etter en grundig psykiatrisk vurdering av tilfellene og etter at Justisdepartementet har innvilget søknader fra de enkelte.

Det var med tilfellet Christine Jorgensen i begynnelsen av 1950-årene at kjønnsskifte-operasjoner ble allment kjent som en behandlingsmulighet for menn som ønsket å bli kvinner — og omvendt. Dette tilfellet var neppe det første av sitt slag i verden, men var det første som ble slått opp i verdenspressen — etter initiativ av pasienten selv. I 1969 publiserte den danske legen Fogh-Andersen som opererte ham/henne sine resultater av kjønnsskifte-operasjoner foretatt på 11 mannlige og 3 kvinnelige pasienter.

Overlege Eskeland: — Våre hjemlige erfaringer med slike operasjoner må karakteriseres som vellykkede. Det kan være tilfelle hvor en kjønnsskifte-operasjon er en høyst nødvendig kirurgi for å skape et levelig liv for pasienten.

På den annen side kan slike operasjoner vanskelig ses på som noen idealløsning, og etter min mening er det riktig å være meget restriktiv når det gjelder å anbefale og tillate slike operasjoner.

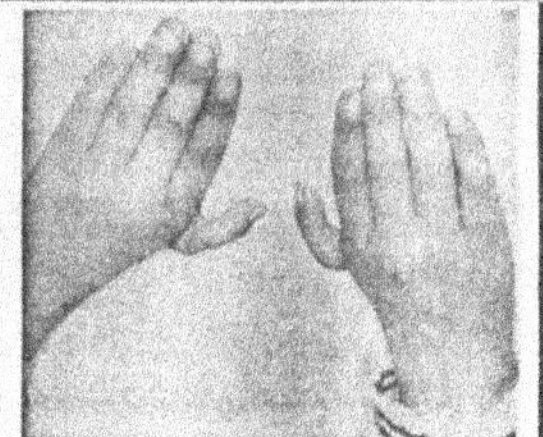

Et barn ble født med underutviklede tommelfingrer som aldri kunne få noen funksjonell verdi. Håndkirurgen flyttet pekefingeren, forkortet den og lagde tommelfinger av den. Resultat: Barnet kan gripe.

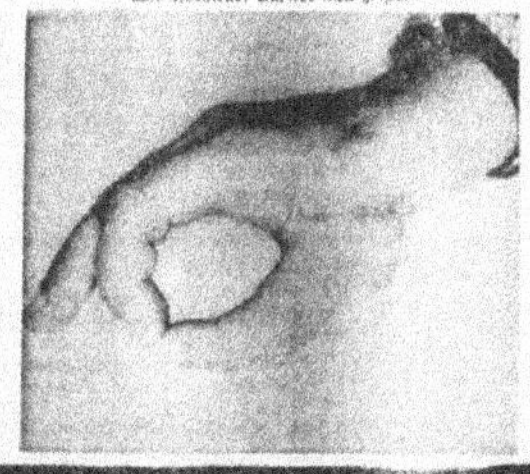

HER ER TO EKSEMPLER

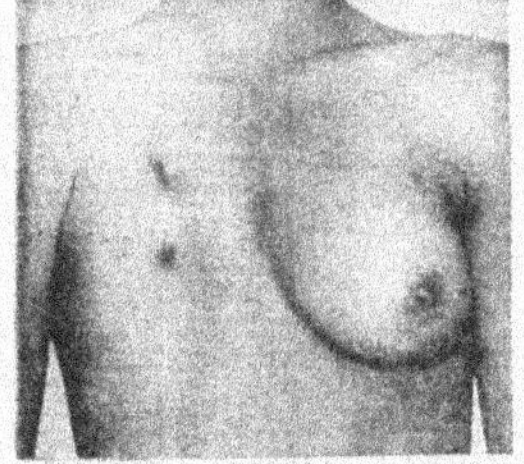

Det ene brystet var uutviklet på denne unge kvinnen.

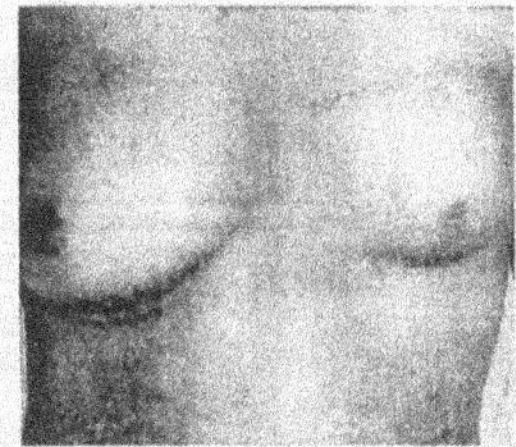

En plastisk kirurgisk operasjon har rettet opp skjevheten.

FIGURE 5.7. (*continued*)

vaginal introitus, the width and depth of the vagina, and asked how well it worked for penetrative sex.[83]

Successful surgery was a combined endpoint of function and aesthetics. Beauty was a composite criterion based on an assessment of overall aesthetic appearance with specific measurements of length, width, and consistency. The goal of combining aesthetics and functionality in sex reassignment demonstrated that there was no clear boundary between cosmetic and curative plastic surgery: It was both.

CHAPTER SIX

Sex and the Binary State

Getting hormones and surgery was one thing; changing one's name and legal sex was quite another. In the postwar decades, all three Scandinavian states implemented castration or irreversible sterilization as a requirement for trans people wanting to change their legal sex status. By tracing the official regulation of name changes and the changing of one's sex in law, the inseparability of clinical reasoning and bureaucratic routines in the welfare state is brought into sharper focus. The state's handling of these requests directly shaped the clinical practice of changing sex in the 1960s and 1970s, even though some trans people and their physicians emphasized how state policies violated trans people's human rights.

The institutionalization of sterilization as necessary for changes of name or sex in the law occurred around the same time as a number of studies highlighted the shadow sides of the welfare state. In the 1967 book *Den ofärdiga välfärden* (The Unfinished Welfare), for example, the authors argued that the Swedish state overlooked the needs of the most marginalized in society, and poor people in particular.[1] The sociologist Vilhelm Aubert's books *Likhet og rett* (Equality and Justice) (1963) and *The Hidden Society* (1965) analyzed the normative elements of the welfare state, and along with other books, such as the criminologist Thomas Mathiesen's *The Defences of the Weak* (1965), this scholarship introduced a new interest in the reality of groups of people living on the margins of society. These books did not criticize the welfare state project per se, but rather how it overlooked certain groups: the solution was *more* welfare state, not less.[2] There was little criticism of the paternalistic ideals of the welfare state and its consequences for minorities such as trans people; nor was there any public attention paid to the state-enforced sterilization practice.

Although not exclusively or explicitly justified with eugenic arguments, there is a line from the state's enforcement of eugenic sterilization policies in the interwar period to the institutionalization of sterilization as a prerequisite for the legal changing of name and sex for trans people in the postwar decades: Both cases demonstrate the state's willingness to prioritize integration and the vision of "the good society" over individual rights. It also shows how the most paternalistic and violent elements of social medicine found new expression in the postwar welfare state.

Different Contexts, Same Outcome

"The greatest challenge concerning the transvestites," noted Knud Sand, chair of the Danish Medico-Legal Council as early as 1954, "is this problem: the change of name." Christine Jorgensen represented a turning point and catalyst. When she succeeded in changing her name after returning to the United States, she "inspired our own transvestites" to request not only surgery but also the "keystone," official recognition of a new name.[3] Previously, Danish authorities had silently allowed trans people to change their names to one similar to the one they had—substituting Kaja for Kaj, for example; Petra for Peter. Official sex status remained unchanged, however, unless the person had undergone castration, and the Medico-Legal Council seldom recommended such operations for trans people prior to the Jorgensen case.

The criterion of sterilization or castration for the change of sex in law was implemented in different ways at different times in the three Scandinavian countries. In Denmark, it likely became routine in 1956, when B. Vestergaard, a thirty-seven-year-old trans man, applied for permission to take a male name. "He was the first man who really understood who I was physically and psychologically, and he was the first person who did not just treat me as a deviant," he said of Christian Hamburger, who had initiated testosterone treatment three years earlier.[4] The gynecologist who examined him and provided an expert opinion was unequivocal: "If permission for a name change is to be granted, sterilization is necessary for social reasons."[5] The Ministry of Justice approved the application after the recommendation by the Medico-Legal Council.

The approval was just the beginning of a control regime intended to last for the rest of his life. As a professor of medicine and director of the Medico-Legal Council, Knud Sand was ideally positioned to pursue his research interest in the biology of sex by requiring applicants to return for follow-up examinations. "I am aware that the professor believes these follow-up examinations should continue in the future and that it is required to monitor my private life as well as my professional activities," Vestergaard wrote to Sand in December 1958. Vestergaard emphasized that the professor was aware that he now was fully employed and leading an independent life, which should be the best evidence that the procedure and change of name had been successful. The repeated questioning about his past and present sexual life and the physical examinations had a profoundly detrimental effect on his mental health, he wrote, leading to prolonged periods of depression after the examination. Vestergaard asked to be allowed to return to an anonymous life like everyone else. "This recurrent control and monitoring of me gives me a strong impression of being a dangerous criminal that society must have under surveillance at all costs."[6]

Norwegian and Swedish authorities also debated what should be the requirements for changing sex status. Sweden was alone in introducing separate legislation for the changing of sex status for transsexuals. In 1972, sterilization was made an obligatory criterion in a separate law regulating sex reassignment, making the National Board of Health and Welfare responsible for all decisions.[7] To change sex status, the law stated, the person had to undergo a long psychiatric evaluation, a "real life test," and provide proof of infertility. In practice, this meant that people had to be sterilized.[8] As members of the commission, the psychiatrist Jan Wålinder and the endocrinologist Rolf Luft played a key role in the wording of the law. In Wålinder's dissertation published a few years earlier, he had already referred to the concept of "gender role," introduced by the Baltimore psychologists, and "gender identity," as developed by Robert J. Stoller, and these concepts profoundly influence the commission, who reasoned that a stable gender identity could not be changed and therefore had to be the basis for legal sex status.[9]

In Norway, the change of legal sex status would remain unregulated. "Under no circumstances" was the person allowed to change his or her name, the Ministry of Justice stated in a letter to the Directorate of Health in December 1953, regarding the inmate at Ila Prison who had requested hormone treatment and permission to wear women's clothing in communal spaces.[10] However, as physicians in the late 1950s and early 1960s prescribed hormones to trans people and even performed genital surgery, it quickly became clear that the official practice had to change. For trans women, the question of gonadectomy was usually not an issue, since all requests at that time involved castration as part of genital surgery. This became the routine: Following castration and genital surgery and the subsequent approval of the Church Ministry, the National Registry changed sex status in the National Register.[11] "Psychologically and socially, she conforms completely with the biological feminization" and "presents herself as a young, very beautiful, and attractive woman," wrote Per Anchersen in one application in October 1970. "From a strictly human point of view—and also from a medical-psychiatric point of view—it would be a catastrophe if the necessary corrections were not made to her sex status in official registers."[12] Henrik Borchgrevink, who had performed the surgery, included a detailed description of the operation, noting that women in the changing room would notice no difference between themselves and his patient postsurgery.[13]

For trans men, changing sex status and name raised more complicated issues. The requirement to undergo gonadectomy to change legal sex status for this group was instituted as a bureaucratic routine supported by clinicians, but not all clinicians agreed. According to Sølve Holm, the Danish Medico-Legal Council believed that patients wanted the procedure to stop menstrual bleeding. Yet none of the men who applied for a name change in the 1950s raised menstruation as a concern. In the years that followed, many people withdrew their applications out of concern about the possible negative health effects of the operation when they realized that gonadectomy was a requirement.[14] In Norway, practice varied. At the Aker Hospital, Jørgen H. Vogt did not follow a standardized routine. In the early 1960s, he advocated for a change of name for his patient Chris Dalen, whom he treated with testosterone injections. Since only his birth name was printed on his identity documents, Dalen was unable to find a job, according to a letter written by Vogt to the Population Register in December 1961, yielding no tangible response.[15] In a letter to the Ministry of Justice two years later, he included two photos of the patient: one taken before testosterone treatment and another taken seven years after. "As you can see from the external appearance, there is no difficulty in accepting him as a man," noted Vogt, "even if the uterus and ovaries are still intact and even if he has no testicles and no penis."[16] Eventually, the Ministry of Justice agreed to the unisex name Chris.

For Vogt, this was a question of good medicine, not a question of conforming to some state principle: The indication for removing the ovaries had to be medical. "This treatment usually stops the menstruation for some time, although somewhat inconsistently, so it is recommended that something more be done to stop menstruation," Vogt told Dalen in 1967. He recommended either X-ray irradiation of the ovaries—"a simple and hygienic method"—or placing a drug-covered tampon in the uterus to "stop the uterine lining from working."[17] Only if bleedings were a "major nuisance" to the patient, or if consistent amenorrhea could only be achieved with high doses of testosterone, did he recommend the removal of the ovaries.[18] For Vogt, the obligatory passage point to request a change of name was the ability to pass oneself off as a member of the sex toward which one was transitioning. "Once a masculine appearance was achieved to such an extent that any uninformed person would accept them as men without hesitation, usually even in bathing suits, they were assisted in obtaining legal permission to take a male name or an ambiguous name applicable to both sexes."[19]

If the patient was over thirty years old, however, gonadectomy or radiation therapy of the ovaries became a standard part of sex reassignment at the Rikshospitalet. "He is treated in one field from the front, one field from the back, both 18.5 × 10 cm, the front field with a lower limit by the upper line of the symphysis. Each field 4 Gy × 3," noted one physician about a patient in 1982.[20] The treatment was repeated three times every other day. Because the surgeon Henrik Borchgrevink did not routinely remove the uterus, a much simpler and less invasive procedure, a primary justification for removing the ovaries must have been to induce infertility and not just to limit vaginal bleeding. Ultimately, there was no single treatment practice or manual governing the treatment of trans men; practice varied from hospital to hospital and from patient to patient.

A New Bureaucratic Convention

In 1964, the Act of Public Registry established a central register of Norwegian citizens. All citizens were assigned a national identity number that also indicated the assigned or legal sex—referred to in Norwegian as *kjønnsstatus*—literally "sex status." Four years later, Denmark followed suit with a similar system.[21] The Norwegian Ministry of Church and Educational Affairs processed applications to change the identity number according to the person's *kjønnsstatus*, and the Ministry of Justice processed applications to change names.[22] A circular from 1969 stated that all ministries—the Ministry of Justice, the Ministry of Church and

Educational Affairs, and the Ministry of Social Affairs (under which the Directorate of Health was organized)—had to be involved in all applications for the change of sex status. If *kjønnsstatus* was changed in the National Registry, it must also be changed in the Birth Registry and in the church book.[23] The Act Relating to Personal Names of 1964 stated that a man could not take a female name, and vice versa. A person's *kjønnsstatus* and name had to correspond. Gradually, a bureaucratic convention developed whereby the Ministry of Justice would allow people—before surgery but during hormone treatment—to take a unisex name, such as Benny, Tore, or Inge.[24] To change a name to an unequivocally male or female name, however, *kjønnsstatus* first had to be changed.[25] But what should be the criteria for determining who had a right to change their state of sex?

Repeatedly, clinicians advocated for allowing their patients to change names or *kjønnsstatus*. For many, it seemed unnecessary to force patients to go through surgery just to change name or sex status. "The name change itself should be the least invasive and least risky part of this treatment," wrote one psychiatrist in a 1969 application to the Ministry of Justice to change the name of one of her patients.[26] In 1972, Vogt sent a new request for one of his patients to be registered as a man. The patient was receiving testosterone injections and had undergone chest surgery and radiation to the ovaries to prevent bleeding. "He is now masculinized to the extent expected, with masculine facial and pubic hair growth, as well as an enlargement of the larynx and a rather pronounced growth of the clitoris," Vogt wrote.[27] But the Ministry of Justice was not convinced: "We assume that NN has a female chromosome constellation and internal organs. Could you confirm whether the X-ray treatment of the ovaries definitely stops the menstruation or whether the sexual organs are still intact? It is especially important to know whether the external sex organs have undergone any changes beyond growth of the clitoris."[28]

Because several state institutions were involved in the process of changing these different personal details—national number, sex status, and name—it happened that people were registered with two names and two different sexes. "The authorities have been deceived," noted a state lawyer in the ministry when, in the summer of 1973, it learned that Chris Dalen, Vogt's patient, had succeeded in getting married to a woman.[29] "It is the duty of the Ministry of Justice to consider whether the authorities should annul the marriage," wrote Ole Herman Fisknes in a letter to Vogt. As a legal adviser in the Ministry of Justice and a lecturer at the MF Free Faculty of Theology, Fisknes was one of the most powerful men in the bureaucracy on church issues.[30] "It now appears, unless I have completely misunderstood you, that if I had simply entered into a marriage without mentioning it to your office, you would not have been informed of it, and I would be happily married today," Dalen wrote in a letter to the ministry. "This is a terrible story, and the way this case has turned out, my life is completely ruined. My motive for the treatment and surgery was to obtain male status, and I or Dr. Vogt must have misunderstood your willingness to help people in my situation. If my status is now to be changed again, then all my stresses and strains from my sex change are completely wasted, and make my situation more difficult than ever," he wrote. "Since the question of my male status and marriage means my entire being or non-being in order to function somewhat normally in society, I will of course never give in to this demand."[31] Vogt also strongly objected to annulling the marriages of his patients who had changed sex status: "This would lead to massive psychological stress, which could have catastrophic consequences for the persons concerned."[32]

Bureaucratic Inertia

As the sex binary dissolved in the hands of bureaucrats, state lawyers looked for new places where sex could be more clearly defined: from chromosomes, gonads, and the shape of the genitals to female and male appearance, sex role, marriage, and sexual performance. Does the patient "present as a man," or is he a "woman with certain masculine characteristics," the ministry asked Vogt. Was it "reasonable or justified" for him to marry a woman? Could he "perform sexually as a man"?[33] Vogt confirmed that it was very likely that menstruation was permanently arrested, emphasizing that his patient looked like a man "with and without clothes." His fiancée even confirmed that he completely fulfilled his sexual duties.[34] The Ministry of Justice now turned to the Office of Psychiatry in the Directorate of Health for their opinion. Christofer Lohne Knudsen, who still headed the office, remained skeptical. Two decades earlier, when confronted with the issue for the first time, he had not been able to stop physicians from providing such treatments for three patients. Now they were witnessing the consequences: "Repeated surgeries and long-term hormone treatments, X-ray castration, etc., then changing the first name to a 'neutral' unisex name, changing sex status in the National Population Register without the approval of the Ministry of Justice and the Ministry of the Church and Educational Affairs, getting engaged and married," he stated. "In doing so, they have brought three more persons into their tragedy—the spouses." These stories were the "worst best proof" of the futility of this type of medical treatment.[35] Once again, Knudsen asked Johan Bremer for his opinion. "I cannot see that the authorities have the appropriate knowledge base to handle these complicated cases—not even for the consent of the change of legal sex," he wrote by way of a response in July 1974.[36] This last sentence in the typewritten letter was underlined with a pen, a likely sign that Knudsen had received the answer he wanted.

As in the case of medical treatment two decades earlier, the solution was to do nothing, but to play for time. The bureaucratic inertia had an explicit goal to protect the sex binary. Still, physicians continued to request changes of name for their patients. "At my first contact with her, I had no problem seeing her as a man—she has a rather deep voice, her stature and facial features are not very feminine. She hides her medium-sized breasts with a tight band," a psychiatrist in an inland county in eastern Norway wrote to the Ministry of Justice in May 1977 with reference to her patient, Magne Indrevik. She had met Indrevik for the first time two years earlier, and now he was requesting that his name be officially recognized.[37] From the time Indrevik was a child, he had only used his male name, and his relatives and family saw him as a man. Having been in a relationship with a woman for several years, they were now engaged and lived with his fiancée's family. Because he still had his female name on his identity documents, however, it was difficult for him to find a job.

In an October 1977 response, the Ministry of Justice insisted that "the permission to adopt a unique boy's or girl's name is granted only after sex change surgery."[38] The psychiatrist therefore referred the patient to the Rikshospitalet where a psychiatrist at the psychosomatic department recommended treatment with testosterone, possibly surgery, and referred the patient onward to the Aker Hospital. There, testosterone treatment was initiated, and a year later Indrevik underwent chest surgery. Again, the psychiatrist contacted the authorities. "For many years, she has lived as a man, using the name Magne in the local community and with strangers, but in constant fear of being exposed. This has meant, among other things, that the client has not dared to look for work."[39] This time, the Ministry forwarded the application to the Ministry of Church and Educational Affairs, which was responsible for registering national identity numbers. "It appears that sex change has only been partially completed, surgically speaking," wrote Fisknes, who had risen through the ranks and was now a senior officer. To change *kjønnsstatus*, the surgery must be completed "for the most part."

Fisknes was no medical expert, however, and he asked the psychiatrist Per Anchersen for advice, "whether there will still be objective criteria for sex when allowing a change of *kjønnstatus* so early in the treatment as in the present case."[40] An absolute prerequisite for a change of *kjønnsstatus*, Anchersen replied, should be "complete sterilization." He had always recommended castration for his patients:

> For women, I have justified the recommendation of oophorectomy on the grounds that without oophorectomy, there is at least a theoretical possibility of being transformed into a menstruating man. And with a simple sterilization (which may not be successful), one can, in the worst case, become a man who can become a mother. Such an (almost imaginary) calamity would lead to unsolvable legal complications and discredit the entire procedure for treating transsexualism.[41]

Torbjørn Mork, who succeeded Karl Evang as director general of health in 1972, agreed. "A change of name to a non-gender-neutral name can only be made when sterility is definitely guaranteed and then in connection with the change of the national identity number. Hormone treatment is reversible and cannot be considered a guarantee of sterility."[42] At all costs, the Office of Psychiatry said, "it must be avoided that people receive sex hormones, change their name, and then do not complete the treatment."[43] Chief psychiatrist Knudsen's skepticism was probably also shaped by a person who asked for having her name changed back to her original female name: "I no longer feel like a man inside. 'God has healed me,'" she wrote in an application for having her old name back.[44]

Legal advisors in the Ministry of Justice argued that Norwegian law was based on the "physical concept of sex," even though it proved impossible to provide a defining criteria that could be translated into legal policy.[45] "I find it somewhat questionable to change a person's *kjønnsstatus* without being sure that they are, in every respect, a so-called 'complete' woman or man," one of the state lawyers noted, echoing the concepts that Magnus Hirschfeld had, forty years earlier, discarded as theoretical abstractions without biological support.[46] The legal advisors and physicians in the bureaucracy desperately needed a clear-cut definition and decided that the removal of the gonads must be the criterion for sex change. "From a legal point of view, this has the advantage of providing an objective criterion," noted one lawyer.[47]

Magne Indrevik's application for name change and change of sex was rejected. This decision formally enshrined gonadectomy as an absolute prerequisite to change name and sex status. Physicians now routinely included letters confirming that the patient had undergone oophorectomy or radiotherapy of the ovaries. Importantly, the state offices established this routine knowing that it would force people to undergo medical treatment to change their sex status. "The authorities must tread carefully on this issue so as not to encourage conversion," one state lawyer wrote.[48] At the end of the document, another lawyer, Elsbeth Bergsland, added her thoughts. "One possibility not mentioned in the case is whether it would be possible to declare the person in question a man with a subsequent change in the National Registry without undergoing surgery. In this case, it would be assumed that the person in question would at least physiologically resemble a man and function socially in a male role," she noted. "Surely, the purpose of the surgery must be to help the person live in his desired sex role, and if that is already working well, it would be a shame to enforce it only for formal reasons. However, I do not have enough knowledge to go into details."[49] This is how gonadectomy became the obligatory passage point of changing legal sex in Norway. The requirement was never written into law, but remained a bureaucratic custom until 2016, when a new law officially allowed trans people to self-declare legal sex without prior castration or sterilization, almost four decades after Bergsland suggested the routine.

The Logics of Social Medicine

The bureaucratic handling of the two policy issues—surgical treatment of transvestism in the 1950s and change of sex status in the 1970s—demonstrates how authorities used inconsistent arguments and shifting logics to justify restrictive policies toward trans people. In the 1950s, state lawyers argued that genital surgery on trans women was illegal, because it violated the Penal Code by removing reproductive capacity. Two decades later, however, when faced with the issue of changing sex status for trans men in public registries, the authorities argued that a man who had been registered as a woman had to have his gonads removed to change *kjønnsstatus*. The dissonance in the way the state dealt with the two issues is striking when one considers that trans women themselves were requesting castration and genital surgery, even if there were at least some trans men who did not want this surgery, but only wanted to have their personal documents changed.

Why did gonadectomy become a requirement to change legal sex? Historian Sigrid Sandal argued that the Norwegian policy of gonadectomy as a prerequisite for changing sex status was about maintaining social order, and, in part, about physicians seeking to protect the integrity of therapeutic practices.[50] Sølve M. Holm has meanwhile argued that the Danish policy was a result of a eugenic discourse aimed at maintaining a genetically and morally "high-quality population." The protection of social order fit into this policy.[51] In order for gender identity to take precedence in a situation where reproductive ability was the most significant indicator of sex, reproductive ability had to be eliminated. According to this argument by Erika Alm, the Swedish law of 1972 was designed to prevent a situation in which a system of gender recognition produced "anomalies," whether it was a person becoming a man after having already given birth, or a marriage that had to be annulled because both partners were registered as having the same sex status.[52]

The Norway case points to a paradox in welfare state regulation of trans life. On the one hand, social medicine as practiced in the Norwegian Directorate of Health sought to limit biomedical solutions to social problems; the psychiatrist Johan Bremer and Directors General of Health Karl Evang and Torbjørn Mork espoused the ideal of avoiding pharmaceutical and surgical interventions for medical and psychiatric problems when social solutions were available. On the other hand, the state instituted a policy that forced people to undergo gonadectomy to change legal sex. At the core of this policy was the binary state and logic of sex. State organs only supported medical interventions that did not threaten the sex binary, and so "partial" sex change must be avoided at all costs. In other words, state physicians at the top of the Norwegian health bureaucracy were willing to sacrifice the bodily integrity of the individual—their reproductive ability—in order not to destabilize a perceived social order. This is perhaps not so surprising, considering that the sterilization laws—which had their impetus in eugenics—stayed active until 1977. Rather than a singular logic of seeking the social causes of health problems, promoting their social solutions, and limiting invasive biomedical procedures, social medicine was a plastic theoretical foundation and framework for state policy. In this version of social medicine, an idea of the common good prioritized the presumed interests of society and the majority over the rights of minorities or the integrity of the individual.

CHAPTER SEVEN

Society as Cause and Cure

In 1975, amid large-scale social movements agitating for sexual liberation and women's rights, a group of lesbian and gay health workers in Oslo founded a peer-support group for homosexual men and women. Three years earlier, sex between men was decriminalized in Norway, but homosexuality remained a psychiatric diagnosis.[1] Medical professionals lacked knowledge and training in lesbian and gay health; many homosexuals were afraid to come out to their physicians, and attempts to convert people to heterosexuality through psychotherapy were widespread. In response to the urgent need for professional health services for the lesbian and gay population, the peer-support group was formalized in 1977 into the counseling service at Oslo Health Council known as Rådgivningstjenesten for homofile. The trans people who very soon turned to the lesbian and gay health care workers for medical treatment found little support, however; inspired by medicalization critique and lesbian feminism on the one hand and psychodynamic theories on the other, health care workers argued that their patients should find solutions outside and beyond medical treatment.

In the postwar welfare state, the health councils—the backbone of the public health system, dating back to the introduction of the Health Act of 1860—were meant to be key vehicles for implementing the health policies of the expanding welfare state. The health councils were directed by physicians (and a community nurse) whose vote counted twice in the council, the highest political unit in the county or city. In the capital, the physician was called *stadsfysikus*, and in the municipalities was known as the district physician. As politicians prioritized financing the hospitals, the vision of a health care system organized around the primary health care centers in the municipalities never materialized.[2] But the Oslo Health Council was

FIGURE 7.1. St. Olavs plass 5, Oslo Health Council, 1969. Photo: Leif Krohn Ørnelund. Oslo Museum, http://oslobilder.no/OMU/OB.%C3%9869/0319, CC BY-SA 3.0 NO, https://creativecommons.org/licenses/by-sa/3.0/no/deed.en.

nevertheless central in hammering out radical responses to health issues in the 1970s. It became a laboratory for the development and large-scale implementation of health expertise, including psychology, sexology, and social medicine.

In 1972, Fredrik Mellbye became *stadsfysikus* in Oslo and medical director of Oslo Health Council, having been chief medical officer at the Directorate of Health's Office of Hygiene. Mellbye was a firm believer in the potential of social medicine to improve public health. While the Oslo Health Council had been housed in an old school since 1887, it moved to a new building in 1969. This brought all the medical departments under one

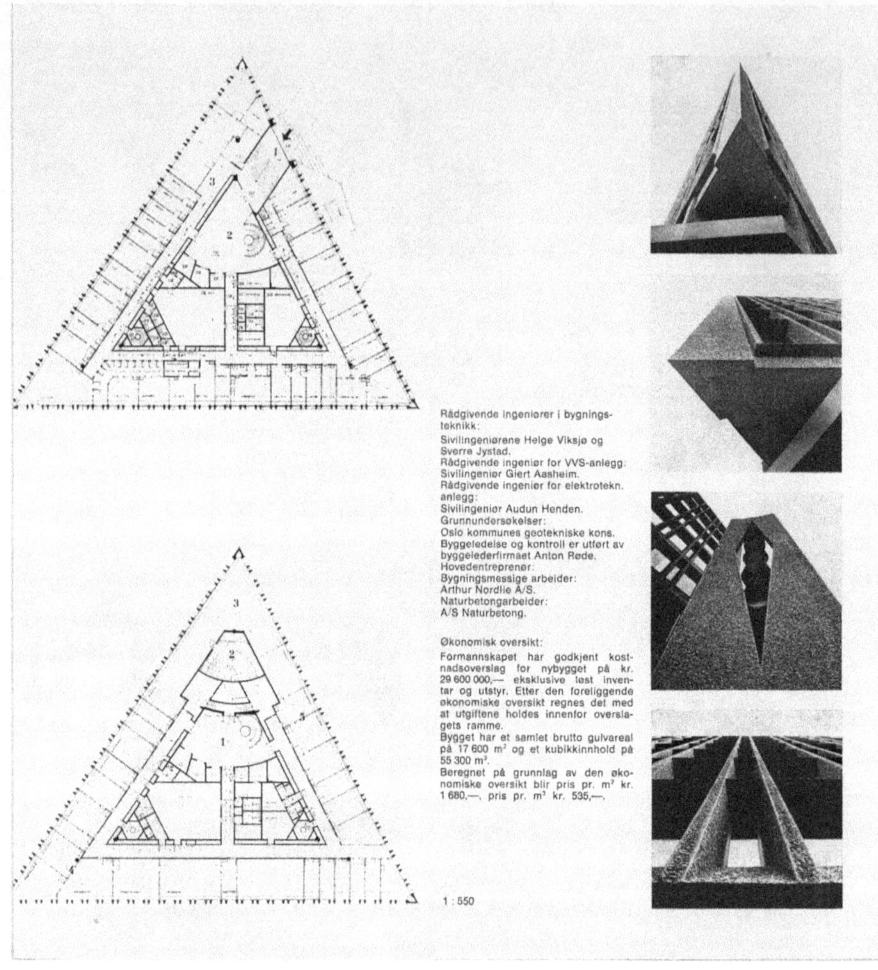

Rådgivende ingeniører i bygningsteknikk:
Sivilingeniørene Helge Viksjø og Sverre Jystad.
Rådgivende ingeniør for VVS-anlegg:
Sivilingeniør Giert Aasheim.
Rådgivende ingeniør for elektrotekn. anlegg:
Sivilingeniør Audun Henden.
Grunnundersøkelser:
Oslo kommunes geotekniske kons.
Byggeledelse og kontroll er utført av byggelederfirmaet Anton Røde.
Hovedentreprenør:
Bygningsmessige arbeider:
Arthur Nordlie A/S.
Naturbetongarbeider:
A/S Naturbetong.

Økonomisk oversikt:
Formannskapet har godkjent kostnadsoverslag for nybygget på kr. 29 600 000,— eksklusive løst inventar og utstyr. Etter den foreliggende økonomiske oversikt regnes det med at utgiftene holdes innenfor overslagets ramme.
Bygget har et samlet brutto gulvareal på 17 600 m² og et kubikkinnhold på 55 300 m³.
Beregnet på grunnlag av den økonomiske oversikt blir pris pr. m² kr. 1 680,—, pris pr. m³ kr. 535,—.

1 : 550

FIGURE 7.2. Architectural plan of Oslo Health Council. Inge A. Dahl and Erling Viksjø, "Oslo Helseråd," *Byggekunst* 51, 6 (1969). © Arkitektur N.

roof, from those dealing with epidemic diseases, housing hygiene, venereal diseases, and food hygiene to those dealing with school health care and those tending to maternal and child health. During the 1970s, the council expanded significantly, adding eight new departments, including general practice, community nursing, physiotherapy, medical genetics, and a support service for parents with disabled children.[3] In addition to providing new premises for the council, the new building was a concrete manifestation of a monumental vision of social medicine and the muscular ambitions of the welfare state.

Built in 1969 at a cost of 29 million kroner, the brutalist structure of natural concrete was designed by Erling Viksjø and Inge A. Dahl. Ten years earlier, Viksjø had designed the high-rise government building in the city center, a stone's throw from the Oslo Health Council. The site of the new council presented problems for the architects, who gave the building a striking triangular shape. The architectural design, floor plan, and choice of materials demonstrated a hypermodern unified vision of architecture, science, and medicine: A small laboratory was set up in the basement, each room was equipped with a sink, and the frames of the building's windows—numbering more than 1,000—were made of solid aluminum.[4] In many ways, the two brutalist concrete buildings—the government buildings and the Oslo Health Council—embodied an ambitious postwar political program. For politicians and physicians alike, the new health council was the symbol of a bright medical future, an expansive public health care system, and a strong belief in the role of social medicine, psychiatry, and psychology within the welfare state. Here, in the late 1970s and early 1980s, psychologists and physicians developed new diagnostic and therapeutic routines for sex reassignment.

Scandinavian Sexology and Social Medicine

Social medicine is a term with different meanings in different places.[5] In the Scandinavian countries by the middle of the twentieth century, it was not merely a subspecialty within medicine nor an academic field closely associated with sociology but played a key role in the creation and reformation of the welfare state after the Second World War. The historian Ida Ohlsson Al Fakir refers to the period from 1950 to 1970 as the "golden age of social medicine."[6] In Norway, one of the most outspoken advocates of social medicine and one of the first to implement its principles was Director General of Health Karl Evang. The state should provide medicine for the public good, he argued, and he promoted a health policy rooted in science and social science. Evang's arguments also found a receptive audience outside of Norway. In its 1946 constitution, the World Health Organization (WHO) defined health as "a state of complete physical, mental and social well-being and not merely the absence of disease or infirmity." Credited by Sunniva Engh as a central architect of the WHO, Evang remained faithful to a concept of health in which there was no separation between people and their environment.[7] Health and disease depended on cultural, societal, and biological factors; a holistic understanding therefore required a bringing together of knowledge of the body, the mind, and of people in society.[8]

Social medicine in Norway, Sweden, and Denmark was firmly rooted in academia, in medical education, and in clinical practice. The establishment of social medicine as a separate subject in the postwar decades was enabled by its separation from the old tradition of hygiene.[9] When the University of Oslo's Institute of Hygiene was divided into an institute for hygiene and an institute for social medicine in 1951, the country's first professorship in social medicine was established. While the Oslo professor of medicine Axel Strøm, in his textbook for medical students, defined social medicine as a discipline concerned with "the reciprocal relationship between society and health," he nevertheless sought to broaden the horizons of social medicine beyond its primary concern with health at the population level to include perspectives and tools for clinical practice.[10] Social medicine in Sweden became a separate subject in the medical curriculum as early as 1954 and had become a proper clinical specialty in Norway by the end of the 1950s.

By the 1970s, social medicine in Norway and Sweden had undergone a radical transformation from having been closely associated in the interwar period with hygiene and the control and prevention of infectious diseases to being increasingly concerned with social benefits and the care of minorities, people with disabilities, and the "outcasts" of society.[11] Experts of social medicine saw this as a shift "from macro perspective to micro perspective"—from the level of the population to that of vulnerable groups.[12] This shift dovetailed with the introduction of a new type of expertise into social medicine: psychiatry. Having previously been dominated by infectious disease specialists and public hygiene, social medicine was now increasingly taken over by psychiatrists. By the 1970s, psychiatrists held all the chairs in social medicine at the University of Oslo.

The reform of social medicine coincided with—and was, in part, spurred onward by—the redevelopment of another way of knowing: sexology. Sexology, of course, was much older. As a field of knowledge, it had roots in the studies of sexual pathology conducted during the late nineteenth century and early twentieth century, alongside the burgeoning discipline of endocrinology and partly in opposition to psychoanalysis. By 1950, researchers such as Alfred Kinsey had increasingly given sexology a basis in statistics and population-based studies, further contributing to its scientification.[13] Not until the 1970s, however, did sexology became a clinical profession (though it was not necessarily officially recognized as such). The International Academy of Sex Research was founded in 1973, followed by the World Association for Sexology in 1978. Sexologists published textbooks, organized conferences, and formed national and transnational organizations. The Nordic Association for Clinical Sexology was founded in 1978. The Norwegian Association for Clinical Sexology was established in 1981, while a Nordic journal of sexology, *Nordisk sexologi*, had its first issue published in 1983. In Denmark, courses in clinical sexology have been offered to medical students since 1978.[14] These national organizations prefigured the formation of the European Federation for Sexology, which was founded in 1988.

As Steven Epstein argued, the linking of *sexual* with *health* allowed for a new discourse on *sexual health*.[15] In 1975, the WHO published a report on the education and training of health professionals in a variety of human aspects, stating: "Sexual health is the integration of the somatic, emotional, intellectual, and social aspects of sexual being, in ways that are positively enriching and that enhance personality, communication, and love."[16] Psychologists and psychiatrists became increasingly concerned with sexual health, particularly among sexual minorities. As early as 1964, the Swedish psychiatrist Lars Ullerstam published a book titled *De erotiska minoriteterna* (The Sexual Minorities) in which he argued for the normalization and destigmatization of sexual minorities. The book was quickly translated into several languages, including Norwegian, and was widely discussed in progressive left circles and the student movement, among medical students, and politically engaged doctors.[17]

In the corridors of the Oslo Health Council, health workers turned to social medicine and sexology to scrutinize the social aspects of their patients' requests. Psychologists and psychiatrists supervised lesbian and gay physicians and nurses who worked on the front lines. The counselors may have been lesbian and gay, but their supervisors were not, nor did they necessarily have any relevant competence in working with sexual or gender minorities. "I had no gay or lesbian friends," says Bodil Solberg, one of the psychologists in the council. She recalls traveling with colleagues from the council to the Nordic sexology conference in Copenhagen: "It was a bit cool and fun and without obligations," she says. "This is a bit embarrassing, but I remember that we walked the streets of Copenhagen so that I could learn to spot a lesbian. We walked around whispering to each other, which shows how new I was in this field."[18]

Among the supervisors was Berthold Grünfeld. In 1976, he was appointed as a sexologist at the Oslo Health Council, where he specialized in information work on contraception and provided clinical sexology services. Under his leadership, Norway's first department of medical sexology was established in the Oslo Health Council in 1979.[19] Born into a Jewish family in Bratislava in what was then Czechoslovakia, his parents were murdered by the Nazis, but the young Berthold was saved by the Nansen Relief and brought to Norway. Having studied medicine and become a specialist in psychiatry, Grünfeld devoted his life to forensic psychiatry and social medicine in which he later became a professor. Throughout his career, he was particularly interested in society's "outcasts" and was a vocal figure in public debates opposing puritanism and sexual moralism. His interest in sexology and social medicine shaped a new model of trans medicine in Oslo from the late 1970s onward.

In Denmark too, sexology was a rapidly developing field of expertise at that time, thanks in no small part to the work of psychiatrist Preben Hertoft, who set up a unit for clinical research in sexology in Copenhagen. Hertoft first started to make a name for himself in 1968 when he defended his doctoral thesis on the sexual behavior of young men in Denmark. The comprehensive study was based on interviews with a total of 400 men in military service and questionnaires from more than 3,100 recruits. This study firmly positioned Hertoft at the forefront of quantified sexology, a new field that had burgeoned after the war, following the publication of the Kinsey Reports.[20] In 1970 he became Denmark's first university-employed sexologist, and a few years later he founded a sexological research department at the Rigshospitalet.

Preben Hertoft stood in a long Danish tradition of the scientific study of human sex and sexuality—from Knud Sand to Christian Hamburger—and he worked to place the role of Danish medicine and Danish physicians at the center of the global history of twentieth-century sexology. In this narrative, Christine Jorgensen and her Danish medical team played a leading role. Hertoft's strategy dovetailed nicely with Jorgensen's own narrative. In the decades following her transition, Jorgensen would continue to emphasize her Danish roots and the importance of Scandinavian culture to her transition, while centering her story in a much bigger narrative of the twentieth-century sexual revolution. She even worked on a Scandinavian cookbook tentatively titled "A Lump, a Pinch, and a Dash Scandinavian Cookery" with hundreds of recipes such as for Danish open sandwiches and how to put up a smorgasbord, all illustrated by "Christine's" color photos.[21]

Therefore, when Hertoft's colleague Teit Ritzau approached her in 1983 about taking part in a film about "Denmark's historical significans [*sic*] in the question of gender identity," emphasizing in his letter that a film without her contribution would not be of much value, since her "experiences have formed the basis of a lot of people's experience of the question of gender identity," she did not hesitate.[22] In the film (whose English title was *Paradise Is Not for Sale*), which premiered in 1984, we see Jorgensen returning to Copenhagen to meet Hamburger in his garden, where they first met thirty-four years earlier. "History cannot be judged while it is happening," Jorgensen had written in a letter to Hamburger some months before. "None of us including the press knew that a new informative era, now known as the 'sexual revolution' was in the making. You and I, dear Christian did not entirely start it but I suspect that we gave it a damn good push."[23] This observation was indeed accurate; however, it left out half the story: It did not mention her own role in promoting this history of the origins of trans medicine, which Hertoft now repeated and once more brought to the public. The book of the same name, cowritten by Hertoft and Ritzau, and the film, shown at film festivals in Chicago and Seattle in 1985 and in Berlin in 1987, placed Denmark and Danish medicine at the center of the birth of trans medicine in general.

Although Hertoft seemed proud of this tradition and the role of his home country and the Rigshospitalet in this history, he remained skeptical about its therapeutic rationale. In 1986, Hertoft expanded the sexological department at Rigshospitalet into a separate sexological clinic.[24] There, he and the doctoral student Thorkil Sørensen systematically studied patients referred for sex reassignment, a patient group that had hitherto been seen by different psychiatrists and had been operated on at the surgical department in the hospital. Soon, this patient group constituted almost 9 percent of all patients at the clinic. Prior to 2017, when a separate unit was opened in Aalborg, the Rigshospitalet was one of two clinics in Denmark (the other in Århus) where trans people could get gender-affirming care.[25] A centralized system was in line with Hertoft's thinking. He saw the future of sexology as a cooperation between different professionals combining knowledge about biology and psychology; a future, however, not as a separate specialty but of centralized sexological units at the university hospitals.[26] At the Sexologisk Klinik, patients met a team of psychiatrists, psychologists, and social workers that coordinated with other departments such as the genetic department, the hormone unit, the gynecological department, the department of plastic surgery, the dermatological department for removal of facial hair, and the speech unit.[27] For Hertoft, an important role for sexologists was to seek solutions other than surgery for their trans patients. Since Christine Jorgensen's treatment in the early 1950s, 110 people, mostly trans women, had requested sex reassignment in Denmark, of whom fifty-six were accepted for surgery, as shown by a follow-up study in 1978.[28] In countries with limited resources, reasoned Hertoft, other patients must be given priority; but in the welfare state, sex reassignment treatment could be used as a last resort measure for a highly selected group of patients.[29] The leading role in this selection process is one, according to Hertoft, that must be played by the sexologist.

The Oslo and Baltimore Model of Gender Identity

For sexologists like Grünfeld and Hertoft—as for Evang—sexuality was a fundamental part of life, and therefore "good" sexual health was an integral part of an expanded concept of health. In his doctoral thesis, Hertoft discarded the idea that "awareness campaigns" would have any effect on people's sexual behavior. Instead, he proposed a broad sex education program beginning in elementary school, similar to how pupils were taught how to read and write. Information was not enough; it must be accompanied by psychological orientation, so that a person was better able to understand "what is happening within himself and others."[30] Hertoft was struck by the emotional intensity with which interviewees talked about homosexuality, an effect he interpreted as being the result of a widespread internalized homophobia.

Grünfeld shared Hertoft's nonmoralizing attitude and a belief in the importance of education and openness about sexual matters. "Sexual health is a resource not only to be preserved but also to be used," wrote Grünfeld in a 1979 textbook on sexuality for laypeople.[31] Sexuality, he insisted, is a "primal force"—the more you try to suppress it, the more it becomes a problem: "Repression dehumanizes it, turns it into something dirty and sordid, something to be ashamed of. Unfortunately, there is still far too much of this destructive attitude toward sexuality in our culture."[32] Grünfeld regarded sexuality and gender identity as products of biological, psychological, and social stimuli, with an ontogenetic model of gender identity formation involving many developmental steps—from genes and fetal hormones through puberty, body image, and environmental factors. This model had strong similarities to the complex model of gender identity and gender role formation developed by the psychologist John Money and his doctoral student Anke A. Ehrhardt in the 1970s. Published in 1972, their book *Man & Woman, Boy & Girl* built on the model of gender role formation developed in the 1950s and 1960s by Money and his colleagues, the psychiatrists Joan and John Hampson, all of whom worked at the Johns Hopkins Hospital with children showing intersex traits.[33]

The framework of the gender role offered a practical medical solution to a perceived problem of ambiguous sex: With the help of hormones and surgery, physicians could shape the bodies of these children so that they would be unambiguously sexed as male or female. This was a precondition for the development of a "secure" role as either a man or a woman. The formation of a "stable" gender role as either a boy or a girl, a man or a woman in these individuals was equivalent to the process in infancy of learning a language. In a series of articles published in the 1950s, Money and the Hampsons hypothesized the existence of a biological program in which gender differences were learned, consciously and unconsciously, in a continuous exchange with the environment. The crucial metaphor, borrowed from ethology, was that of *imprinting*. The plasticity of gender role formation was limited to the first eighteen months of an individual's life, after which the window of imprinting closed and the gender role became fixed.[34]

As Sandra Eder has pointed out, Money's ideas in the 1970s shifted from emphasizing primarily the role of nurture (as opposed to nature) in gender role formation to ideas including the effects of fetal hormones on the brain. This was inspired by new research on prenatal brain organization and the concept of "core gender identity" coined by Robert J. Stoller, psychoanalyst and professor of psychiatry at the University of California, Los Angeles. Though Money and Ehrhardt increasingly incorporated the notion that gender role or gender identity formation is shaped by the brain's exposure to hormones during fetal development, they never abandoned their early insistence on the importance of nurture to the formation of gender role and identity.[35]

Between the 1950s and 1970s, the Baltimore model of gender role formation circulated widely between the United States and Europe, shaping clinical practice in different contexts.[36] It is unclear whether Grünfeld corresponded with the Baltimore team. Years later, when a tabloid newspaper discovered that he had thrown patient records in the trash at his summer home, the ensuing scandal led him to destroy his entire archive.[37] We know that Grünfeld was very familiar with their work and cited it in his writings. In fact, Grünfeld's flowchart of gender development was a

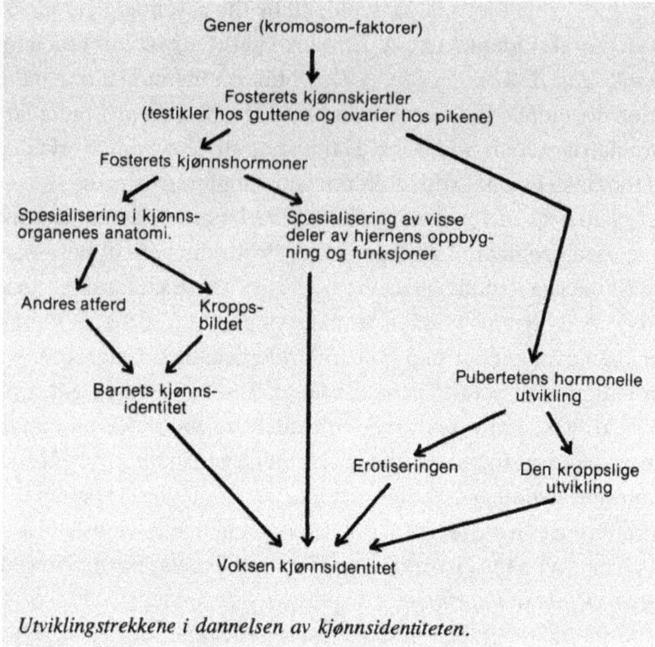

FIGURE 7.3. Flowchart of gender identity formation. Berthold Grünfeld, *Vårt seksuelle liv*. Oslo: Gyldendal forlag, 1979. © Gyldendal forlag.

carbon copy of a chart printed in *Man & Woman, Boy & Girl* published a few years earlier.[38] While it appears in retrospect that there was a conjoint model for the management of intersex conditions, Sandra Eder has problematized this notion of a uniform model, claiming instead that the theories and practices emerging in Baltimore circulated between clinics nationally and were modified in local contexts.[39] This is also what happened in Oslo and Copenhagen.

According to Grünfeld and Hertoft, there only existed two *kjønn/køn* (sex). "Gender roles, gender identity, hormones, and other factors go together and gradually create man and woman, the adult psycho-sexual personality," wrote Grünfeld in his textbook.[40] In a 1984 book on gender identity, transsexuality, and transvestism, Hertoft wrote that Hirschfeld's theory of sexual intermediates had not been confirmed, at least not from a biological point of view: "Biologically seen, these people are like everyone else."[41] In a similar vein, Money and Ehrhardt argued that "alone among the diverse functional systems of embryonic development, the reproductive system is sexually dimorphic. So also, in subsequent behavioral and psychic development, there is sexual dimorphism."[42] Human "dimorphism" was of paramount importance to this learning process: For the child to develop a "secure" gender identity as a woman or a man, mother and father had to be *distinguishable* from each other, as manifested in anatomy (external genitals), physique, voice pitch, and so on. By this reasoning, trans people who had undergone hormonal and surgical transition could become parents as long as their "outward appearance" was easily distinguishable as female or male and they fulfilled traditional roles as wife and husband.[43] As Jennifer Germon has pointed out, Money situated the intersexed body as "unfinished and finish-able"; medical technologies promised to restore what nature had not completed: "When perceived as unfinished—or disordered—in a context where interventions are possible, medical science is compelled to make an intervention, to make things right, to finish what nature failed to do, to bring order."[44] The models of gender identity formation thus perpetuated a dichotomous understanding of sex.

The model of gender formation in the handling of children with intersex conditions was extended into a biosocial framework for gender development more generally: The multifactorial ontogenetic model of gender identity allowed for the possibility that gender identity might not always develop congruently with the sex assigned at birth. Gender identity, which Grünfeld defined as "psychological sex" (*kjønn*), could differ from "biological sex" (*kjønn*): "Usually, there is congruence between the biological and the psychological sexes (*kjønn*). But sometimes there is conflict between the two."[45] These ideas went hand in hand with Stoller's model of the formation of "core gender identity," developed in the early 1960s at UCLA's Gender Identity Research Clinic, which opened its doors in 1962.[46] By gender identity, Stoller referred to the inner conception of the self as either a man or a woman. Stoller also based his concept of gender identity on case histories of patients with intersex traits. He agreed with Money's emphasis on the physical appearance of the genitals and the importance of nurture, as opposed to nature, in the formation of gender role. For those few patients whose gender identity did not develop in accordance with the assigned sex or the external genitals, however, he argued that a "biological force" must be the defining factor.[47]

One of these patients was Agnes, who came to him in October 1958 when she was nineteen. She explained that she had been born and raised as a boy but had, in puberty, developed "all the secondary sex characteristics of a girl, including full breasts, feminine-appearing waist and buttocks, female hair distribution with an absence of facial hair, peaches and cream complexion, etc., all in the presence of a normal-sized penis and testes." Ever since she was three years old, her "conscious fantasy life consisted completely of playing at being a female." Reasoning that Agnes had an intersex condition, which demonstrated the importance of the biological force in the development of a core gender identity, Stoller acceded to her requests and at the age of twenty she underwent genital surgery. When Agnes returned to Stoller seven years later, however, she revealed that the development of her teenage body in a female direction had been caused by her taking synthetic estrogen in the form of her mother's stilbestrol pills. This was a major blow to Stoller's model of a biological force underpinning gender identity formation. Agnes now talked more openly with him, and he was also allowed to interview Agnes's mother. Based on the new information thus obtained, Stoller revised his theory about the importance of the biological force: Agnes was no longer a case of intersexuality but of transsexuality in a boy, and her development demonstrated the importance of nurture on gender identity formation. Agnes and her mother had a symbiotic relationship, wrote Stoller; an absent father and a domineering mother had extinguished Agnes's masculinity. The "successful" development of a male identity in boys depended on a breaking away from a "profound identification" with the mother.[48] In Stoller's view, gender identity was not fully developed until the end of adolescence, but the formation of the core gender identity was complete before the phallic phase. From emphasizing a biological force, Stoller's model of gender identity now became a traditional psychoanalytic model, anchoring gender identity in the oedipal triangle.

"If it is true that there are no biological explanations—and all the evidence points in that direction—then it must be the interaction between the individual and the environment—parents, family, and society—that conditions gender-identity conflicts," Hertoft wrote in 1984, agreeing with Stoller's theory but emphasizing the role of society.[49] The goal, he insisted, was to prevent these conflicts from developing in the first place. Therefore, since a gender identity that did not match the sex at birth did not fit into his concept of normal and healthy sexuality, there is no surprise that Hertoft was very restrictive in accepting people for sex reassignment surgery. He saw himself, in fact, as part of a generational change in attitudes to castration for trans people: While sexual biologists such as Knud Sand had seen castration as directed at the "root of the evil"—the biological basis of an abnormal sexual instinct, in other words—this view, reasoned Hertoft, completely overlooked the phenomenological aspects: "Today one is to a greater degree prepared to observe the biological angle, the individual intrapsychic angle, and the interaction of the psychological and social aspects on the surroundings."[50] The paradox of this putatively more multifactorial model is that it provided less leeway for listening to and taking seriously what trans people said about their bodies.

When it came to the social aspects, Money, Stoller, Hertoft, and Grünfeld nonetheless differed significantly in their ideas of where the "social" was to be located. In the Baltimore model, the social was limited to the socialization—or "imprinting"—of the child within the family; a model of psychosexual differentiation heavily influenced by behaviorist thinking.[51] For Stoller, the social was limited to the dynamics within the nuclear family, "funneled through the mother."[52] For Grünfeld, as for Hertoft, the role of society in the formation of sex roles and identities was of "colossal importance," whether at school, among friends, in sports, at home, in the mass media, in books, movies, television—"everywhere": "The signals are constantly there giving messages." Grünfeld constantly sought to integrate a societal dimension into his approach to sexuality and psychological sex: "The role pattern is clear and unambiguous. It tells you what you can and cannot do. Breaking the laws will be punished."[53]

Grünfeld bemoaned society's restrained and conservative norms about sex and sexuality, which required boys to be tough and to never show feelings of sadness, and girls to be polite, sweet, and proper. School textbooks reproduced stereotypes of mothers as weak but kind housewives and the strong and strict father as the family's breadwinner: "The message is that women are more helpless, weaker, and inferior to men. This seeps into the child's mind like hot wheat bread with jam sliding down the throat of a hungry child. And it stays there in the mind, it is stored."[54] The importance of society and social relations—at least, according to Grünfeld—in the formation of psychological sex in social medicine would have a profound impact on the approach taken at the Oslo Health Council to trans people seeking medical treatment.

Standardizing Sex Change

One of the striking features of the Agnes case, which historian John Forrester identified as having played a decisive role in shifting an understanding of sex, at least within psychoanalysis—from the Freudian understanding of the polymorphous perverse child to a binary gender system *à la* Stoller[55]—was the lack of agency attributed to the protagonist herself: Gender identity was always something that was imposed on the child, either by biology or by a controlling mother, but was not something you could play and experiment with yourself. The immutability of gender identity nevertheless became an axiom and fundamental justification for sex reassignment in the first place: The medical model of variation in sex-gender as "development gone awry" and the notion of gender identity as fixed—after childhood—were fundamental to the treatment regime that grew out of the gender clinics in the United States and in the Oslo Health Council in the late 1970s.

Confronted with people requesting medical transition and by their own lack of clinical experience, the professionals at the Oslo Health Council saw the need for more formal procedures for this type of treatment, including a government-supported team or clinic with a formalized administrative routine: The patient group was small, the diagnostic considerations were difficult, the evaluations required time and expertise, and optimal treatment outcomes required strict selection criteria. In a letter to Mork in early 1979, Mellbye therefore suggested that a centralized clinic for this patient group should be established.[56] But Mork rejected the idea of a national service under government control: Formalization would lead to more people seeking treatment, people who were already taken care of by an informal network of physicians and psychiatrists with an interest in this kind of medicine. Mork's argument reflected the general policy of social medicine in the health administration dating back to Evang's time, which promoted social solutions to health problems and a limiting of biomedical, pharmaceutical, or technological alternatives. Mork feared that once such a service was formalized, numerous legal issues would have to be resolved regarding such matters as sterilization, castration, and name changes; these issues had hitherto been dealt with satisfactorily on an ad hoc basis. Finally, in such a peculiar and marginal area of medicine, efficient care depended, wrote Mork, on clinicians with "personal commitment, interests, and attitudes." To ensure the medical autonomy of the experts, Mork preferred doctors to continue to handle these issues independently in "an informal professional forum."[57] In other words, the health authorities maintained the status quo, and the question of formalizing and institutionalizing a specialized service was set aside. In the absence of official support for the establishment of a centralized clinic for gender identity issues, the Oslo team sought other ways to ensure the legitimacy of diagnostic and therapeutic practice. The professionals met with the Norwegian authorities, and they consulted with the Scandinavian experts Jan Wålinder and Preben Hertoft. This cooperation led to the formulation of guidelines for sex reassignment, which were introduced in 1979—the same year that the Harry Benjamin International Gender Dysphoria Association published the first edition of the *Standards of Care*.

According to the guidelines, gender identity was "irreversible" after puberty; the "conflict" therefore had to be resolved by means, usually, of hormones and surgery.[58] Not everyone who applied for medical transition would benefit, however. Criteria had to be strict, and treatment would be granted only to "well-selected cases." Two criteria were fundamental to treatment: A person had to be diagnosed as a "true transsexual" and there could be no contraindications. For the diagnosis of transsexuality, the Oslo team used the same diagnostic criteria that Wålinder characterized in his dissertation, and these criteria mirrored those of the ninth edition of the *International Classification of Diseases*, published in 1978: Since childhood, patients had to have had the experience of "belonging to the opposite sex" and "feelings of disgust" with their "own biological sex," as well as a desire to be recognized as the "opposite sex" and a desire for hormonal and surgical therapy to align their bodies with their gender identity.[59]

An important part of the diagnostic evaluation was to establish the chronology for the development of individual gender identity and to distinguish gender identity from sexuality. After seeing a patient three times, a gay physician at Oslo Health Council noted: "Psychosexual development: First sexual experience at the age of fourteen with a friend of the same age. Since then, now and then, but always with men. Does not consider himself homosexual." Then followed some reflections on sexuality: "Says he has little desire for sex, but always enjoys it. When he is with men, he imagines himself to be the woman. Masturbates occasionally, and he then often fantasizes about the men he meets with, but as a woman. Never wears women's clothing for sexual stimulation. Occasionally uses women's clothing, but with the social intent of being recognized as the woman that he feels he is."[60] The diagnostic thinking used here reflected the distinction established two decades earlier between transvestism as symptom and transsexuality as nosological entity. In a broader sense, this represented a growing recognition within scientific discourse of sex, gender, and sexuality as interrelated yet discrete human phenomena.

One way to protect the integrity of the treatment decisions was to distribute decision-making among a diverse group of specialists. A team of experts—psychiatrists, experts of social medicine, sexologists, endocrinologists, social workers, psychologists, and plastic surgeons—were all involved early in the diagnostic process, and all therapeutic decisions were based on their collective expertise. This was a very different approach from that of the American sexologist Harry Benjamin who left recommendations for surgery to the psychiatrist.[61] Making decisions as a team allowed each case to be scrutinized from biological, psychological, and social perspectives. For example, a physician or surgeon would conduct a thorough clinical examination that included all aspects of the clients' "somatic *kjønn*" to exclude genetic, hormonal, or genital "incongruities." The diagnostic process took six to twelve months, during which time no hormonal treatment was initiated.

In rare cases, the Oslo guidelines argued, the request to change sex was caused by psychosis. While signs of psychosis were a clear contraindication to hormonal or surgical treatment, a diagnosis of depression was not. Depression, the experts reasoned, could be caused by negative life experiences, by having hidden a part of oneself, and by having waited a long time for treatment.[62] While psychiatrists and psychologists linked depression to societal factors such as stigma, they recognized that health was inextricably linked to the negative health effects of marginalization and ostracism, and they treated psychotic symptoms as a separate disease entity unrelated to minority stress. The hierarchy of contraindications, between psychosis and depression, probably reflected a much longer psychiatric tradition of distinguishing between severe and milder forms of mental illness and of attributing lower self-awareness and impaired decision-making capacity to people with psychotic symptoms.

The likelihood of successful "adjustment" after "conversion" increased if social stability was good before treatment, according to the guidelines.[63] Social factors were not treated as individual and separable, but as interwoven and co-constitutive. A good surgical result was of little use if the social situation was unstable, and the patient had to be psychologically robust enough to cope with the changes following hormonal and surgical therapy. The numerous contraindications to treatment also highlighted the importance of social aspects in the decision-making process: Lack of resources, poor social situation, or lack of support from friends and family were important contraindications, as were complicating sexological aspects, "advanced age," and "unsuitable physique."[64] At each level, the patient's identity and desire to transition were seen in relation to society. In a referral to the psychologist for one patient, for example, the physician provided the following details: "Tall, slender 21-year-old male, feminine appearance and mannerisms, although the style of clothing is not particularly feminine. Sparse facial hair, otherwise normal sex characteristics." He noted that the patient seemed "balanced and well-considered" regarding the decision to "change sex" and about her life in general.[65]

From the outset, the patient's appearance, voice, movement patterns, clothing, and physical characteristics were evaluated in relation to a perceived idea of successfully "passing" after treatment. The criteria of a "suitable" physique or physical characteristics could be seen as signaling underlying stereotypical and normative notions of what constitutes a man or a woman. Kirsti Malterud—a feminist, lesbian activist, and general practitioner who was part of the team—confirms this: "One of the criteria for sex change, which was very strict, was that you had to be able to pass as the other *kjønn*. That's why tall men didn't get treatment, people with big shoes," she says in an interview. "I remember very well how this passing criterion was talked about. Talk about cultural production of masculinity and femininity, and what is right and wrong and normal and abnormal! It's very strange to think about today."[66]

Medicalization and Nordic Feminism

Internalized homophobia, relationship problems, repressed sexual fantasies, coming-out processes, and the health consequences of stigmatization—all were familiar topics for lesbian and gay counselors at the Oslo Health Council. Confronted with the clinical problem of sex reassignment, they brought this knowledge to bear on the issue, raising further questions: What role did societal norms, and especially homophobia, play in the desire to transition? How did norms of what men and women should look like and how they should behave become mixed with a desire for changing sex? What was gender identity and what was sexual orientation, and how did their clients envision their sex lives after treatment? Having themselves experienced the importance of intimacy and sex for good health, the homosexual counselors were particularly concerned with exploring these aspects in patients seeking sex reassignment.[67] Jan-Henrik Pederstad, a physician at the Rådgivningstjenesten, recalled in an interview that he was skeptical about the desire to remove healthy body parts.

"It was like you were discarding your ability to have an orgasm," he says. In his view, the importance of sexual pleasure was often downplayed in treatment decisions. "I remember trying to find other solutions for my clients," he says. "Even if you were born in the wrong body, my goal was to try to help people appreciate and enjoy what their bodies were already capable of doing."[68]

For some people, however, it was only after surgery that they were able to truly enjoy sex. Janni Christin Adeen-Wintherbauer was in her early thirties when she was assessed by the team at Oslo Health Council. Adeen-Wintherbauer grew up with her mother in Brøttum, a tiny village 170 km north of Oslo on the shores of Mjøsa, Norway's largest lake. Fathered by a German soldier, she was mocked as *tyskerunge*, or a "German brat," one of the most stigmatized cohorts in postwar Norway. When she was thirteen, she left Brøttum for Oslo, where she survived by selling sex. Since she had worked as a sex worker, the team at Oslo Health Council was skeptical about recommending surgery, reasoning that her social situation was too unstable. When she finally got the surgery she wanted, she was for the first time able to truly enjoy sex. "The first time I could see, touch and feel inside with a finger—I have no words. It was incredible. It was like a tumor had been removed and I was finally complete," she recalled in a book about her life. Although it would be some time before her vagina was wide and elastic enough for her to be able to enjoy penetrative sex and have orgasms, she still remembered, of the first time, that "the feeling that a man could penetrate me in the normal way, as it should have been from day one, was extraordinary."[69]

The Oslo Health Council's position in the public health system provided its counselors with opportunities that colleagues elsewhere could only dream of. Their counselors traveled to sexology conferences around the world to keep abreast of the latest research and to bring the latest findings back to Norway. These were certainly extraordinary events, not least because of the high cost of travel. When Georg Petersen and Kirsti Malterud from the council traveled with 1,600 professionals to the fourth world congress of sexology in Mexico City in 1979, it was reported in the lesbian and gay press and in the national press under headlines such as "Let's talk about sex change."[70] This was a very different policy from two decades earlier, when the health authorities tried as much as possible to prevent public discussion of the issue, fearing that it would lead to more requests for treatment. In the late 1970s, sex change was something physicians openly discussed in public. The fear of "social contagion" was not gone, however. To the contrary, sex reassignment was now problematized from a sexological perspective as yet another example of the "medicalization" of social and political issues.[71]

On their return from the conference, Petersen and Malterud reported that sexologists were increasingly approaching sexuality as a dimensional—not as a binary—human phenomenon. The categorization into two or three types—heterosexuality, homosexuality, or bisexuality—reflected a "repressive society with a tendency to polarized grouping."[72] Keynote speakers at the conference included John Money and Richard Green, who presented their research on children with "extremely atypical gender role behavior," where all eventually identified as homosexual males. Gender roles, gender identity, and sexual orientation must be considered together, Petersen and Malterud stated, arguing for a careful approach to people seeking help to transition.

Some of the professionals at the Oslo Health Council were intensely engaged at that time in feminist critiques of medical patriarchy and the rapidly growing literature on the difficulties of turning social issues into medical problems. The concept of "medicalization" was first introduced by the American sociologists Jesse R. Pitts (1968) and Irving K. Zola (1972).[73] Medicalization was a flexible concept that could be deployed against any form of medical definition or treatment, be it hyperkinesis or sex reassignment. The harshest criticism came from Janice G. Raymond in her 1979 book *The Transsexual Empire*. The "transsexual empire," she argued, was a "medical empire" based on a patriarchal medical model that had "medicalized moral and social questions of sex-role oppression," depleting the questions of their actual meaning.[74] This reasoning resonated with some experts at the Oslo Health Council. When, in 1983, Kirsti Malterud and Bodil Solberg published an article in the national medical journal about their experiences with people seeking sex reassignment, they referenced Raymond's book as well the book *Sex by Prescription* by psychiatrist and medicalization critic Thomas Szasz. In 1961, Szasz famously declared mental illness to be a "myth" in the sense that "mental illness" was a response to social issues and life problems, and that psychiatry had become little more than an instrument of social control. According to Malterud and Solberg, critics such as Raymond and Szasz had shown how "sex change" raised "big ethical and political dilemmas."[75] At the time, Malterud was concerned with medicalization in medicine more generally, such as the tendency to turn menopause into an illness in need of medical treatment. "Being a woman is not an illness," she said in a February 1982 newspaper interview. "The pathologization of women's lives has many negative effects," she said. "It makes our life cycle invisible, almost something forbidden. It makes women more dependent on the healthcare system and robs us of our independence. This dependence is unfortunate and hinders women's liberation."[76] Today, she recalls thinking that these dilemmas were not given enough attention in the medical handling of people requesting sex reassignment. "There were strong ideological objections to everything becoming disease and having a medical solution," she says.[77]

Sexology is a "political science," wrote Malterud in an essay in a left-wing journal in 1980.[78] She was deeply skeptical of paternalistic traditions in sexology that reduced sexuality to individual problems, divorced from political issues such as class struggle and women's liberation. For Malterud and Solberg, "sex change treatment" cemented the gender norms they sought to crush. "Our clients describe a strong despair of living in a society where women are not allowed to look like men and vice versa," they wrote. Medical treatment was a "quick fix" for structural and societal problems, reflecting a much older skepticism about the use of pharmaceuticals and about the "medicalization" of life problems in social medicine and the Norwegian health bureaucracy.[79] "For our clients, these limitations are so narrow that only surgery can resolve the conflict between the expectation of others and one's own behavior."[80] The title of their article—"Clinical Experiences with Sex Change Clients"—suggested in itself a medicalization critique: People seeking sex reassignment were not patients but "clients," shopping for health services. "It was an expression of a fundamentally essentialist attitude," recalls Malterud. "Something is wrong, and we can change it. We are not going to change it a little bit; we are going to change it properly." This is also why they referred to it as *kjønnsskifte*—sex change. "The idea that the person's sex (*kjønn*) was changed through treatment was very strong," she says. "This concerned not only genitals or hormones, but the person had become a man or a woman."[81]

As lesbian and gay activists, many of the counselors at the Rådgivningstjenesten had invested a lot of effort into challenging stereotypical gender norms. Malterud recalls that many lesbians at the time—herself included, though she referred to herself at that time as a homosexual woman—wanted to break with traditional norms of femininity or beauty, for example, by wearing large watches, good shoes, comfortable clothes, and no makeup. "They were not supposed to pass as just any other woman," Malterud says of the trans women she met at the Oslo Health Council. "It was Marilyn Monroe." She remembers feeling alienated by these ideals. "It was a kind of *outré* form that I reacted very strongly to."[82] Some of the physicians and psychologists feared that people would want medical interventions to transition as a "cheap solution" to internalized homophobia. Malterud recalls that it was much harder to be openly lesbian or gay back then. "If you could disguise it with surgery and hormones and clothes and social roles, that was actually more attractive for some people," she says. "We thought we knew quite a lot about sexual orientation, so in some of the people we talked to, we concluded that he is gay, do not pursue this project, sex change is not the solution to this." But Malterud also recalls that they lacked an understanding that people requesting sex reassignment were not necessarily heterosexual. "I don't think anybody believed at the time that trans people, or 'sex change clients' as we used to call them, could be anything other than heterosexual," she says. "It was part of the definition that if they wanted to become the opposite sex, then they wanted a partner of the same sex as the one with which they were registered at birth. It was almost a requirement."[83]

In the end, the team at the Oslo Health Council approved only one individual for treatment, and this was Isak, whom we first met in chapter 5. Writing to the team in 1981, he expressed himself as follows: "I have taken the step towards a full life. The responsibility for this and for what happens to me next is mine alone. I will never choose any other solution. I can say this because I feel in my bones that I have made the only right decision and that I am a man, as I have always known. Therefore, I categorically refuse any offer of treatment to live as a woman." Forestalling any objections, he continued: "Why on earth would I attempt such a thing? Renounce my entire reality, which is good, and embark on a totally unknown path towards an undesirable goal, just in the event that my lifelong experience of myself may have been wrong." He rejected any such "hypothetical thought experiments" as being of no interest. "I encourage the working group to consider what responsibility you will assume in my place if you deny my reality as true and good for me. You have had the opportunity to challenge me against what science has to offer. This has not shaken my experience or my conviction," wrote Isak. "The final word and full responsibility must rest with me."[84]

Grünfeld began treatment with testosterone injections in cooperation with an endocrinologist at the Rikshospitalet and eventually referred Isak to Borchgrevink for surgery. As we will see, however, Isak's case proved to be a tipping point for the team at the Oslo Health Council, wherein disagreements concerning the handling of trans people came to a head. Soon afterward, the team disbanded. We do not know what happened to those whose applications were rejected. Some probably managed to access hormones through friends or contacts with physicians, others traveled to Brussels. Very few had enough money to travel to Morocco for genital surgery at Georges Burou's clinic in Casablanca. For those who did make this journey, there is very little documentary evidence in the state archives for what became of them subsequently, but we know that—for at least some of them—their primary care physicians continued hormone treatment after their return to Norway and helped them apply for changes of their legal sex status.[85]

By the late 1970s, sex change had undergone its own transformation: From a marginal, albeit specialized endeavor for dedicated physicians—psychiatrists, endocrinologists, and plastic surgeons—it was now at the center of professional debates within sexology, social medicine, and sexual health. From originally being a question of diagnosing and treating what medicine and psychiatry labeled as sexual abnormality, sex reassignment increasingly became a social issue and a question of social justice. This opened medical practice to criticism from the outside. Invoking medicalization theory, Nordic feminism, and lesbian and gay liberation, some health professionals at the beginning of the 1980s—inspired by trans-exclusionary radical feminism—sought to restrict access to medical transition for trans people. As social medicine shifted in the 1970s and 1980s from focusing on large-scale hygiene and public health to instead focus on minorities, its paternalism also changed: An emancipatory service created by health activists within the public health care system directed at the lesbian and gay populations, and established as an explicit critique of a heteronormative and paternalistic version of social medicine, now became the main obstacle for trans people seeking to access transition-related health services.

CHAPTER EIGHT

Draw Your Sex and I Will Tell You Who You Are

The reform of trans medicine in the 1970s and 1980s coincided with another shift in the welfare state: the expanding role of psychologists in the health care system. Already in the late 1950s, the Oslo Health Council had opened a social-psychiatric department for outpatient services, providing prevention work, acute psychiatric care, and follow-up care for patients discharged from psychiatric hospitals. By the mid-1970s, the Oslo Health Council had expanded to the point that it was responsible for coordinating all mental health services in the capital. Psychologists gained a foothold and became increasingly involved in the diagnostic assessment of trans patients.

The entry of psychologists into trans medicine was a sign of an increasingly assertive profession expanding into areas previously dominated by physicians and other medical professionals. By contrast with the United States and in Sweden—where psychology was dominated by Wundtian psychophysics, behaviorism, and statistics—psychoanalysis in Norway held a strong position among psychologists from the very beginning.[1] The first professor of psychology, Harald K. Schjelderup, was a psychoanalyst and also became a mentor to a group of medical students who would play an important role in medicine and psychiatry; this group included Karl Evang, the future director general of health.[2] After the Second World War, psychology got a proper foundation in the universities and shifted from a theoretical to a clinical training.[3] In 1973, an act on the authorization of psychologists ensured that only people with a degree in psychology from a Norwegian university could be licensed to practice as psychologists. "Psychologist" became a protected title. The role of this profession in the diagnostic assessment of trans people in the 1970s was thus part of a more general history of its involvement in the reformation of the welfare state.[4]

None of the professionals at the health council had any previous experience with the health needs of trans people, and teaching about gender identity was still not part of university curriculum. "I don't know if I had even heard the word 'transsexualism.' It drew a complete blank," recalls Bodil Solberg.[5] She had just graduated from the university when she started working as a psychologist at the council in the late 1970s: "At that time, I had just become a clinical specialist, and I was very committed to my work. From the records, I see that I have written 'based on this' and so on"; her suggestion is that this more formal way of writing—and of backing up her decisions with facts and research—reflected that she was new to the profession and took her work very seriously: "In a way, the work is conducted in a very orderly manner."[6]

Until then, treatment decisions regarding sex reassignment were left to the physicians—to the psychiatrists, plastic surgeons, and endocrinologists. As psychologists entered the picture, they asked new questions: What were the unconscious motives for seeking medical treatment? Was the request for medical treatment to transition an expression of an underlying psychological conflict? To provide answers to these questions, psychologists used projective instruments, most commonly the Rorschach method and the Draw-a-Man test.[7] On the other side of the Atlantic, these tests were in daily use in postwar psychiatry, and they quickly became popular in the gender identity clinics that opened in the United States in the late 1960s and 1970s.[8] The state-of-the-art book on transsexualism, published in 1969 and edited by the leading experts in America—Richard Green and John Money—recommended the use of projective instruments such as the Rorschach test and the house-tree-person drawing test as part of the assessment.[9]

In Scandinavian countries too, these tests were part of standard psychiatric practice. In the 1950s, Swedish psychiatrists were already using projective testing in their assessments of trans patients, both at Karolinska Hospital in Stockholm and at St. Jörgen Hospital in Gothenburg.[10] At St. Jörgen, for example, Jan Wålinder used a battery of tests in the assessment of trans patients; these included personality questionnaires for masculinity and femininity, the Marke-Gottfries Attitude-Interest Questionnaire for emotions and interests, the Minnesota Multiphasic Personality Inventory for personality traits and psychopathy, a nonverbal test (the FDCT), and a body image/self-perception test (Ira Pauly). In addition to the Draw-a-Man test, he also introduced the Frank Drawing Completion Test consisting of thirty-six unfinished drawings that patients are asked to complete.[11] At a meeting in Oslo in December 1979, Wålinder shared his experiences with the professionals at the Oslo Health Council, prompting them to start using these tests there as well. In Norway, it was not routine to use psychological tests in decisions about sex reassignment. At the Oslo meeting, Per Anchersen said that he based decisions on "clinical judgment."[12] But this was not entirely true. In 1954, he conducted a diagnostic examination on a patient in her late twenties at the psychiatric ward for men at Ullevål Hospital. During the stay of just over a week, she underwent extensive psychiatric and psychological testing, including the Rorschach test, the Wechsler-Bellevue Intelligence Scale, electroencephalography, and the Thematic Apperception Test (TAT).

The TAT test consisted of black-and-white drawings of ambiguous situations on the basis of which the test subjects were asked to tell a story. Based on the narrative, the examiner would interpret the speaker's conscious and unconscious motives and personality structures. At the Ullevål Hospital, the patient was asked to look at twenty drawings. One drawing depicted two men in fine clothing, a younger man staring into the distance and an older man looking at him. "He looks like he is telling a sad story to his father," said the patient. "It might have been a story about a crime, something he has done. Maybe he went to jail, served his sentence, and was released. It was a very sad time for him. The father listens and tries to understand what his son experienced during a sad time. He tries to understand as best he can."[13] Another drawing depicted a person huddled on the floor next to a couch with the back turned to the viewer: "It looks like that boy; it looks like there is a bed there. Maybe he just got up from the bed, maybe he was sleepwalking, perhaps he fell asleep again next to the bed, maybe he had a bad dream, something special since he left his bed. It looks like he's just sitting there sleeping."[14] According to the examiner, the test responses revealed an "unfriendly" experience of the world that exposed the test subject to "aggression," "exclusion," "dominance," and "physical nuisance": "The most evident is the need to <u>avoid and escape discomfort</u>, almost like <u>fear</u>. There is also a very strong <u>need for passivity</u>. There is also a need for <u>self-humiliation.</u>"[15] This compulsion to escape, the examiner reasoned, demonstrated that the patient often chose "the path of least resistance."

The examiner went on to conduct a Rorschach test, scoring the locations, determinants, content, and responses to the inkblots, examining perceptual, cognitive, and emotional personality structures. "According to the Rorschach, he tends to make superficial, simplistic generalizations. His ability to think critically is less developed, and his field of association is quite narrow," wrote the examiner.[16] The patient had a "restricted register," with an emotional life that was "undifferentiated," "immature," and "unstable." The examiner concluded that the patient was "intellectually and emotionally" underdeveloped and was therefore likely to have difficulty "coping with the problems his peculiar apparition will present."[17] The subject herself clearly found the testing meaningless. Her only desire was to have sex surgery to transition. "There is nothing wrong with me," she said.[18]

The Paperwork of the Self

Psychologists believed they knew the language of the unconscious: They were trained in psychological development and had the clinical skills to probe into people's unconscious drives and personality structures. To investigate the dynamics of "the desire to change sex," wrote Kirsti Malterud and Bodil Solberg at the Oslo Health Council in 1983, the clinical investigation had to be extensive and prolonged.[19] "The most important thing we saw," says Solberg looking back at that time, "was underlying conflicts. You could see a lack of resources."[20] The psychologists reasoned that in-depth projective testing was a way to explore the "forces and motives" unknown to the patient.[21] The psychologists believed that the tests enabled a "look behind," or, as Rebecca Lemov put it, enabled a way to "regularize the observation of all that otherwise eluded ordinary visibility."[22] By combining projective tests such as the Rorschach and Draw-a-Man tests with clinical interviews and observation, psychologists sought to create a window into the deepest layers of the patients' identities, much like an X-ray of the self.

Originally developed by the Swiss psychiatrist Hermann Rorschach and published in his book *Psychodynamik* in 1921, projective testing, he argued, provided unique access to the inner human world. The inner world organized psychic structures, interacted with the test material, and produced affective responses in the test situation, he reasoned, and the meanings projected onto the plates reflected underlying personality structures that organized perception and cognition. Rorschach's primary interest was not in *what* people saw; he was not concerned with people's imagination or fantasies. He was interested instead in *how* the test subject saw the inkblots. The Rorschach instrument tested perception and apperception, as evidenced by the number of responses, the response time, the shape and part of the blot upon which the patient focused, the color of the blot, and so on.[23]

According to Lorraine Daston and Peter Galison's argument, the Rorschach test exemplifies a new ethical-epistemological scientific ideal of trained judgment emerging in the early twentieth century that complemented the older ideal of mechanical objectivity. While mechanical objectivity valorized objective observations free from human interpretation, the ideal of trained judgment appreciated the subjective and unconscious self of the researcher in the production of knowledge: "Objective" access to the subjective went through a trained expert.[24] A good example of this regime, they argued, is the Rorschach test, unifying two seemingly contradictory scientific demands: the desire and need for objectivity, and the desire to probe into subjectivity.[25] The inkblot test was an instrument for the measurement of the modern self, and as an X-ray of the self, in Galison's words, "it both reflected this new interiority and, more actively, provided a powerful assessment procedure, a universally recognized visual sign, and a compelling central metaphor."[26]

While researchers in the postwar period increasingly criticized the Rorschach test for its lack of validity and reliability, many psychologists praised it for its usefulness in clinical practice. According to a compendium on the Rorschach method used by psychology students at Norwegian universities in the 1970s, it offered a unique insight into structural and dynamic individual characteristics that could not be obtained by other means.[27] The work was challenging but stimulating. "I had a lot of fun," says Solberg in an interview. "It was exciting. Clinically, it was new to me." This was in the period just before the implementation of diagnostic Structured Clinical Interviews (SCID) and before the attendant third edition of the *Diagnostic and Statistical Manual of Mental Disorders* (DSM-III) was imported from the United States. It was an era of clinical implementation and exploration. "It was extremely exciting to be a psychologist at that time, before the advent of standardized patient tracks, bureaucratization, and statistics," says Solberg.[28] Malterud also remembered the 1970s as a time when clinical judgment played a major role in clinical psychiatric assessment. "Back then, in the late 1970s, it was not like in psychiatry today, where you cannot talk to someone without first finding a scale or a questionnaire," says Malterud laconically in an interview today. "Most things, even in areas other than what we are talking about now, were much more informal, much more based on clinical judgment."[29]

Combining clinical interviews, interpretation of transference and countertransference, and projective tests such as the Rorschach, psychologists made recommendations for trans patients. "There is a doubleness in him," one psychologist noted of a patient. "On the one hand he feels and wants to be obliterated and castrated, on the other hand, he is fundamentally afraid of this. He desperately wants to be something." Seeking medical treatment was just a way of avoiding inner conflicts, the psychologist reasoned. "To be a woman would confirm to him his own self-contempt, a position so pitiful that he would be unassailable where nothing could be expected of him. It would free him from conflict, save him from anxiety and challenges, save him from ridicule and further castration." Based on clinical interviews and the Rorschach test, the psychologist was convinced that medical treatment was not the best solution. "A sex change would confirm his negative self-perception and make him, in his own unconscious, the most pathetic thing there is, as if it would free him from demands. He would not gain a deeper understanding of himself nor develop a psychologically stable identity." Patients were often unaware of the deeper motives for their demands, the psychologist reasoned. "It would go against his psychological nature, and this would prevent further psychological growth." The conclusion was clear: "This rejection will most likely be a great disappointment to him. He seems to be part of the community to such an extent and is monomaniacally fixated on the sex change. Nevertheless, I will not give this primacy."[30]

If psychologists were unaware of their own biases of pathologizing human experiences of sex that did not fit into the binary biological model of sex as unchangeable and given at birth, however, we must question how open-ended and exploratory were projective tests in practice? The Rorschach was not a "magical" or "omnipotent" instrument, wrote Bjørn Killingmo, a psychoanalyst and later professor of psychology at the University of Oslo, in a widely used Rorschach textbook published in 1980.[31] The great advantage of the method was that it allowed for an unstructured but standardized way of observing the patient, but data from the testing situation had to be translated into theoretical concepts and had to fit into clinical reasoning. "In practice, clinical knowledge works largely in an intuitive way, as a trained judgment (*skjønn*)—or as a 'feeling,'" wrote Killingmo. "The experienced clinician catches the ego deficit directly. He simply 'sees' it as immediately given. Seen in this way, his way of working is close to that of the artist." Killingmo long considered becoming an architect before turning to psychology, and it is as if you can see the aspiring architect in his writing: "A painter can suddenly feel that his painting is organizing itself, *becoming* something, assuming a valid form. It is a total reaction—like an answer to a series of implicit questions."[32] But an intuitive way of working would quickly become unsatisfactory for the development of theory. For Killingmo, the Rorschach test combined clinical observation with theoretical insight and development: Less a test in the traditional sense, it was more a method, a creative process involving the psychologist's theoretical position and clinical experience.[33] This is also how the test was used in the assessment of trans patients: Clinical observations were fit into a preconceived theorization about the reasons for patients' requests for medical treatment.

Psychological Conflicts and Pencil Strokes

We see more clearly how clinical material was molded to fit a theoretical framework by looking at another projective test in action: the Draw-a-Man test. Developed by the American psychologist Florence L. Goodenough in the 1920s, the Draw-a-Man test was originally designed to test intelligence and cognitive ability in children, but soon psychologists and psychiatrists also started using it in the assessment of adults.[34] In a book published in 1949, the American psychologist Karen Machover argued that the test could be used to assess personality: "Wide and concentrated experience with drawings of the human figure indicates an intimate tie-up between the figure drawn and the personality of the individual who is doing the drawing."[35] The subject was given a paper, "preferably letter-sized," a pencil, "medium-soft lead," and an eraser, and was then asked to draw a person. In the process of drawing a person, reasoned Machover, the subject projected "himself in all of the body meanings and attitudes that have come to be represented in his body image [. . .] whether he realizes it or not."[36]

Drawing a person was a "vehicle for self-expression," for consciously or unconsciously projecting one's body image. Machover observed that the drawing's form and structure, its size, the shape of the line, whether the test subject used "long, continuous lines or short, jagged ones," whether the figure had an aggressive stance, whether its posture was rigid or fluid, the proportions of the body, incompletions, and shading, were all more regularly associated with personality structure than with content in itself.[37] This was a thinking in line with that of Rorschach: Structure, not content, provided access to the inner nature. After completing one drawing, the subject was instructed to draw the opposite sex, and according to Machover, people tended to draw their "self-sex" first: "From the standpoint of sexual identification, it is assumed to be most normal to draw the self-sex first," and therefore, if the subject drew the opposite sex first, it reflected a "degree of sexual inversion."[38] Although a review article some years later rejected this hypothesis, researchers and clinicians continued to use it to "test" for homosexuality, and in the 1970s, it was increasingly used to test the "true sex" of trans people.[39] In a 1979 study, for example, researchers compared the drawings of "gender dysphoric adults" with a control group. The majority of those classified as "biological males" and "biological females" drew women and men, respectively, in contrast to the control group who tended to draw the "self-sex" first; this made the researchers conclude that the sex that was first drawn provided clues to a person's sexual identity. These findings, according to the researchers, underscored the usefulness of projective testing in examining ambivalence in the patient group, including defense mechanisms such as repression and denial, dissociation, conversion, reaction formation, and seductiveness.[40]

Because the drawing exercise was seen as a probe into the deepest layers of a person's personality, everything about the setting mattered. According to Machover, the clinician must pay attention to every detail: "The size of the figure, where it is placed on the sheet, the rapidity of the graphic movement, the pressure, the solidity and variability of line used, the succession of parts drawn, the stance, the use of background or grounding effects, the extension of the arms toward the body or away from it, the spontaneity or rigidity, whether the figure is drawn profile or front view are all pertinent aspects of the subject's self-presentation."[41] This is also how the Swedish psychiatrist Jan Wålinder used the drawing test on his trans patients. The examiner, stated Wålinder in his dissertation, should pay attention not only to the sex of the figure but also to the overall impression, such as clothing and anatomical details.[42]

When patients met with the psychologist in the Oslo Health Council, they were given a pencil and paper and asked to draw a person. No further instructions were given, since the disorganized, nonstreamlined procedure was believed to provide rich "raw" material for analysis of the personality structure. "All of my drawings look so angry," one person said according to the psychologist's note. "I don't want them to look like that. The other day, I made a drawing for my mother. She said she could recognize the drawings from when I was seven. This is a woman."[43] Looking at the drawing today, it is hard to categorize it as either male or female; the "hesitant" pencil strokes and contours convey a "blurred" expression. Based on the examination of the drawing and the clinical interview, however, the psychologist concluded that "the patient displays strong suppression of depression."[44] In other words, the angry drawings were a sign of repressed anger turned inward, leading to depression. In another patient, the psychologist made a similar observation. "The client's pattern of denial and repression creates a fundamental emptiness (latent depression). His actual self-esteem is very low. He creates an illusion that he knows what it is to be himself, but behind this illusion, he provides a clear description of how helpless and depressed he is," the psychologist noted. "In many situations, the client manages to maintain a balance, that is, he sustains his defenses. This gives him a false sense of security, and to protect his defenses, he wants to be released from his inner tensions. On his own premises, he wants to be free of tension."[45] According to the psychologist, the examination revealed a "denial of inner chaos." But anything below the "tip of a chaotic iceberg" remained unknown to the patient.

On a crisp pink sheet, one patient had drawn several small matchstick men around a larger, more elaborate pencil drawing of an elderly woman in traditional dress looking a bit like a stereotypical grandmother. "To be a bit plump, there is little of a 'good woman' in her," Solberg says examining the drawing today.[46] Similarly, in a chapter on psychological tests of trans patients published in the book *Transsexualism and Sex Reassignment*—the authoritative source on trans medicine at the time—a psychologist stated that the drawings "often resemble high-fashion clothing models, [who] gaze at themselves in mirrors, or pose for photographers, and much attention in the drawings is paid to dress, jewelry, and hairstyle."[47] According to this reasoning, the drawings by trans women reproduced archaic and even regressive ideas of femininity that often contradicted the political positions of the psychologists. For their part, however, the psychologists did not seem to reflect on a fundamental bias of their testing technology: How easy was it for people who were not trained as artists to draw men and women without resorting to stereotypical representations of sex?

The test exercise made Isak very upset. The psychologist had called him by his birth name in the waiting room, and people had looked at him in surprise. Even before medical treatment, he passed in public as a man. "I had the feeling that she was a feminist who did not like this, who thought that women should speak up for themselves and not live like men."[48] He felt that the Rorschach test and the Draw-a-Man test were given too much weight in the process of deciding who should receive treatment. "I remember that I was supposed to draw a man, but I can't draw, so I drew a girl, since I find it so much easier to draw a skirt than pants. I remember that the psychologist made a big deal out of the fact that I drew a woman when I was asked to draw a person."[49] After the first session, the psychologist analyzed the test and told him to go into psychoanalysis and "get rid of the problem," according to Isak's recollection, "or at least give myself time to see if things would change."[50]

This idea that projective tests provided a deeper truth about the motives for medical treatment to transition, truer than what people explicitly revealed about themselves, became the biggest obstacle for those trying to access hormones and surgery. "It strikes me how inadequate words are, especially how impossible it is to explain the obvious. Then, my goal is not to make you fully understand this, something only I and other transsexuals can do," wrote Isak in a letter to the team in the early 1980s. "The case is that it is only the physique and some bodily functions that separate me from other men. Can any of you give an account and explain why you are certain that you are either a man or a woman?"[51] In the end, patients were confronted with an impossible task: How could they possibly prove that their reality was true?

The Politics of Transformation

For both patients and psychologists, transformation was the goal, but transformation meant different things to patients than it did to psychologists. Projective testing and the meeting with the psychologist enacted two profoundly different versions of the future. For psychologists and psychiatrists, personal development and growth meant getting to the bottom of psychological conflicts, repressed feelings, and defense mechanisms. "There is reason to believe that the client's desire to change sex is rooted in the hope of further reducing psychological tensions. What the client really wants is to be sexless, free of desire and sexual tension," wrote one psychologist. "The client imagines that after the sex change, he will be able to unfold in the world. His desire for transformation, however, is more an expression of a pattern of denial and repression in which he seeks to change the world outside himself to avoid chaos in his own emotional life." Therefore, the psychologist would not recommend medical treatment. "A sex change will only strengthen his defenses, which will cause him to live more defensively for the rest of his life."[52]

For those among the psychologists and physicians at the Oslo Health Council who were most critical of sex reassignment, the main fear was that hormonal and surgical treatment would "cement" underlying psychological conflicts. "If the desire for a sex change has similar motives as in our seven clients, fulfilling the desire would mean psychologically depriving the person of the opportunity to resolve the inner conflicts in their original sex," wrote Bodil Solberg and Kirsti Malterud in 1983. "Surgical intervention hinders a potential maturation and cements the attitude of keeping the feelings away."[53] After personality tests, the Rorschach test, and the Draw-a-Man test, one psychologist concluded of a twenty-two-year-old trans woman that "it is possible that the client will adapt to change sex. The main objection, however, is that this will deprive him of the opportunity for potential genuine growth and maturation, i.e., the opportunity to integrate unconscious forces in the future development of his personality."[54] The professionals saw transitioning as a bereavement that would lock patients in a "defensive state" for the rest of their lives. Looking at one of the files in an interview, Solberg explained that "the Rorschach test makes impulse pressure visible," and that the patient was not in touch with these unconscious forces: "He has to keep much sealed off, which leads to the external defensive patterns that become the whole personality."[55]

At the heart of the clinical reasoning was an irresolvable paradox. On the one hand, only a complete "sex change" was possible: Hormones should only be started if genital surgery was planned. The goal of treatment, as expressed in its definitions such as *kjønnsskifte* (sex change), *kjønnskorrigering* (sex correction), and *kjønnsbytte* (sex shift), was to effect a transition from a "before" to an "after." On the other hand, psychologists were often skeptical when clients expressed a desire for a complete "transformation": Transition should not become a "quick fix" for other psychological problems. Expectations of treatment should not be too high, as psychologists witnessed how some patients imagined a completely different life, if only they could get medical treatment. An eighteen-year-old trans woman, for example, had drawn a feminine woman in meticulous detail with far more precise pencil strokes than others. "This is my current ideal," the patient said, according to the psychologist's note. "Seems so clean. Lives in Paris, the widow of Sukarno, who was overthrown in

Indonesia." For the psychologist, this was a warning sign, an unrealistic dream of escaping the deeper psychological conflict: "The idea of becoming a woman is a way of protecting yourself, but you do not experience this yourself."[56]

But not all psychologists shared the same skepticism toward treatment or the same negative attitudes toward trans people. Astri Eidsbø Lindholm, another psychologist at the Oslo Health Council, recalls finding herself in a challenging position. "I found it very hard to say no," she says. "Who was I to make decisions in their lives?"[57] She recalls having many conversations with her clients about choosing a name. "It was important to me that they should not become too *outré*, too over-the-top. Often, they wanted names like Sonja and Silvia, but I tried to convince them to choose names closer to what they already had."[58] The desire for traditionally gendered, royal names probably expressed a longing for certain gender ideals at the time, and these names gestured to a flamboyant kind of celebrity, to a femininity far removed from the patients' social class and everyday life.[59] From a professional perspective, however, Lindholm felt it was important to facilitate the integration of the patient's past into the clinical setting and treatment goal. "I wanted them to keep some of their identity," Lindholm says in an interview. "It was a bit sad if they thought it was going to be a complete transformation. I found that they already had a lot of value in them, that they wouldn't become completely new."[60]

The psychological framework of transformation, shifting, and changing sex nonetheless overlooked the fact that many people requested treatment primarily to modify their bodies to be more in line with an experience of self. "For me, the term 'sex change' is not a fitting description of what I have always been and what I am about to go through," wrote Isak to the team, after he was asked to describe his biography and reasons for seeking treatment. "I have always been male, and that has always felt right. What I am doing now is simply telling people and living as the person I am. I do this with the firm conviction that this is the right decision, and therefore with peace in my heart. The whole thing is so undramatic and natural to me that I don't even know what belongs in a report like this."[61]

Paper Technologies and Trained Judgment

Based on the analysis of these psychological tests, trained judgment seems to have been less about talent or intuition—echoed in clichés such as the "art of medicine" or the "clinician as artist"—but much more a mundane task with pencil and paper: taking notes, transcribing the notes, and synthesizing the findings into a conclusion. After all, trained judgment hinged on practice and repetition—on *training*: it was something the psychologist had to learn through repeated action. When psychologists conducted a Rorschach test, they plotted observations on the so-called localization sheets. These were standardized forms with small symbols representing all the inkblots in the test. During the test situation, when the patient talked about what they saw in the drawings, the psychologist circled elements of the blots on the localization sheet. Next to the Draw-a-Man sketches, or on the back of the paper, the psychologists also scribbled comments, notes, and interpretations during testing. After the consultation, the psychologist transcribed the notes and turned them into a report. This may seem unimportant; note-taking, after all, is a part of so many mundane tasks. But note-taking was not just a tool for remembering; it was part of clinical reasoning and hypothesis testing. It was a tool, in other words, in the production of knowledge in the psychologist's office.

Take, for example, the patient records. The consultation was not a verbatim transcript of the clinical interview but was written as a dialogue between the psychologist and the patient, with statements and responses. "The examination shows that you have many strong forces within you that you are actually afraid of, whether you are aware of it or not," wrote the psychologist, recording a statement put to an eighteen-year-old trans woman. "Yes, I know," the patient responded. "For example, last night, I was afraid that someone would come and cut my face." The psychologist continued:

> You spend a lot of energy trying to keep these forces away and trying to create a universe where they don't exist. You are deeply insecure about who you are, and you are easily captured by your surroundings.
>
> —Yes, if I meet a stranger on a street that I've heard is dangerous, I get scared. I get a pressure in my head and a throbbing headache when I return home.
>
> Yes, you are an easy target.
>
> —Yes, didn't I tell you that I see so many scary things?

The psychologist went on to explain the test.

> The examination shows that becoming a woman won't solve anything for you, even if you think this yourself. It seems to be a way of protecting yourself, in a

way you think it will become less dangerous for you if you are a woman . . . You love what is beautiful. You say that poverty outrages you. Yes, isn't that also a way of avoiding what is painful? . . . Jewels, furs.

—That is my life-lie. I invest thousands in skin products every year.

You are deeply insecure about who you are. You can get stuck thinking about becoming a woman. That closes off the possibility of growth inside in you. Every 18-year-old is interested in developing. So are you.

"Becoming a woman is not the best thing for you," the psychologist concluded, despite the patient's protestations to the contrary: "But that is not part of my life-lie. On this, I disagree with you." The psychologist's concluding laconicism, "That's interesting," suggested that they were decidedly uninterested in prolonging their discussion.[62]

The case stories were not written like medical anamneses, with prefaces to what the patient had said along the lines of "she claims," "according to the patient," "in his view," "she describes," "he suggested," and so on. Rather, these dialogues were written retrospectively as dialogues. Importantly, the patient's own views and experiences contradicting the psychologist's hypotheses were not omitted. To the contrary, disagreements were put on paper. Projective testing was not an objective science with a definite answer, but was a pragmatic way of testing hypotheses and negotiating insecurities with what can be defined as trained judgment. As "inscription devices" for probing into human nature, the dialogic interview was an "epistemic thing," a prerequisite for psychological reasoning.[63] Looking back on these files today, however, it is impossible not to be struck by the team's failure to address how their own prejudices—their norms of "healthy" psychological development and their ideals of the sex binary—prevented the professionals from acknowledging what their patients knew about themselves.

Traces of the Future

As Cornelia Vismann argued, archives can have the beguiling effect of seeming like "chambers of the real."[64] But as all archivists know, archives have as much to do with destruction as they do with preservation; the decisions are about what to keep and preserve and to store for the future, and which stories are to be discarded.[65] To document the "lingering impact of past traumas that continue to shape, and sometimes haunt, queer lives across time," argues Heike Bauer, it is the historian's task to account for the "unspoken acquiescence alongside overt forms of resistance."[66] In the diagnostic regime of projective testing, which narratives were acknowledged and which were disavowed? While most patients made drawings of people who could easily be identified as female or male, some refused to be categorized. One drawing showed a person with an androgynous face, long hair, and a hairy chest. "I don't recall that being addressed in any way," says Bodil Solberg, reflecting on whether nonbinary sex and identity was discussed at the council.[67] "I'm sure that some of the people we saw could today be defined as nonbinary," says Kirsti Malterud, looking back on that time. "But that was not how they saw themselves, or how they referred to themselves."[68] The question then becomes how a model of care in which people were asked hyperconsciously to articulate what they feel, believe, and desire in seeking care could possibly fit with an exploratory, open-ended approach to human phenomena? Paradoxically, the psychoanalytic scheme of transness as repressed conflict, latent depression, or a quick fix for unresolved identity issues—alongside the use of projective instruments to operationalize this framework—prevented an open-ended psychodynamic exploration of human sex embodiment and identity.[69]

Perhaps we can see the androgynous drawing as what Carlo Ginzburg has called "sometimes-infinitesimal traces" that open a "deeper reality that would have been impossible by other means"?[70] Perhaps we can approach it as a "trace of the future," a reality that was "not allowed to come into being"?[71] At the time, there were no terms in the clinic, in psychological testing, or in clinical textbooks for human experiences of sex transcending the binary. When such experiences were recorded, they were interpreted as statistical noise—and were erased from the analysis. Jan Wålinder, for example, included the Draw-a-Man results of forty-three patients in his dissertation: "Five drawings had to be excluded from the subsequent analysis because it was not clear what sex was intended."[72] In the model of gender identity as implemented in the Oslo and Gothenburg clinics, the nosological framework of transsexuality and the practice of projective testing human experiences outside of a sex binary were impossible; any evidence for human experiences outside of a sex binary was dismissed simply as noise.

Out of seven patients assessed by psychologists in the late 1970s and early 1980s, the team of professionals at the Oslo Health Council reluctantly accepted only one for medical treatment—as disclosed above, this one patient was Isak.[73] The psychological examination and projective testing were the main reasons why so few patients eventually received treatment. This story has some similarities with the assessment of post-traumatic stress disorder (PTSD) among Vietnam War veterans in the United States, where a critical goal for health professionals, as Allan Young has argued, was to explore the temporal-causal relationship between etiology (the trauma) and symptoms. Only if the disability was "service-connected" would the soldier be eligible for compensation and care within the Veterans Administration medical system.[74] In both cases, patients came with a clear request for medical treatment or a diagnosis. This was different from other clinical situations in which the patient presented with a symptom or a clinical problem, and the physician made a diagnosis and selected a treatment. Gatekeeping was a way to reduce costs and limit compensation.

"In meeting the most extreme human behavior or the greatest mysteries of social and cultural existence during these years," wrote Rebecca Lemov, referring to the mid-twentieth-century United States, "the logical step to take, it seemed, was to administer a scientific test or a battery of tests."[75] Although projective instruments required interpretation and lacked many of the features of more standardized instruments such as IQ tests, they at least provided an experimental-like control over complex and chaotic human phenomena. An important difference between the history of PTSD diagnosis in the United States and the diagnosis of trans people in Sweden and Norway, however, was the role of health professionals in the public health system. In the case of PTSD among war veterans, gatekeeping was a way of limiting compensation payments. In the case of sex reassignment, gatekeeping was about professional judgment: Is medical treatment the right solution in this situation? But this is only half of the story. In the early 1980s welfare state, the implementation of projective testing also responded to the objective of social medicine to enable the "common good." Psychologists willingly pursued this goal as they entered the public health care system, introducing new theories and testing technologies into trans medicine. Although they claimed to speak the language of the unconscious and to promote autonomy through introspection, this vision of individual growth or the good society did not reserve a space for people to transition.

CHAPTER NINE

Community Care and Scientific Activism

One day, in the autumn of 1966, Anette Hall, Gunnel, Betty, and Yvonne met at Norrköping Central Station in eastern Sweden. To be able to identify each other, each was carrying an issue of *Transvestia*, an American transvestite magazine.[1] It was Hall who had taken the initiative to set up this meeting, having returned from the United States where she met Virginia Prince, founder of the American transvestite organization FPE, known as Phi Pi Epsilon for the name in Greek. Hall was thinking about creating a Scandinavian branch of FPE and was surprised to learn that it already had a few members in the Nordic countries. When she returned to Sweden, she contacted the other members in Denmark, Norway, Finland, and Sweden. After meeting at the central station, the group of four traveled to Hall's summer home where they started to lay out the plan for a transvestite organization in Scandinavia.

On November 17, 1966, FPE-NE was founded at Fjädern restaurant in Stockholm. Those who wanted to join were required to come as women, but those who needed a place to get ready were invited to use Hall's motel room before the dinner. In the invitation, Hall encouraged people to bring their wives. She, for one, would bring her wife, and it would be a shame if she was the only one. FPE-NE started with only eleven members—eight from Sweden, one from Denmark, one from Finland, and one from the United States—but it grew rapidly in the following decades, and by the mid-1980s, it counted 350 people in its membership.[2] Gradually, FPE-NE transformed from a social club for self-defined heterosexual transvestites to become an organization for trans advocacy and activism.[3] Its most important aim was to get rid of the psychiatric diagnoses pathologizing their community, and to achieve this goal, they invited medical experts into their secret world: The transvestite community invited psychiatrists to their parties, and they exchanged ideas in their own members' magazine. These circles of knowledge shaped nosological categories and the rationales for treatment used in trans medicine.

The histories of transvestite communities have often been neglected in trans activism and trans historiography. According to Susan Stryker, the self-understandings, knowledge, and practices of transvestite communities did not fit with the "revitalized transgender politics" emerging from the gay liberation and feminist movements of the 1970s and the queer movement of the 1990s. This renewed political climate occasionally cast Virginia Prince as the "fusty, feisty old spinster in the [transvestite] community's attic—a rigid and somewhat irrelevant anachronism of a less enlightened era."[4] But having her own PhD in pharmacology, Prince contributed original, but overlooked and underappreciated, writings in sexology and trans philosophy.[5] As transvestites were creating connections across Scandinavia, physicians and psychologists with research interests in the psychology of gender identity and outcomes after sex reassignment were at the same time establishing networks and organizations to exchange ideas and plan research projects. They attended conferences, discussed the latest findings, and published in the same journals. In 1971, the three-day Second International Symposium on Gender Identity was held in Elsinore on the northeastern coast of Denmark (the first was held in London two years earlier). Among the participants were researchers from the United States such as Anke Ehrhardt and Ira Pauly, alongside Scandinavian experts Poul Fogh-Andersen, Georg Stürup, and Jan Wålinder.[6] International conferences led to new friendships. The Norwegian psychologist Thore Langfeldt recalls that, at these conferences, he met the Danish psychiatrist Preben Hertoft, and from the United States, both the psychiatrist Richard Green and psychologist John Money; all four shared research interests that led to lifelong friendships. In 1978, their professional network was formalized as the Harry Benjamin International Gender Dysphoria Association, connecting researchers and clinicians across the Atlantic.

As historian Jules Gill-Peterson has suggested, an acknowledgment of the role of trans individuals in the production of knowledge and shaping of practices in trans medicine should extend beyond the importance of key figures; it is also a matter of interpretative practice.[7] The importance of trans individuals and communities in the establishment and development of trans medicine has long been underappreciated: In both the United States and in Scandinavia, the formalization and specialization of trans medicine would not have been possible without medical experts gaining a foothold in the trans subcultures, contacts that gave them unique insights into the communities that were the object of their research.

Forms of Expertise

Unlike in the United States, where trans people have been at the center of a broad queer liberation movement since the Stonewall uprising of 1969, the gay and trans communities in Scandinavia have long waged separate struggles. The first homosexual liberation movements in Denmark and Norway were founded in 1948 and 1950 respectively, but it would be a long time before activism acquired a public face. According to anthropologist Hans W. Kristiansen, 1950s activists stayed low and hid "behind lawyer titles and pseudonyms," but by the mid-1960s, some lesbian and gay activists were coming out.[8]

It took longer for trans people to come out in public. FPE-NE was a secretive society. All members chose female pseudonyms and were given a personal code in the organization register: the first letter for the country of residence, a second letter for the real surname, and a registration number. Meetings were held discreetly in hotel rooms throughout Scandinavia. When Aase Schibsted Knudsen joined the organization in the late 1990s, FPE-NE used the premises of a BDSM (Bondage & Discipline, Dominance & Submission, Sadism & Masochism) club in Oslo. Two torches at the entrance signaled where to go, and she used a secret knock to enter.[9] In addition to planning events, FPE-NE members published a magazine, *Feminform*.[10] Traveling to other countries was not always an option, but the magazine made it possible to stay connected, offer support, share advice, and exchange medical information and research findings. The first issues of *Feminform* were printed in an A4 format with stapled pages and double-sided printing; in the beginning, it was more of a pamphlet than a real magazine. Bjørn Ellen Brudal, the first editor of the Norwegian edition and later the Scandinavian editor, recalls that *Feminform* was a lot of "copy and paste."[11] *Feminform* published letters and questions from its members, texts about transvestite-related issues in various formats; it reproduced translated articles from the American magazine *Transvestia* and included advertisements such as for wigs and breast prostheses. The

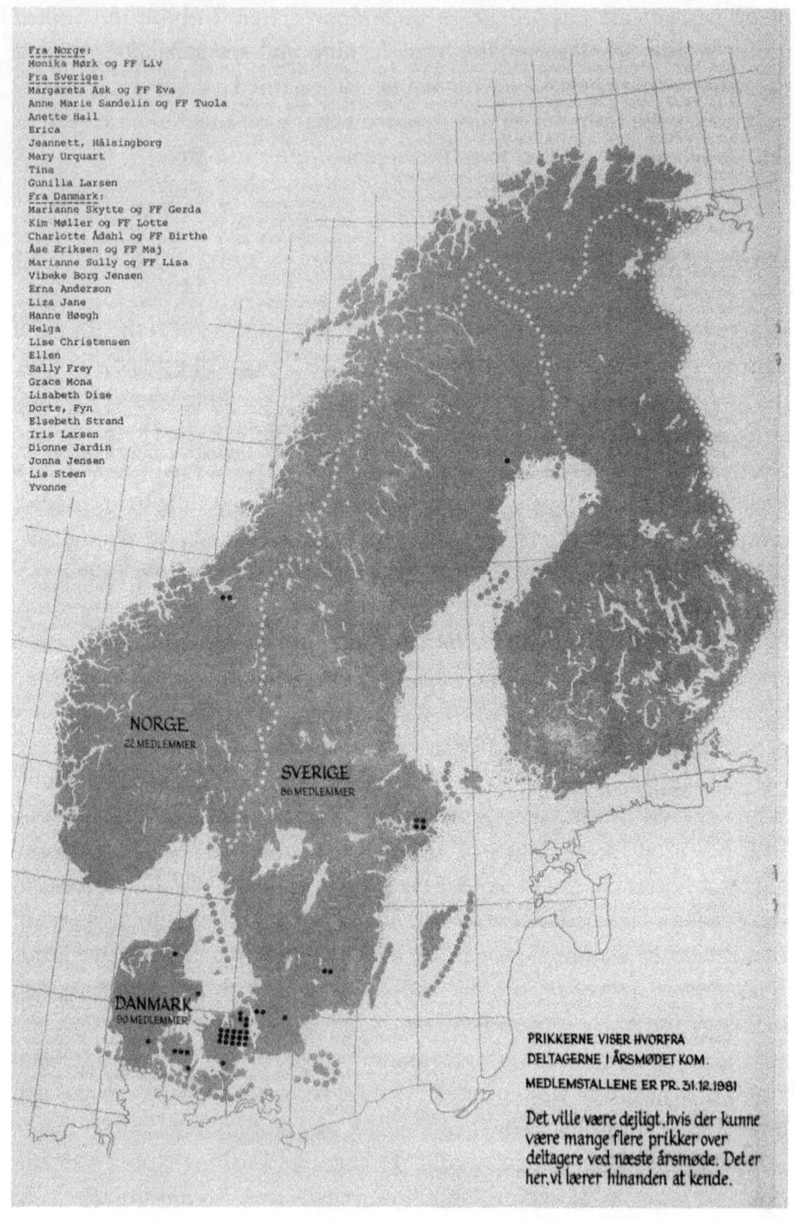

FIGURE 9.1. Map showing the FPE-NE members in December 1981. *Feminform*, no. 87 (1982). Skeivt arkiv.

articles were written in any one of the Scandinavian languages, with translations of words that non-native speakers would not understand.[12]

Feminform also served as a platform to contest theories and concepts pathologizing transvestite communities and to bring medical professionals and sexologists into direct conversation. Even in the early 1970s, the Danish transvestite community invited the sexologist Preben Hertoft and his colleague Thorkil Sørensen to attend their meetings and parties at a club in Frederiksberg outside of Copenhagen. As Sørensen later acknowledged, his doctoral research on transsexuality would not have been possible without these meetings and the hospitality of the transvestite community.[13] At that time, Hertoft and the physician and filmmaker Teit Ritzau were collaborating on a film, and Ritzau was also invited to FPE-NE gatherings. "When you come from the outside with a superficial textbook knowledge, you can be shocked," Ritzau later wrote in *Paradiset er ikke til salg* (*Paradise Is Not for Sale*), published in Danish in 1984, accompanying the movie release. "I had to accept things as they were, put aside my prejudices, and acknowledge that I was in the minority." Occasionally, someone would remove their wig, revealing thin hair growth, and Ritzau noted that he was impressed by how the community could look at themselves with "humor and distance."[14] For the sexologists, it was a chance to gain insights into a secretive community. Similarly, the transvestite community had many reasons to establish and maintain ties with the clinic and leading sexologists. This way, they maintained a direct channel to change how sexologists wrote about them, they could encourage research that would benefit the community, and they could advocate for medical treatment for those who needed it. Hertoft recalled that people often shared secrets with him at these meetings. The unwritten agreement was that you give something, and you get something in return.

It is understandable, therefore, that when Hertoft published a textbook on clinical sexology in 1976 in which he defined transvestism as a "gender identity disorder," many FPE-NE members were upset and disappointed.[15] They felt betrayed. "Are people now going to be taught that TVism—the phenomenon of TVism—is itself an illness—an identity disorder?" asked Erna in *Feminform*, using the internal abbreviation for transvestism.[16] Hertoft had clearly missed the true nature of transvestism, in her opinion: "Is it still true that someone who hasn't experienced this phenomenon themselves can only imagine it as deviant, as something outside of the normal, something that the non-TV has in his head and that he tries to follow to the letter??"[17] Åse Eriksen compared the so-called "expert" to a blind person describing colors.[18] The publication of Hertoft's book showed the other side of the coin of maintaining too close a relationship with medical experts. Medical terms might once have benefited the trans communities by providing scientific legitimacy and by steering the language around trans experience away from sin or moral condemnation or even persecution. But it was time to ask who was speaking for whom. "A man born blind does not know colors—a man born deaf does not know sounds," wrote Erna. "A man who is not a TV cannot know what TVism is."[19] Transvestism was a talent to be celebrated. "It is so wonderfully delightful sometimes to wrap oneself in female clothes and lose oneself in reverie," wrote Britt-Eva in *Feminform* a few years later. "Why does everything that is good have to be taboo? Is it against the law to be happy?"[20] Like a prince in a fairy tale whose treasure was stolen from him as a baby but later reclaimed, "the TV has found," according to an opinion piece in *Feminform*, "a part of himself which was taken away from him as a child, and he certainly has no intention of giving this part away."[21] Society was sick, not the transvestite.

Hertoft was invited to respond. He had not used the word "disorder" in a pejorative way, he clarified. His intention was not to suggest that transvestism was a mental disorder. "I have been in contact with Phi Pi Epsilon for seven years now, and it has taught me more about transvestism than anything else," Hertoft wrote. "And I hope to continue to be considered a friend of Phi Pi Epsilon."[22] FPE-NE's openness and inclusiveness, he wrote, were a prerequisite for his having any understanding at all of the phenomenon. The fact that Hertoft responded, and that FPE-NE found his opinions worthy of publication in their magazine, is probably a sign that both parties saw the value in maintaining a good relationship. The exchange bore fruit. In the second edition of his textbook, Hertoft no longer categorized transvestism and transsexuality as gender identity disorders but as "variations of gender identity" (*kønsidentitetsændringer*). In essence, the community succeeded in shaping medical concepts by insisting that knowledge about trans communities was impossible if based solely on a bookish understanding that did not take the embodied knowledge of trans people as its starting point.

In the United States too, formal and informal bonds between communities and medical professionals shaped trans medicine, such as those links forged, for example, by Reed Erickson, a trans man and philanthropist who funded Harry Benjamin's book *The Transsexual Phenomenon* and numerous gender identity clinics.[23] The welfare state context, however, provided a vastly different landscape for members to navigate and coordinate actions in Scandinavia. FPE-NE members worked through the channels and networks of academic medicine and public health institutions. While activists in the United States, as Erika Alm has noted, often wrote off state institutions as a support and vehicle for change, Swedish activists insisted on the importance of the state for transformative action.[24] The history of Danish sexologists and FPE shows how informal connections and exchanges *outside* of clinics and state institutions could reflect back and inform the work carried on within these institutions. Knowledge emerged through personal connections and intimate relationships, similar to how Benjamin learned from relationships with his patients, such as that between himself and Carla Erskine.[25]

Given the ambivalent history of the relationship between transvestites and medical experts, it may seem paradoxical that FPE-NE never abandoned the term transvestism. Perhaps it carried too much historical weight, was too personal, or maybe it served its purpose by bringing people together? Much was at stake when Hertoft, Erna, and Åse Eriksen debated in the columns of *Feminform*: Both parties realized that they had something to lose, and they used their personal connections to bridge the gap of understanding between them, hoping that the other party would understand and even ask for forgiveness.

The Homosexual, the Transvestite, the Femiphile, and the Transsexual

Before FPE-NE there was Transvestia. Transvestia was the first pen-pal and social club for trans people in Sweden, and it welcomed transsexuals, transvestites, drag queens, fetishists, and lesbians.[26] The club was founded by Eva-Lisa Bengtson and Erik Sjöman, after Bengtson found a personal ad by Sjöman in a Swedish porn magazine. Thanks to activist and scholar Sam Hultin, who interviewed Bengtson and got access to her archive after she died, we now know about this very early unique effort to create a community organization for trans people in the early 1960s.[27] When Bengtson and Sjöman placed an advertisement in the Swedish newspaper to recruit new members, they used the headline "Transvestia seeks members" to circumvent the newspaper's policy of not publishing anything that directly addressed—or used the word—transvestites. On March 8, 1964—without deliberately choosing International Women's Day—Bengtson opened the club's post office box.[28] Transvestia had an open and inclusive membership policy, while FPE-NE was from the beginning an organization for self-identified heterosexual transvestites, almost exclusively people assigned male at birth.[29] Anette Hall, in fact, was a member of Transvestia before she founded FPE-NE. Since Bengtson identified as a lesbian, however, she was not eligible for membership in the new organization.

Homosexual transvestites were generally not accepted in FPE-NE, and the rationale behind their exclusion was to protect the marriages of its members. Already in the second issue of *Feminform*, Hall stated that the policy was implemented to remain an organization that was respectful "to our FFs, wives, parents, children, and friends."[30] It was believed that excluding homosexual transvestites would help to secure the respect, understanding, and tolerance of physicians, lawyers, and the police. "Under no circumstances will there be any sex parties in our meetings!" wrote Hall.[31] Some members were critical of the exclusionary policy, however, arguing that it turned FPE-NE into a "half-fossilized" organization for the elites. In a 1978 letter published in *Feminform*, Eva-Karin Rydberg argued that the FPE-NE should instead look to the United States, where "old moss-covered sisterhood organizations" were replaced by broader, more inclusive, and more active organizations that welcomed queens, fetishists, and transsexuals.[32] Rydberg's letter was met with much criticism. In her response, Åse Eriksen wrote that "Behind most FPs stand FFs"—referring respectively to transvestites and "genuine/genetic girls"—there were family and loved ones to consider: "In short, we are not members of the FPE alone, but together with our fiancées or wives."[33] An editorial in the next issue of *Feminform* stated that "we must not forget our responsibility to our members and their wives and fiancées."[34]

Although homosexual transvestites were not admitted into the organization, many of its members, including the leadership, believed that transvestism and homosexuality were not necessarily fixed identities but developmental processes. For both heterosexual and homosexual transvestites, the transvestic process began with the arousal of wearing women's clothes, shoes, or makeup. In an article on sex and FPE-NE, Hall wrote, "From the beginning, we have all associated our TV-drive with sex," hereby reflecting the views of Prince and the policy of the American organization.[35] In an article translated into Danish and published in 1978, Prince wrote that the homosexual transvestite feels like a woman when he sees himself in the mirror dressed as a woman. However, homosexual transvestites, like all homosexuals, have no "special love" for women.[36] For homosexual transvestites, cross-dressing was satire—a sign of a "negative relationship" to women, wrote Åse Eriksen.[37] For homosexuals, cross-dressing was an easy way to escape narrow norms of masculinity and femininity, to assimilate heterosexual relationships. But in cases where cross-dressing was for more than just sexual arousal, the transvestite could take a different developmental path. "This usually happens, when he for some reason, keeps the clothes on after sexual arousal," Prince reasoned. "Up to that point, the pleasure and arousal have been exclusively erotic. But if he remains in women's clothing for half an hour or more, he may begin to feel 'the girl within.' "[38] It took time for this girl within to come alive. The crucial developmental step occurred when the action pattern of cross-dressing became a pleasure in itself, Prince argued. "It's that experience that transforms a person from just being a transvestite, i.e., a cross-dressing person, a TV person, if you will, to an FP person, a femiphile, someone who loves the feminine."[39]

For many self-identified transvestites, the female identity was not always positively present but something that had to be brought into life. Dressing up was one way of making the latent *patent*, of bringing alive a hidden aspect. "It is you—and yet it is not you. It is your other self—your other half. The girl looks at the boy from within the mirror—and the boy looks at the girl in the mirror," wrote Erna in 1978. "She is flesh of your flesh and blood of your blood, brought to life by your imagination in your brain. You have created her, literally painted her on your skin," she noted, referencing the creation of Eve from Adam's rib in Genesis:[40] "This is now bone of my bones and flesh of my flesh; she shall be called 'woman,' for she was taken out of man."[41]

Several members of FPE-NE could relate to the experience of moving in and out of roles and identities. "Even though I feel like Egil and like a man deep inside, I want people to see me as Evelyn when I am dressed as a woman," Egil/Evelyn told Norwegian lesbian and gay magazine *Fritt Fram* in 1979, in the first open interview with a transvestite in the Norwegian press.[42] This resonated with Hans/Hanne Rasmusen's experience. Rasmusen, who along with Christine Jorgensen was one of the protagonists in Hertoft and Ritzau's combined book and film project, said that the members of FPE-NE referred to their two selves as brothers and sisters. When they dressed as a woman and wanted to talk about their male self, they referred to their "brother," referring to their "sister" when the roles were reversed.[43] Erik/Eva, for example, a "happily married family father in his thirties," told a major Norwegian newspaper some years later that he thought of Eva as "another person"; "she even has her own room in our house." For Erik/Eva, transvestism was almost cyclical. "He can be Eva for up to two weeks at a time. This is followed by an equally strong need to be a man for a while—often up to months."[44]

This temporal-developmental self—beginning with cross-dressing and depending on the path, script, and repetition, differentiating into homosexual desire or femiphilic transvestism—was very different from the nosological category of transvestism in psychiatric and sexological discourse. For example, the ninth edition of the *International Classification of Diseases* (ICD) published in 1977 defined *transvestism* as "a sexual deviation in which sexual pleasure is derived from dressing in clothes of the opposite sex." According to the ICD, there was "no consistent attempt to take on the identity or behavior of the opposite sex." *Trans-sexualism* on the other hand was defined as "a sexual deviation centered around fixed beliefs that the overt bodily sex is wrong. The resulting behavior is directed towards changing the sexual organs by operation, or completely concealing the bodily sex by adopting both the dress and behavior of the opposite sex." The DSM-III (1980) followed a similar distinction between two mutually exclusive diagnoses. In other words, one diagnosis was about arousal and sexual perversion, the other about sexual deviation related to identity. "Closely related to fetishism is *transvestism*," wrote Grünfeld in a 1979 book on sexuality for the layperson. Transvestism was the experience of sexual arousal and satisfaction by dressing in the clothes of the opposite sex. "Some feel an intense urge to do so, an urge that is almost impossible to resist, while others can control the need relatively freely."[45] At the same time, even though FPE-NE rejected this definition of transvestism, its policy of sharply distinguishing between homosexual and heterosexual transvestism, between cross-dressing as arousal and the femiphile as identity, and the policy of excluding homosexual transvestites from the organization, were signs that many in the community—or, at least, its leadership—often thought within, and evoked, the same medical categories that they rejected as pathologizing. This was also the case with transsexuality.

"If a real TS is heterosexual, this means that this person has felt from birth that she has a female soul in a male body, and that due to her physical appearance she could not receive a man as a sexual partner, but that surgery could make this possible," wrote Åse Eriksen in a 1978 issue of *Feminform*, using the internally accepted abbreviation for transsexuals. Another member asked what would happen to a member who underwent surgery and "changed sex." If this person was attracted to women and was now a lesbian, would she then be kicked out of the organization? In her response, Eriksen noted that the question was strictly theoretical: transvestites were by definition heterosexual, and the person would not be a member in the first place. "The surgery should allow for 'normal' intercourse with a man," she wrote, but surgery would make "her" a "fugitive," insofar as she was now sexually attracted to women with whom she "could not have intercourse."[46]

Due to the strict membership criteria, people sometimes had to sign up for identity categories that they felt did not really fit. Jan Elisabeth Lindvik, who today identifies as trans and nonbinary, applied for membership in 1991. Before being allowed to join the organization, Lindvik had to meet for an "audition" and recalled showing up nicely dressed at Krølle Kro, a pub in Oslo. "In the corner, a bald person in a black suit, white shirt, and a tie waved me over, and that was the leader of the FPE-NE," says Lindvik. "I almost had to swear that I was a heterosexual transvestite, but I hated the word 'transvestite.' And as for heterosexual, well, it felt very strange to sign off on that."[47] Again, the strictness with which this policy was enforced probably had to do with protecting marriages and maintaining respectability. According to Esben Esther Pirelli Benestad, who had joined FPE-NE a few years earlier, the leadership was particularly afraid of jeopardizing their members' marriages. "The members were mostly people who were assigned male sex at birth, with wives and children, and it was extremely important that the wives were assured that they were not transsexual, that they did not want to change sex," says Pirelli Benestad in an interview.[48]

At the same time, it should be noted that self-descriptions might change over time and depending on the context. For some, it started with clothing and gradually developed into a full-time life as a woman. Ragnhild, who was born in the late 1940s, has this to say in an interview: "When I was seven years old, I remember that there was a lot of garbage in the area where we lived where people threw away their clothes, women's clothes and men's clothes. I felt as if I was in heaven." She remembered trying on the clothes under the tarp. "I had such joy in these girls' clothes, it felt natural to me."[49] When she was in her twenties, she wrote to the local hospital asking for help, but in response, the doctor recommended that she stay in her job and get a family. "When the desire to be a girl was too strong, I packed the clothes from my room into the car and drove to a remote place to dress up as a woman."[50] She kept her job in the military and wore women's underwear under her uniform. For Ragnhild, clothing became a way to survive, to preserve her identity. Only now, in her late sixties, has she begun the process of medically transitioning.

Jeanette Solstad, who joined the FPE-NE in 1986, has never been comfortable with the term "transvestite." Today, she identifies as a gender incongruent trans woman. From early childhood, she knew that she was a girl, even though, as she puts it, she was born with a boy's body. At the age of four, she told her mother that she wanted to change her name and dress in girls' clothes, but soon felt that others around her did not see her the way she saw herself. She continued to live a double life. "I continued to do this for years, taking every opportunity to put on girls' clothes because I was a girl in my mind, but I continued to play the role of a boy, and I often overplayed that role to show how tough I was. It was a time filled with anxiety and fear of being exposed." As an adult, she grew a beard and pursued a career in the Navy as a submarine captain. A very masculine appearance became a "shield" to protect herself, she recalls. "Psychiatrists and the Norwegian Directorate of Health classified transvestism as a mental illness, so if people had found out, they would have fired me immediately. The Navy could not have a mentally ill submarine captain! This was during the Cold War. I was responsible for twenty men, patrolling the coast of Norway, the Barents Sea, and parts of the Atlantic with warheads and torpedoes." She survived by always carrying women's clothes at sea. The submarine vault kept her and NATO safe. "I stored the clothes in the safe along with documents labelled NATO Cosmic Top Secret. These were top-secret documents that told us where we would be deployed if Norway was attacked and what sectors we would operate in," she says. "Almost no one knew about these documents, so I put my clothes in the same safe and only I had the combination." Before going to bed, she used to tell herself "I am a woman." In this way, she carried her identity into her dreams. "Even though I had to dress as a man to survive in the society at that time and in the job, I had chosen: I was a woman in my head and in the person I was."[51] After leaving the Navy and becoming a state sea pilot, she joined FPE-NE.

Ragnhild's and Jeanette's life stories have many similarities: both felt different from an early age, both used clothes to express themselves and live out their identity as women when they were alone, and both live as women today. However, while Ragnhild has chosen to transition with the help of hormones and surgery, Jeanette has not. For a long time, there was in FPE-NE an outspoken skepticism about hormonal and surgical treatment. "We all know that a number of transvestites (heterosexual) have been deceived by the TS mirage in order to get surgery," wrote Åse Eriksen, already in 1978. "There are doctors, doctors who want to make money, and there are people who are trying to escape from various problems that they think will disappear if they just get an operation."[52] This demonstration of skepticism was also a strategy of protecting their marriages and of retaining social respectability. When researchers began to interview members in detail, however, a more complex picture emerged. This research project was initiated from the inside by Esben Esther Pirelli Benestad.

Crossing Boundaries

Unable to remember the exact year, Benestad—some time in the late 1980s—saw an advertisement for FPE-NE in *Dagbladet*, a left-liberal newspaper. At the time, Benestad was practicing as a general practitioner in Grimstad, a little town bordering on the Skagerrak, a strait running between Norway and Denmark. Having cut out the ad and kept it in their wallet for a half year before contacting the FPE-NE, Benestad met Berthold Grünfeld at one of the first meetings, while the latter was still working as a sexologist at the Oslo Health Council. Like the Danish suborganization, the Norwegian transvestite community invited the country's leading sexologist to its meetings, parties, and anniversaries. The sexologist sometimes brought his wife, Gunhild Grünfeld. In an interview, she recalls meeting members of FPE-NE for the first time at Larkollen, an idyllic village on the fjord just outside of Oslo. She says that people at the party joked that they would not be hurt if she didn't recognize them on the street the next day, an anecdote pointing to a shared experience of giving and receiving.[53] Grünfeld was invited into a secret society; by bringing along his wife, he shared something personal in return. When people made jokes about themselves, it was also a sign of caring to include a nontransvestite in their secret world. Not being able to recognize someone on the street, after all, was a powerful thing, but the joke rendered it harmless.

The meeting between Grünfeld and Benestad was the beginning of a clinical and scientific collaboration. Based on questionnaires and interviews with people in FPE-NE, they wrote scientific papers together. In an interview, Benestad says that Grünfeld had a genuine interest in the group and cared about the community. In return, the community welcomed Grünfeld. The meeting also became the impetus for Benestad's career in trans medicine. At first, Benestad was worried that Grünfeld would be skeptical of their entering a field in which he was the leading expert. "We developed a good working relationship. He was happy that he was not the only one who cared about this group," says Benestad.[54] They traveled together to sexology conferences and to meetings in the WPATH, which was founded in 1979. After a while, Benestad began to refer patients from their general practice to Grünfeld, but also started administering hormone treatment independently. "As a general practitioner, you prescribe birth control pills and give testosterone after mumps. I wasn't unfamiliar with prescribing hormones and knowing what to look for. Of course, at first it was a little scary to start in this field, but then I realized that it is not complicated."[55]

When FPE-NE members were interrogated about their desire for medical treatment, they received more varied reactions than the open hostility of some members that had been expressed in *Feminform* a few years earlier. "Our transvestites desire and dream of body modifications such as removal of male hair growth, breast development, and wider hips," they wrote in an article published in the Norwegian medical journal in 1986. Thirty-one of the fifty Norwegian members returned the questionnaire. "The general goal seems to be to stay within a frame that allows the possibility of playing both roles."[56] They also published their findings in *Nordisk Sexologi* and *Feminform*, targeting both Scandinavian sexologists and the transvestite community. For most respondents, the desire to cross-dress began before puberty with a "fetishistic/masturbatory character" associated with strong feelings of sexual arousal. Only later in life, after marriage and children, did cross-dressing develop into a "social" phenomenon, but the sexual excitement never completely disappeared. "After having settled down and gained the respect of himself and often others, he experiences himself like both man and woman. To a certain extent, this depends on the situation—depending on how he is dressed, but he can also experience this psychological hermaphroditism without the stimulus of clothing." Only one was unhappy about their life as a transvestite. Even so, they wrote, "He values his transvestism very much."[57] Nearly half of the responders saw themselves as women when cross-dressing and the other half felt themselves to be both woman and man.[58] "It is therefore reasonable to assume that the boundaries are quite fluid between people who get sexual arousal and satisfaction from putting on a single piece of female clothing, i.e., a fetishistic transvestite, to cross-dressers with an androgynous self-image to transsexuals," wrote Benestad to their community in *Feminform*.[59]

The transsexual autobiography, according to Kadji Amin, narrated around the transsexual "self-actualization" made possible by medical treatment, is "a diachronic narrative form that retrospectively bestows an illusion of teleological progression upon the aleatory chaos of life experience." For some, this form has created experiences of healing; for others, Amin argues, of fragmentation and invalidation.[60] The responses by FPE-NE members complicate the sharp boundaries of transvestism and transsexuality as introduced in the nosological classification, as practiced in the selection of people for sex reassignment, and as enforced in the FPE-NE membership rules. Self-descriptions and experiences varied throughout life and were even dependent on time and context, but in the nosological framework of transsexuality and transvestism, there was little space for these nuances. A praxiographic analysis of the negotiation of identity labels and experiences between trans communities and medical experts provides a historical counterexample to the nosological master narrative.

CHAPTER TEN

Epidemiological Dreams and the Operationalization of Regret

In the 1980s, the political organization and structure of health care services in Norway underwent large shifts as the centralized and physician-controlled political model of social medicine came to an end. This happened in a changing political context in which the social democratic hegemony was put under pressure by a neoliberal push to reform the welfare state. In 1981, a conservative coalition won a majority in the parliamentary elections and the Conservative Party formed the government. Two years later, the Directorate of Health was separated from the Ministry of Social Affairs and became a purely specialist and administrative institution. This meant that the directorate lost its dual role as ministry and a specialized directorate, and the director general of health lost much of the political power formerly associated with that role. Enacted the following year, the Municipal Health Act dissolved the institution of the state-employed district physician, breaking the axis with the director general of health, and replaced it with the less powerful role of the municipal physician.

This was the beginning of the decline of the centralized physician-controlled political health care system, and it was initiated by a broad coalition of politicians from across the political spectrum who wanted more democratic control of health policy, a more socially just health service, and better distribution of health services in rural areas. Owing to this reorganization, the Oslo Health Council closed its doors for the last time in 1988 and many of its departments were transferred to the Oslo municipality. Still, the municipal physician represented the public, albeit at a decentralized level, and the result was not less but more public control over the health care system.[1] The neoliberal turn in the 1980s did not lead to a direct attack on the universalist principle of the welfare state, but, as anthropologists of the welfare state have argued, the decentralization of state services and increased municipal responsibility ushered in an indirect dismantling of the welfare state: Political power remained at the central level; the fact that the municipalities were economically unable to provide these services ultimately paved the way for privatization.[2]

The changes in trans care in the 1980s were not a direct result of this larger political shift, but we can nevertheless see here the final acts of an old tradition of social medicine and attempts to adjust to a new political reality. In 1981, the disagreements between the experts in the Oslo Health Council came to a head. In a letter to the *stadsfysikus*, three health professionals articulated these concerns as follows: "We question the provision of health services that in the long run only reinforce traditional gender roles, stereotypes, and expectations of conflict resolution."[3] Berthold Grünfeld, by contrast, saw it as his duty as a doctor to help trans people, based on a "subjective experience of their situation."[4] Grünfeld was aware of the asymmetrical distribution of power in these decisions and the tendency of doctors—as he noted in a book a few years later—to revert to paternalism "disguised as so-called medical reason."[5] The disbanding of the team led to a situation where a new routine emerged for the handling of trans people in the two decades that followed. Grünfeld, who later became a professor of social medicine, led a team of physicians consisting of a surgeon, an endocrinologist, and a psychiatrist, a group that collectively referred to themselves as the Working Group for Transsexuals/Transsexualism. The diagnostic approach shifted from psychodynamic reasoning and projective testing to one based on sexology and social medicine, and the focus shifted from concealing supposedly repressed psychological conflicts to anchoring diagnostic and treatment decisions in society and the patient's milieu. This was a shift in focus from the micro to the macro level, from internal psychological issues to external societal factors.

Internationally also, there were changes underway in trans medicine. Colleagues working with the same patient population in other countries were increasingly concerned that patients might come to regret the treatment they received for transitioning. It was unknown how many regretted the treatment. While clinicians did not have data on the long-term effects, the issue of regret was, of course, not completely new. Already in the 1920s when the first genital operations were performed, surgeons addressed the possibility of regret. In outlining the practice of Harry Benjamin and the urologist Elmer Belt in the United States in the 1950s and early 1960s, Beans Velocci describes a key consideration as being the avoidance of future regret on the part of the trans patient and the possibility of their taking legal action or seeking personal revenge on those who performed the operations. Benjamin and Belt thus set strict selection criteria for treatment.[6] It was not until the 1970s and 1980s, however, that medical researchers began to study regret more systematically. The epidemiological framework in Scandinavia, which included national identity numbers, population registers, and centralized health facilities, provided optimal conditions for studying the long-term effects of treatment. The study of regret now became "scientific." This had direct implications for clinical practice: Research findings were translated into prognostic and predictive tools. At the same time, in an attempt to protect the legitimacy of treatment decisions or even to reduce the risk of regret, clinicians were also increasingly using techniques with which they were already familiar. One of these tools was the active use of *time* in diagnostic assessment and therapeutic routines. Both technologies of time and the social aspects of transitioning shaped social medicine and trans care in the 1980s and 1990s.

Time as a Tool

The all-male working group functioned as a "team as a whole," with members contributing their expertise in addition to their regular jobs.[7] "In each case," the surgeon Henrik Borchgrevink said, "the team decided together on the indication before starting treatment for a final sex change."[8] All the specialists in the group personally examined the patients, after which each case was discussed at irregular meetings—often in the doctors' private homes after work hours—where all members were present. "It was a mix of socializing and professional matters," the plastic surgeon Knut Skolleborg recalled in an interview.[9]

Usually, Grünfeld and another psychiatrist saw the patients first. The process, which often took more than a year, included "extensive psychosocial observation, selection and treatment."[10] For many people, this waiting time was particularly tough, because it added to the time already spent hiding a fundamental part of themselves before coming out and daring to seek medical help. When they finally managed to find the few experts who dealt with this type of medicine in Norway, they had to convince the team of their gender identity and their determination to fully undergo the process. When Grünfeld first met with Isak, he explained that the team needed time and a thorough evaluation before coming to a decision. "We also emphasized that we could not be guided by her current decision to share her plans and goals with the outside world. The patient is in full agreement with our expectant and observant approach towards the situation," noted Grünfeld. "She also expects that there will be tensions and conflicts between her desire for a quick conversion and our more watchful and cautious approach, but this is something we have to live with, the patient says."[11] This waiting period was followed by another eighteen months to two years of endocrinologic and psychosocial treatment and evaluation, including an "extensive laboratory screening, personal consultation with each specialist, and evaluation in regular meetings where the whole team discusses the cases."[12] Isak remembers being impatient: "I had no problem understanding that the process had to take time. When the experts agreed to destroy a healthy body, they had to be sure of their decision," he says. "It felt like I was waiting forever, but it only took two years. In retrospect, it was not a very long time. But I was impatient. You constantly felt you had to defend yourself and explain yourself."[13]

Once the working group determined that a person was ready to begin the medical transition, the preoperative "conversion" took place on several levels concomitantly: hormonal, mental, and social. Often, it was Grünfeld, the specialist in social medicine, who prescribed the hormones and who gave patients testosterone injections in his office. The following phase of preoperative transition lasted another six to twelve months and required that the team closely monitor and support the patient through major physical and life changes. From this point, the working group required that the patient live in their "new" sex on a full-time basis. The fact that Grünfeld provided the testosterone injections in the office highlights that the expertise of social medicine extended beyond purely social considerations, but incorporated somatic and biological factors, and psychological treatments. By following the practices of social medicine and examining how the thinking of social medicine was translated into clinical action, we get a very different view of social medicine from the image provided in historiography, with its focus on ideology, grand ideas, and large-scale politics: The meeting with trans patients and the tinkering with ways of responding to their requests changed social medicine from within.

The waiting time for sex reassignment unfolded in a step-by-step pattern. For patients it was much like climbing a mountain: Just when you think you have reached the supposed peak, you realize that there is another peak behind it. Clinicians in Denmark, Sweden, and Norway all implemented variations on this stepwise approach. Bengt Lundström and Jan Wålinder, who followed a traditional psychiatric approach to transsexuality, required that patients be observed for at least two years before starting treatment. For at least one year, moreover, candidates for transitioning had to "have lived and behaved in the contrary sex role" to show "suitability and capacity for coping with a life in this role."[14] At Sexologisk Klinik at Rigshospitalet in Copenhagen, Hertoft and Sørensen—whose approach differed from the Oslo model of social medicine—saw transsexuality as an intrapersonal identity conflict that could best be helped by nonmedical interventions, but in some cases the conflict was so "rigid" that they conceded to hormone treatment and surgery. Their team followed patients for a least one year before hormone treatment and then another year before surgery.[15]

The protraction of time was not entirely new to the trans medicine of the 1980s and 1990s, however. "I had been going for several years and was getting impatient because nothing was happening," recalls Janni Christin Adeen-Wintherbauer in an interview. She met the counselors at the Oslo Health Council in the late 1970s. Patients were frustrated not only by the time dragging on but also by the opacity of the diagnostic process and therapeutic reasoning over which they had no control. For her, as for many other patients, it was an encounter with a Kafkaesque system. A few medical professionals had absolute power over her future, but the diagnostic process was opaque. "One year goes by after another, everyone else around me is having surgery," she told her physician, a gay man at the Oslo Health Council. She knew several people who had surgery. She remembers her physician saying that even if she had gotten hormones, it did not automatically mean that she would get surgery. "By that time, so many years had passed that I had developed breasts and all," says Adeen-Wintherbauer.[16] She describes how, at one point, she was so upset and frustrated that she threw one of the typewriters in the clinic against the wall. Eventually she was referred to Borchgrevink for surgery.

These stories point to a dissonance in the experience of physicians and patients in terms of what was considered "reversible" and different experiences of time and tempo. Both were waiting, time passed, but the experience was radically different. Some have the luxury of being patient, of slowing down; others do not. This is nothing unique to trans medicine. Physicians have a long history of using time as a diagnostic and therapeutic tool. In cases where the clinical picture was unclear and the diagnosis uncertain—cases such as atypical forms of headache, unclear gastrointestinal symptoms, or diffuse muscle pain—physicians have used time to track spontaneous development. Physicians have also used time to keep an eye on conditions where treatment may do more harm than good. In prostate cancer in the elderly, for example, doctors often choose to carefully monitor its progress without treatment, a practice called "watchful waiting." The rationale is that more patients are likely to die *with* cancer than *from* it. This cautious form of medicine—doing enough but not too much, perhaps what might be called a "noninterventionist" form of medicine—was probably particularly well suited to being widely adopted in the welfare state. In 1974, the government introduced the lowest efficient level of care (LEON) principle as a cornerstone in the Norwegian health care system. It stated that health care services should be provided by the lowest possible level of the health care system to reduce costs.[17]

When clinicians extended the time that would be allocated to transitioning in the 1970s, it was often because they were not familiar with the patient group and did not know how to deal with the requests. While clinicians such as Grünfeld had two decades of experience by the 1990s, their patients nevertheless still had to wait months after first contacting the working group before they received a response. Jonas's doctor referred him to the group in the early 1990s, but as he never heard back, he wrote a letter directly to the group explaining that the lack of response was making his already difficult life situation even worse. He was then given an appointment with Grünfeld, at which it was explained to him that "extending the time, not responding or waiting was a way to test people's seriousness; it was a way to filter out some of the patients."[18]

The tiered system was implemented to ensure that people took the time to make informed decisions. "In a sense, we used time as a yardstick. If we were in doubt, we used more time," says psychiatrist Dahl about the approach in the working group. "Patients had to show consistency, had to turn up for check-ups, comply with medication, and demonstrate their conviction." Many patients were denied treatment. But this "clinical no" was a "time-limited no." People kept coming back, "nagging about their problems," as Dahl put it. "Some patients kept coming back to the working group, and then perhaps in the end they had a positive decision."[19] In the regime of "watchful waiting," the door was left ajar. Patients kept coming back, begging for help, hoping that the working group would find the moment to be right. For the group, it became a question of *timing*.[20] The routine bore similarities with what anthropologist Todd Sekuler, writing about sex reassignment in France, defined as a "prudent regime of care." The goal of prolonging time in diagnostic decisions, he reasoned, was to reduce so-called "fragilities" or "embodiments of a perceived future regret in the present." Even so, the approach of "moving forward little by little"—the "politics of the 'not-yet' "—was ever undermined by the constant anticipation of regret.[21]

Mobilizing Society

The meetings with Isak and other patients in the late 1970s shaped Grünfeld's diagnostic thinking. For example, Grünfeld insisted to Isak that his decision to come out to his family and colleagues would have no bearing on the ultimate therapeutic decision. As Grünfeld gained more clinical experience, however, the process of coming out and social transition gradually became a key condition for the working group to agree to treatment. "They were very afraid that you would regret the decision. If you were a heterosexual woman like me, things were fine, but if you were a lesbian, it was not, then they would not operate on you," says Hanna in an interview. She consulted with the Working Group in the mid-1980s, and she met Grünfeld every other month in his office and talked about what had happened since the last time. "To exaggerate, if you had said you wanted a spouse, two kids, a family car, a villa, and a dog, the team would have been thrilled."[22] At first, Grünfeld concluded that Hanna was an effeminate gay man because she dated gay men. But at the time, Hanna did not care if the guy was gay or straight, and she says it was much easier to meet men in Oslo's queer scene. "But when I told Grünfeld that I dated straight guys, went to the movies, and we had a glass of wine, he asked me why it stopped there. I had to explain to him that I had not yet had genital surgery or developed breasts."[23]

The requirement to end in a heterosexual relationship was not so much about enforcing medical norms of a respectable life, but rather the belief that social support was essential for optimal treatment outcomes. As a result, patients were expected to come out to their families and colleagues early in the process. "One of the things the doctors stressed, the only thing they cared about, was whether your family was supportive. Even the surgeon asked what my mom, dad, and siblings said," recalls Hanna. "There was no way around it. Grünfeld wouldn't even have considered me as a patient if I hadn't come out." In fact, Grünfeld only began to support her application after she decided to come out to her colleagues in her new job, and Hanna remembers that Grünfeld was very surprised by her decision and by her acting upon it. "I finally got the hormones. He said I had to run to get to the pharmacy before it closed. There and then, he created an atmosphere that made me want to give him a hug," says Hanna. "I had the impression that he did not want to be held responsible if my family distanced themselves from me if I had not tried to do something about it," she says. "The social aspect was of very great importance to him. During our conversations, he kept coming back to what I had done with my family, holidays with my siblings visiting. He emphasized your social life as a woman to an extreme degree. That is why he was probing how I managed my life, my daily routines, and my free time."[24]

When the first patients came to the council in the mid- and late 1970s, social workers were already helping to facilitate social transition in the workplace, even arranging for vocational rehabilitation or relocation to a new job. In one report by a physician and a social worker at the Rådgivningstjenesten, they describe how "the client is in a very complicated situation psychologically and socially," with reference to Janni Christin Adeen-Wintherbauer when she requested treatment in the late 1970s. Since adolescence, she had sold sex to earn a living, but now she wanted a more stable life. "Since childhood he has had the need to dress and appear like a woman. For this reason, he has not found a foothold in a work situation or a situation that has given him general social status/acceptance."[25] Having a stable work life was seen as a prerequisite for successful sex reassignment treatment, however, and the counselors suggested that the patient's options for vocational rehabilitation or finding a new job should be explored first before any consideration of hormonal or surgical treatment. In other words, the integration into work life was as a crucial part of the therapeutic process.

Since society and the milieu played such an important role in realizing optimal treatment outcomes, it was also logical for the working group to help patients have children after treatment, even if it meant using high-tech medicine such as IVF treatment. Private clinics began experimenting with artificial insemination as early as the late 1920s, but it was not until after the Second World War that it attracted much interest. For the Norwegian health authorities, artificial insemination fit well with the goals of social medicine. According to historian Eira Bjørvik, the technology became a tool for the state to promote the goal of population growth in the postwar period, and this went hand in hand with the expanded concept of health in social medicine, which included presumptions about women's "biological need" for children.[26]

This also coincided with Grünfeld's view. For him, infertility was a disease, and he strongly defended access to IVF technologies, also for trans people, at least those he approved for sex reassignment.[27] But it was not easy to find doctors who were willing to offer IVF in these situations. "The doctor states that she cannot, for reasons of conscience, carry out the task in this case," wrote a professor of gynecology at the Rikshospitalet, in response to Grünfeld's request for insemination treatment for Isak and his wife.[28] "Despite some of the concerns one might well have in a case like this," Grünfeld wrote to colleagues at the Ullevål Hospital, it would be a pity if the couple had to travel abroad when they could get help from the health care system at home. "I am quite convinced that they will be able to take care of a child in a satisfactory way."[29] This thinking resonated with the approach taken to trans patients on the other side of the Atlantic at Johns Hopkins Hospital. Some years earlier, John Money and Anke A. Ehrhardt noted that "it does not even matter if the father (by adoption) is a female-to-male transsexual, provided his hormonal and surgical masculinization have given him the outward appearance and voice of a man, and provided he relates to the child's mother as her lover and husband—irrespective of how they actually perform coitally."[30] What mattered was a person's ability to fulfill the role of mother or father and to meet certain standards for the "phenotype" of a woman or a man.

The Oslo working group also selected patients for treatment based on bodily phenotype. "It was a very conservative view of what a woman should look like and be like," says Hanna. "I know of several people who were stopped because they looked a little too masculine. Those who were too tall were stopped."[31] The selection of patients based on physical appearance mirrored Harry Benjamin's practice two decades earlier. His approval of sex reassignment surgery depended largely on whether he envisioned the person as a "successful woman," based on physical appearance and the impression of the personality. "A heavy masculine build, a height of six feet or more, and a strong, dark beard were causes for worry and doubt," wrote Benjamin in *The Transsexual Phenomenon*.[32] In an article published in 1973, Robert J. Stoller reached the same conclusion: that in the absence of proper follow-up studies and prognostic criteria, only "those males who are the most feminine" should be offered medical transition.[33] For the Oslo working group, however, height and build were not per se determinants of who received treatment. A psychologist and colleague of Grünfeld's at the Oslo Health Council in the 1980s recalled that a patient who was 1.9 meters tall and had a shoe size of 14 was accepted for treatment.[34] This statement is consistent with Daisy Hafstad's experience, but in an interview with Skeivt arkiv, she recalled how hurtful Grünfeld could be: "I understand that he had to be strict, you had to be one hundred percent sure that you really wanted it, because there was no way back," she says. "But you can't say to a person, you're not very pretty as a man, how do you think you'll look when you are an ugly old lady." She said that Grünfeld looked not only at how she dressed or whether she put on makeup, but also at how she behaved. "I understood that things go deeper, that a dress and makeup are not enough to convince a sexologist." She recalls thinking it must be important to have a supportive environment, so she applied for work at a local high school and gradually built a network. Finally, she was approved for treatment and underwent surgery in July 1989.[35]

Although phenotypic criteria reflected traditional ideals of gender, they also expressed a founding principle of social medicine. The therapeutic goal was not—as it had been in the 1950s and 1960s—to reproduce stereotypical gender phenotypes per se. We can see the contrast, for example, between Per Anchersen telling his patient Sophia in the late 1960s to take care of her "gracile" body after treatment and the situation two decades later, when the working group saw the goals of treatment in relation to society: After treatment, the patient had to adapt, adjust, and integrate into society. Sigrid Sandal argued that this shift reflected a more complex understanding of sex (*kjønn*) that included gender identity and a turn from the importance in the 1950s of "gender aesthetics" to what was technically possible in the 1970s.[36] However, it was not so much that "gender aesthetics" disappeared from diagnostic considerations but that the medical concept of gender, including its aesthetics, had become inseparable from the socio-medical ideal of the patient's integration into society. Because the expanded concept of health in social medicine relied so heavily on seeing health and illness as the interaction of biological, psychological, and social factors, this also meant that clinicians had to imagine how a person would integrate into a society with narrow norms of gender expression. In other words, the goal of the diagnostic regime was not primarily to determine who could fulfill narrow physical norms of masculinity or femininity, but to protect a socio-medical norm of gender that implied conformity to society. Importantly, the site of intervention remained the individual, who had to absorb, adapt, and transform social norms. This points to a paradox in

Grünfeld's thinking, since, beyond his own clinical practice, he was otherwise publicly critical of stereotypical gender norms, moralistic attitudes to sex, and heteronormative attitudes to sexuality. A potentially subversive technique, in other words, was predicated on reinforcing narrow gender norms. Yet health workers knew that social norms were slow to change, and they probably regarded their therapeutic responses as pragmatic responses to medical interventions raising complex ethical questions.

Regret in Research

Until the 1960s, most research on the hormonal and surgical treatment of trans people was limited to case reports. Some of these cases involved patients who had been followed for years. In 1966, Harry Benjamin published results based on his patient series. Out of seventy-three MtF patients who had undergone surgery in Europe (in Italy, the Netherlands, or Denmark), in the United States, in Mexico, or in Casablanca, and whom he had observed for at least three months, he found the outcome to be good or satisfactory in sixty-two patients, doubtful in nine, and unsatisfactory in two. He based his assessment on the "overall result," factoring in physical health, emotional state, surgical result, and sex life. Of the twenty FtM patients who underwent surgery, he concluded that all but one had good or satisfactory results.[37] Although Benjamin did not attempt to identify prognostic factors based on his series, further follow-up studies were published in the following decade, and these studies prompted researchers to identify factors associated with good outcomes.[38]

In pursuit of data on the factors most likely to yield positive outcomes, researchers in Scandinavia saw an opportunity to use the infrastructures of the welfare state and public health care systems. As Todd Sekuler has pointed out, the Swedish National Board of Health and Welfare helped to subsume regret into a clinical tool: prognosis.[39] National registers in all Scandinavian countries—archives of medical records, public offices, specialized clinics, and epidemiological infrastructure—in fact provided a comprehensive archive of clinical data for medical researchers to construct prognostic tools. Centralized health facilities and comprehensive patient registries provided unique opportunities to collect data on everybody who had ever undergone sex reassignment. National identity numbers and extensive population registers brought together science, citizens, and the state, and the epidemiological infrastructure and state-science nexus represented a treasure trove for the study of prognosis and regret. “Scandinavian epidemiology,” noted Susanne Bauer, was “an epidemiologist’s dream.”[40]

“Through the existence of special registration procedures in Sweden it is possible to obtain from social registers information as to how far any person has availed himself of social security benefit or has become known to the authorities for alcoholic excess or the commission of crime,” noted psychiatrist Jan Wålinder and psychiatric social worker Inga Thuwe in a follow-up study published in 1975 of twenty-four persons who had undergone sex reassignment.[41] To conduct the research they received special permission to extract data on their patients from these registers, in addition to information on sick leave and disability pensions from the records of public insurance funds. The minimum observation period was three years, but the actual mean observation period was seven and a half years for the trans masculine group, and a little more than six years for the trans feminine group. None of the trans men regretted undergoing the procedure, while for two in the trans feminine group, the procedure was deemed a “failure.”[42] Wålinder and Thuwe concluded that the study was too small to allow for statistical analysis, but they nonetheless tried to draw prognostic conclusions based on the two cases considered to be a failure. There were numerous factors, however, complicating the situation of each of these individuals: a diagnosis of personality disorder with immaturity and emotional instability, a history of criminal behavior, alcoholism, and poor abilities in long-term planning. Both had suicidal thoughts before and after surgery. Physical characteristics such as heavy bone structure, height, and body hair led to difficulties in social adjustment. Difficulty in maintaining regular contact with the team after surgery and poor cooperation in the postoperative period led to excessive scarring and contracture in the surgical area, which made sex complicated. Although both were said to regret the procedures, importantly, both continued to live as women.[43]

These two cases inspired the Swedish researchers to initiate a more comprehensive study of prognostic factors. Three years later, they published their findings in the *British Journal of Psychiatry*. The study compared five “regretters”—including the two from the earlier study—with nine “very satisfactory” controls, and from these fourteen cases, the research team developed a list of thirteen prognostic factors. Valuable though this kind of research was, the study had many limitations from the very beginning: The researchers used data from only five individuals; they hand-picked the controls; three of the control patients had not

Table

Occurrence of factors suspected to be prognostically unfavourable, division by individuals

Factors suspected to be prognostically unfavourable	Repentant cases (1–5)					Satisfied cases (6–14)									P*
	1	2	3	4	5	6	7	8	9	10	11	12	13	14	
1. Psychotic reaction ..	−	−	+†	−	−	−	−	−	−	−	−	−	−	−	NS
2. Mental retardation ..	−	−	+	−	−	−	−	−	−	−	−	−	−	−	NS
3. Unstable personality ..	+	+	+	+	−	−	+	−	−	−	−	+	−	−	·06
4. Alcoholism—drug addiction	+	−	+	+	−	−	−	−	+	−	−	+	−	−	NS
5. Criminality	−	+	+	+	+	−	−	−	−	−	−	+	−	−	·02
6. Inadequacy in self-support	+	+	+	+	+	−	+	−	+	+	−	+	−	−	·06
7. Inadequate support from family	+	+	+	+	+	−	+	−	−	−	−	+	−	−	·01
8. Excessive geographical distance	−	+	−	−	−	+	+	+	+	−	−	+	+	−	NS
9. Physical build inappropriate to the new sex role	+	+	+	+	−	−	−	−	−	−	−	−	−	−	·005
10. Completed military service	−	−	+	+	+	−	−	+	−	−	−	+	−	−	NS
11. Heterosexual experience	+	−	+	+	+	−	+	−	−	−	−	−	−	+	·06
12. Strong sexual interest ..	+	−	+	+	+	−	+	+	−	−	+	−	−	+	NS
13. Age at request for intervention	26	26	36	37	32	22	34	21	16	23	21	26	19	20	

* P is determined by the Fisher Exact Test (one-tail).

† Repeated psychotic reactions in phases of abstinence after alcohol—also drug addiction.

FIGURE 10.1. Table of "prognostically unfavorable" factors for sex reassignment. From Jan Wålinder, Bengt Lundström, and Inga Thuwe, "Prognostic Factors in the Assessment of Male Transsexuals," *British Journal of Psychiatry* 132 (1978): 16–20.

yet undergone surgery; and in three cases, they relied on medical records rather than interviews. Despite the limitations of the methodological design and the lack of conceptual clarity, however, the team used statistical tests to characterize generalizable factors for a bad prognosis, which they enumerated as follows: a history of alcoholism or criminality, "inadequacy in self-support," inadequate family support, a body type "inappropriate to the new sex role," and old age.[44]

After the publication of the study, the predictive criteria started to circulate in the international research community. One of the researchers, Bengt Lundström, was excited to share the results of his study with his colleague across the Atlantic, Harry Benjamin, and over the years, he sent copies of his papers to the latter. "Let me thank you ever so much for your letter of April 2, and the copy of your lecture," wrote Benjamin in 1979. "I am looking forward to reading it as soon as I have been fitted with the proper glasses following cataract surgery."[45] Three years later, he expressed his gratitude again: "I am nearly blind and cannot read for myself anymore, but part of your study was read to me, and I was very much impressed with your findings."[46]

Standardizing Regret

The research project sought to conjoin a surgical procedure with an emotional state—one of satisfaction or regret—yet at no point did the team provide a definition of regret. The prognostic criteria they had proposed, nevertheless, soon took on a life of their own—even without scientific validation. The Swedish studies pioneered research on prognostic factors, and medical researchers continued to cite them in subsequent publications.[47] Gradually, research on prognostic factors and the goal of predicting regret found its way into clinical practice, and physicians and psychologists in Sweden and Norway started to implement prognostic factors in the process of selecting patients for treatment.[48] "The most crucial issue in transsexuality is whether sex reassignment was a mistake," wrote three Swedish psychiatrists in a 1986 follow-up study of sex reassignment.[49] In an article published in the main national Swedish medical journal in 1977, the psychiatrist authors stressed the importance of adhering to the prognostic factors from the Swedish studies in treatment decisions: "There is ample evidence that certain factors signal a less favorable prognosis. The more of these factors are present in the individual case, the more reluctantly one should proceed. Otherwise," they wrote, "there is an imminent danger that the transsexual will regret the extensive measures, or be profoundly disharmonized with a less successful outcome." The making of prognosis became the most important task for the physician in the evaluation of who should receive surgery. "Among the most important measures in the change of sex in transsexuals," noted the Swedish researchers, "seems to be to conduct a prognostic evaluation."[50]

In Norway, the working group for transsexuals/transsexualism also introduced the prognostic criteria when selecting candidates for treatment. Astrid remembers that the psychiatrist showed her the criteria when the group initially decided to reject her for treatment. They explained that their reasons for withholding treatment were that she still hadn't as yet come out to her mother, and she was struggling with anxiety and depression due to her difficult relationship with her mother. "They had ticked off the criteria of Wålinder and Thuwe, and I didn't get the operation because of my social situation," says Astrid. "It was almost like a lottery ticket. You didn't get 12 or 11, you got 10, which didn't win you a prize."[51]

Researchers also translated prognostic factors into a standardized scoring format for postoperative assessment of sex reassignment surgery in clinical and research settings.[52] For example, when psychiatrist Alv A. Dahl in the Oslo working group in the late 1980s initiated the first long-term follow-up study of trans patients in Norway, he applied Wålinder and Thuwe's prognostic criteria. The Oslo researchers managed to include sixty-three of all the sixty-five patients (forty MtF and twenty-three FtM) who had undergone surgery at the Rikshospitalet since the first operation in 1963, at an average eight years after surgery (ranging from one to twenty-six years). The uniquely high enrollment rate, partly due to the centralized public health care services and comprehensive patient register, made this study one of the most comprehensive of its kind. Because very few Norwegian physicians had experience with trans medicine, patients remained in contact with the team after the most important surgical interventions for "various minor operations to improve the functional and cosmetic results of their treatment," the team stated. "This contact made it relatively easy to obtain the patients' cooperation in the outcome study."[53] The results were very encouraging. Physical and secondary sex characteristics were "acceptable for passing as the other sex" in almost all cases (MtF 96%, FtM 87%). The patients displayed a stable social and economic situation (MtF 96%, FtM 79%), and they remained in a reliable relationship with the team in a very high proportion of all cases (MtF 96%, FtM 92%).

Henrik Borchgrevink performed the surgical examination, while the examination of social, psychiatric, and sexological factors was conducted under the supervision of Dahl and Grünfeld. This evaluation included several questionnaires and psychometric instruments—SCL-90 for psychiatric symptoms, Global Assessment of Functioning (GAF) scale, the General Health Questionnaire (GHQ), Derogatis's Sexual Functioning Inventory, Cloninger Temperament and Character Inventory, and Bond's Defense Style Questionnaire—and Structured Clinical Interviews for DSM disorders (SCID axis I and II interviews). The team also conducted clinical interviews to assess development, treatment satisfaction, sex life, relationship status, substance use, and mental health problems, and based on this information, the researchers made diagnoses of mental illness and scored functioning levels based on instruments such as GAF and Hunt & Hampson's standardized psychosocial postoperative format.[54]

The team never published their results. Dahl says in an interview that he was not confident with statistical calculations. In May 1993, however, he nevertheless wrote to the team that, based on the preliminary calculations, the results in terms of psychopathology, sexual adjustment, family reactions, diagnoses, indication for surgery, and general assessment of function were all very good.[55] "For patients with a definite diagnosis of transsexualism and with a high motivation," concluded the team, "sex reassignment surgery is a successful treatment on all outcome measures."[56] In addition to demonstrating good treatment outcomes, the study also seemed to confirm the usefulness of prognostic factors and thereby reified the use of prognostic criteria in clinical practice.

The introduction of prognostic factors in clinical practice in the 1980s was an attempt to operationalize "objective" criteria in support of treatment decisions. Inevitably, such criteria flattened life stories and homogenized individual nuances. Outcome became a dichotomy: good or bad, a matter of success or regret, as if an operation or a whole life could be equated with one emotional state. Yet underneath "regret" were stories of stigma and lack of social support, dissatisfaction with surgical results, broken marriages, and lost jobs. No wonder that surgeons wanted to be sure that they were doing no harm before operating on people, "destroying healthy bodies," as Isak phrased it. For good reason, physicians sought to identify prognostic factors to ensure the legitimacy of therapeutic practices, to make sure their patients fared better after treatment. But in the process, it is as if they forgot that not even physicians can predict the future.

The approach to trans people in 1980s and 1990s reflected how social medicine itself was changing: In place of the focus on physical norms of femininity and masculinity from the 1950s, it was now the social aspects of transition that became the most important consideration in diagnostic assessment. This implied a shift in focus from body parts to social norms, but clinical assessment still centered on the individual. Clinicians used the months or years of waiting to assess the suitability for treatment and to ensure that the timing was right. At the same time, the turn to prognostic criteria to prevent transitioners' regret meant a shift from the individual to the large number, from the single case to the mean. Social medicine in the late twentieth century, as practiced in the encounter with trans people, was thus marked by a paradox: On the one hand, it manifested as an individualized form of care rooted in a social understanding of health; on the other, as a mean-based and generalizable concept of health.

CHAPTER ELEVEN

Bureaucratizing Medicine

When Bjørn looks back on his first experience of the Rikshospitalet in Oslo in early 2000, he does not have good memories. That was the first time he met with the psychiatric team that assessed all patients requesting hormonal and surgical sex reassignment. Bjørn knew from a very early age that he did not feel like a girl. As a child, he told his family that he did not want to have a high-pitched female voice when he grew up and asked if there were other pronouns for him apart from "she" and "he." He recalled the cold winter day when he first entered the gates of the hospital in the city center of Oslo. There Bjørn encountered a psychiatric diagnostic regime that he experienced as aloof, distanced, and judgmental. "I felt desperate. I felt completely rejected, as if I was being looked at like a creature under a microscope," Bjørn says. "There was no empathy. I was devastated."[1]

Before the consultation, he had already completed nearly eighty pages of questionnaires, with diagnostic instruments and psychiatric interview questions probing his past and his current situation. These forms, which were not discussed during the interview at the hospital, were kept in his medical record for use by the psychiatric team in diagnostic assessment and clinical decision-making. Although Bjørn eventually succeeded in obtaining hormonal and surgical treatment, the process was to prove lengthy and complicated—largely because, at that very time, the diagnostic and therapeutic services for individuals requesting sex reassignment were being reorganized, and a specialized and centralized gender identity clinic under psychiatric control was being established. If Bjørn had applied for treatment a few years earlier, he would have encountered a completely different system.

National Specialist Services

In the 1980s and 1990s, several European welfare states, including those in Scandinavia, were under pressure to liberalize. In Norway, such liberalization of the health care system was allowed to take place, but the process was kept under tight governmental control: Some services were decentralized, some were privatized, and the financing of public hospitals was reorganized to stimulate productivity.[2] Within this shifting landscape, medical services for trans people were negotiated and recreated; in an increasingly liberalized but regulated public health care system, it was guidelines, standards, economic planning, and reimbursement based on diagnosis-related groups that took center stage. As the twentieth century ended, *kjønnsskifte* (sex change)—and therefore, all that is denoted by *kjønn* (sex)—ultimately became intertwined with, and inseparable from, the politico-bureaucratic processes of reforming the public health care system. In this period, the purpose and practice of sex change was shaped by the bureaucratic processes involved in the government's decentralization of some health care services and the centralization of others. To understand the discussions about the formalization, institutionalization, and governmentalization of a national sex reassignment service in the 1990s, however, we must look to the context of the more general reorganization of the health care system that took place in the 1970s and 1980s.

While the organization and planning of the hospital structure in Norway had been on the agenda of medical authorities since the 1930s, it was not until the Storting—the Norwegian parliament—passed the first Hospitals Act in 1969 that county municipalities were given formal responsibility for the future organization and development of hospitals. The exception was the single state-owned hospital, the Rikshospitalet. In the decades following World War Two, public health officers, politicians, and experts began to distinguish between local and central hospitals. Politicians and bureaucrats were inspired by developments in other European countries, such as in England and Sweden, and the Hospitals Act was part of a more general policy change in health care administration that sought to centralize and regionalize the country's hospital structure.[3] At this time, a third regional hospital level was introduced, as indicated in a report to the Storting.[4] According to the official strategic document, which for the first time laid out a plan for the future organization of hospitals, certain specialized tasks should be carried out only in one single location.

Following convention, the organization of these functions had been decided by "the special interests of particular departments and doctors," but, in the future, this had to be better coordinated and was to include the centralization of certain health care services to control costs and better meet the needs and demands of the population.[5]

The question of centralizing specialized medical services was increasingly debated among Norwegian politicians from the late 1980s, partly as a response to the World Health Organization (WHO) initiative to put the infrastructure of health care services in member countries on the political agenda. In a report to the Storting about future health policy, the government underscored the fact that, even if medical and technological progress demanded that the organization of regional and national functions remain relatively dynamic, it was still important to control the distribution of highly specialized services, mainly because such services were extremely expensive and labor intensive.[6] These discussions were crucial to the adoption of the concept of *landsfunksjon*, a "national, countrywide function," which the report defined as a "highly specialized medical activity that from a total appraisal of the magnitude of healthcare need, degree of difficulty, and treatment costs should only be built at one, possibly a few locations in the country."[7] Three prerequisites had to be fulfilled before a service provision could be considered for *landsfunksjon*: the disease had to be rare; the service had to require expensive equipment, technology, or labor-intensive care; and, finally, the clinical expertise had to be concentrated in a single location to guarantee that knowledge and services remained up-to-date and of a high quality.[8] These stipulations would become crucial to the future organization of health care services for trans people.

This was not an uncontroversial policy shift in a country with a long, narrow landmass and a scattered population but with strong traditions of local democracy.[9] Although the health care administration was centralized, Norway had a long tradition of decentralizing political power. Choices about where to build new hospitals were always hotly contested, with the closure of hospitals mobilizing much resistance. Local communities wanted to keep the small hospitals in their districts while politicians were looking to cut costs by limiting numbers and centralizing services. Politicians therefore sought to balance the demands of local communities while creating space for more government regulation of regional hospitals. The Storting ultimately gave the Ministry of Social Affairs authority to allocate specialized services to different hospitals. As a part of this rearrangement of Norway's health care services, health authorities appointed a committee to provide an overview of psychiatric services and assess the need for centralized national functions. The committee was composed of health care professionals, mostly psychiatrists, one of whom was part of the informal team of doctors who decided at that time on requests for medical transition. In 1995, when the committee delivered its report, it recommended only three such national psychiatric services: a department for the assessment and treatment of "transsexual patients," a psychiatric epidemiological register, and a department for deaf people with psychiatric illnesses.[10] Transsexualism was so rare, the committee reasoned, and the patients in need of such highly specialized competence, that the service should remain centralized.[11]

This suggestion was in line with the prevailing opinion at the Rikshospitalet. At the beginning of the 1990s, the Directorate of Health had begun to collect reports on regional and national services, and hospitals were asked to list the number of patients treated yearly, including their demographic data and treatment results, how the services were financed, and whether they could be regionalized. The Rikshospitalet report specifically identified the following three specialized practices for which it had national responsibility: the treatment of transsexualism, the treatment of people with intersex conditions, and the embolization of cerebral and spinal arteriovenous malformations.[12] The Rikshospitalet department of plastic surgery concurred that the surgical treatment of "transsexual patients" was indeed "very special." Among the two to three patients treated annually, there were equal numbers of men and women. From 1995, gynecologists from the women's clinic at the Rikshospitalet regularly performed endoscopic oophorectomies as part of the procedure on trans men. Patients would undergo masculinizing chest surgery with the subsequent option of phalloplasty, the construction of a penis. As an alternative to flap surgery, so-called metoidioplasty—that is, using the clitoris, which had become enlarged through testosterone therapy, to construct a functional penis—became an option in the 1990s. The surgery was complicated, time-consuming, and expensive, with the average patient spending 43 days in the hospital. Since these extensive operations were performed partly on the recommendations of colleagues, head surgeon Henrik Borchgrevink explained in the report that it was reassuring to know the other team members through continuous, direct collaboration. The treatment required a "particular routine," not only surgically and technically, but also among the nursing staff; the nature of the condition and the procedures required the caregivers to demonstrate certain "human abilities."[13] Due to the special nature of this surgery, in short, sex reassignment was to be offered in only one location in Norway—at the Rikshospitalet.[14]

Formalizing a Health Service

This new position on having centralized trans medical provision was very different from the position adopted just over a decade earlier, when clinicians and representatives from the Directorate of Health discussed the organization of these health services in the Oslo Health Council in the late 1970s. While *stadsfysikus* Fredrik Mellbye argued that more formalized procedures for sex reassignment were needed, including a government-supported clinic, not all clinicians agreed.[15] According to Harald Frey, an endocrinologist and professor of medicine at the Aker Hospital, most surgical procedures for trans people, at least masculinizing chest surgery, were "so simple" that they could be performed at "a number of hospitals." The same was true for the endocrinologic treatment. "This is very easy and requires no expertise," he said. "Any internist should be able to perform such treatment."[16] Torbjørn Mork, the director general of health, rejected the idea of setting up a specialized clinic for sex reassignment. Formalization would lead more people to seek treatment, he reasoned—people who had, until then, been taken care of by individual clinicians. Mork feared that the moment such a service was formalized, several legal questions would have to be settled regarding issues such as sterilization, castration, and name changes; this collection of problems had hitherto been satisfactorily addressed on an ad hoc basis. Finally, he maintained that efficient care, in such a peculiar and marginal field of medicine, depended on highly committed clinicians with "personal engagement, interests, and attitudes." To secure the medical autonomy of experts, he preferred that physicians should continue to handle these issues independently of the Directorate of Health in an informal professional forum. "Until now," he observed, "this has been carried out and worked well on an informal basis."[17]

When the question of formalizing health care services for sex reassignment was put back on the agenda at the highest political level in the 1990s, legal matters were once again at the forefront of arguments for and against. Writing to the Norwegian Board of Health Supervision in December 1994, the psychiatrists of the Working Group for Transsexuals/Transsexualism, Berthold Grünfeld and Alv A. Dahl, raised concerns regarding the lack of capacity to follow up with patients after treatment.

Patients after surgery were routinely left to fend for themselves or left to their primary care providers. To follow up with patients, they argued, a nurse or a social worker should be hired, and this also presented a good opportunity to formalize and institutionalize diagnostic and therapeutic procedures. For the first time, the working group had received a formal complaint from a person who was denied treatment. Until then, patients who were denied treatment had nowhere to direct their complaints and no avenues for obtaining a second opinion; given the constant stream of patients seeking sex reassignment, the psychiatrists argued, future complaints could be expected. Judicial aspects, which had previously been handled on an ad hoc basis, now had to be addressed.[18] In a further letter to the chief county medical officer in Oslo, Grünfeld explained that the question of a second opinion for people whose applications had been rejected "might be worth discussing," but, he added, "We have never encountered this as a problem up until now, even though we have assessed approximately one hundred cases since the mid-1970s."[19]

Grünfeld's concerns reflected a much broader development in medicine and society. In the 1970s and 1980s, several grassroots, patient, and activist movements emerged internationally, as well as in Scandinavia, that challenged hierarchical power structures in medicine and contested medical knowledge that was "detached" from lived experience. Activists demanded rights and control over their bodies and a seat at the table in scientific and clinical decision-making.[20] In medicine and society at large, patients' rights were increasingly discussed and acknowledged. As one professor of law put it when looking back on that time, "'patients' rights' became a term with strong positive connotations with major argumentational power in health politics."[21] The historian Anne-Lise Seip characterized this moment as a change from a collectivist state to the rights-based state.[22] In Norway, several laws aiming to increase individual rights were implemented: an Organ Transplantation Act (1973), which required donors to consent before biological material could be used in the treatment of others; a law allowing abortion on demand (1978); the Physicians' Act (1980), which introduced patients' rights to information, access to their medical records, and participation in treatment decisions; and the Municipal Health Services Act (1982), which stated that everyone staying in a municipality had the right to necessary health care. In a 1992 bill, it was recommended that patients have the option of obtaining a second opinion in treatment decisions, but provision of the service was not mandatory.[23] These shifts culminated in the Patients' Rights Act of 1999, which included a paragraph on the right to a second opinion. This is the backdrop against which we must interpret the question of the right to a second opinion for patients denied sex reassignment, and the working group's increasing awareness that patients had thus far been unable to exercise this option. In an interview, Dahl recalled that the working group discussed legal developments in other European countries.[24] In 1989 the Council of Europe recommended that member states introduce legislation to enable people legally to change their name or sex and protect their private life in the case of "irreversible transsexualism." The European Court of Human Rights condemned France in 1992 for not giving a trans woman the option of a legal sex change.[25] More and more, patients' rights had to be taken into account when planning health services.

The complaint to which Grünfeld and Dahl referred in their letter was not described as the first such complaint *ever* made; rather, they called it the first *formal* one: "We have recently had the first formal complaint on the rejection of sex change." The complaint in question concerned a trans woman who had changed her name after approval by the county governor. Her endocrinologist had initiated hormonal treatment, but the working group rejected her application for genital reconstruction surgery.[26] The doctors noted that she had had several "heterosexual relationships" and did not fulfill the criteria for a "true transsexual." In addition, her "older age" was an important contraindication for the operation, and her height and weight and other pronounced "phenotypical traits" would "not give a very convincing representation in the role."[27]

The fact that the working group, as we have seen, shared experiences with colleagues in neighboring countries means that second opinions could have been sought in cases of doubt. The idea of a right to a second opinion, however, seems to have remained irrelevant to medical thinking and so physicians never pursued this option prior to the 1990s. Only then did the question of institutionalizing health care services for medical transition as a national function activate the matter of patients' rights, in a medico-political context where patients' rights had gained traction, including the right to a second opinion on treatment decisions. By the turn of the century, the Rikshospitalet had reached the conclusion that "if the working group rejects a patient, we recommend that the Nordic centers for sex conversion can offer a second opinion."[28] This never happened.

The co-production of patient autonomy and formalized health care services to secure patients' rights had unintended consequences. Paradoxically, while the debates about patients' rights propelled the institutionalization of a clinic for trans medicine, the establishing of that clinic in the early 2000s cemented the monopoly over such medicine without increasing patient rights, such as the right to a second opinion; nor did it improve the transparency of therapeutic reasoning and decision-making. When the Directorate of Health in 2020 published new guidelines on the organizational structure for this type of treatment, the lack of a second opinion was one of the justifications given for the breaking up of this very monopoly by regionalizing some of the services.[29] The inevitable effect of the processes leading up to the formalization of the clinic was that the doctors' leeway was constrained. The nondisciplinary functioning of the team was rooted in the experts' medical knowledge and professional ethics, not in institutionalized guidelines or protocols. "We have always followed a policy that we should be free to make the decisions in this type of treatment, regardless of who referred the patient for examination and assessment," wrote Grünfeld, defending their restrictive approach.[30] Since physicians handled these requests in addition to their usual practice, and because they lacked support from a formalized public structure or institution, the pressure to make the right decision was increased.

Mobilizing the Public

When looking at the formalization of the service for medical transition in the early 2000s, one question to be addressed is this: Why did it take place now but not earlier, when similar attempts were made? Twice, the director general of health—first Karl Evang in the 1950s and next Torbjørn Mork in the 1970s—blocked attempts to formalize an administrative routine for sex change or for a health service for medical transition. The health authorities succeeded in their strategies because the media was not aware of these matters and there were no activist groups to represent patients' interests. By the end of the century this situation had changed, and sex reassignment now assembled and engaged a range of actors and institutions: physicians, patients, the Norwegian Board of Health Supervision, the Rikshospitalet, and the Directorate of Health. No fresh resources were set aside, however, for the formalization of the working group or for the establishment of a clinic. Even though the question of formalizing the service continued to circulate among ministries, directorates, and hospitals throughout the 1990s, the issue did not seem to gain enough traction to lead to any actual change.

Matters took a radical turn in 2000, however, as—for nine months—all diagnostic assessment and treatment of patients seeking medical transition was abruptly stopped. Ninety-five patients waiting for help suddenly had no health care service, and many wrote desperate letters to the Rikshospitalet, to the health authorities, and to the minister of health. Even the ombudsman of Norway's second largest county weighed in, arguing that the abrupt closure had provoked deep psychological crises among patients.[31] The reason for the closure was as follows: That same year the Rikshospitalet relocated from the center of Oslo, where it had been since 1883, to a new site in a residential area at the edge of the large forest surrounding the city. In the relocation process, the medical needs of individuals requesting sex reassignment were forgotten—or at least neglected—as the hospital administration did not keep this patient group in mind when planning the new department of plastic surgery.[32] According to one surgeon in the working group, the chief of the department did not see sex reassignment surgery as a priority and originally wanted to shut it down.[33] In addition, major conflicts between the plastic surgeons and the chief of the department had broken out; owing to professional disagreements and dissatisfaction with the head's leadership style, all but one surgeon left the department between 1990 and 2000. Eventually, the only surgeon who had been responsible for genital reconstructive surgery also left.[34] Finally, the capacity of the working group's psychiatrists, who had been serving on a voluntary basis, had long since been exceeded, meaning that when Grünfeld retired an impasse was inevitable.

Economic factors also played a part. In a letter to the Ministry of Health and Social Affairs, the Rikshospitalet administration stated that the plan had been to formalize a team and service for this treatment, but the hospital lacked money.[35] In the years leading up to the crisis, the model for funding public hospitals had undergone major changes. In 1997, a share of these hospitals' sources of funding was replaced by an activity-based system where the state reimbursed hospitals for costs according to a set of diagnosis-related groups (DRGs).[36] A major portion of the budget of the Rikshospitalet—at the time, still the only state-owned hospital, as public hospitals were operated by the county councils—was based on yearly block funding, but neither framework grants nor activity-based funding took into account the need to establish new health care services or to expand and redevelop old ones. "Reimbursement via the system of activity-based funding does not cover the establishment of such services," the hospital administration wrote.[37] No specific DRG rates existed for such surgical procedures, the hospital administration explained to the government, and "based on experience with other specialized procedures, the disparities between the DRG reimbursement and the actual costs per procedure are huge (20,000–50,000 kroner)."[38] Expensive and specialized genital surgery therefore exemplified an inherent problem with hospital funding that had paradoxical consequences: When the hospital used its own budget or external funding, such as grants, to research and develop new diagnostic methods or advanced forms of therapy, the public's expectations for these same interventions and therapies grew. Put differently, the Rikshospitalet took part in the creation of "new" groups of patients, the needs of which it lacked the capacity and funding to meet. "When this becomes established treatment," the hospital administration reasoned in reference to new, advanced forms of therapy, "the patients have already lined up."[39]

Concomitantly, and partly as a response to this complex situation, activists began to organize. As in the United States, the 1990s witnessed a rise in trans activism and visibility in media and society.[40] FPE-NE had remained a secretive society, but in 1994, Esben Esther Pirelli Benestad came out in public as a transvestite, "identifying as both man and woman," first in a popular TV show, and then on the front page of the most-read tabloid in the country. In the newspaper article, Pirelli Benestad was photographed twice sitting on the same barstool—in one photo as Esben, and in the other as Esther.[41] In 2001 they published a scholarly book about gender and sexology, *Kjønn i bevegelse* (Gender in Motion), that received much media coverage, and with the release of their son's 2002 film *Alt om min far* (All About My Father), they became a national celebrity.[42]

By contrast with the United States, where trans activists had made common cause with lesbian and gay activists since Stonewall in 1969, there was no similar communal activism in Norway.[43] The first advocacy organization for people seeking sex reassignment—the Landsforeningen for Transseksuelle (LFTS; National Organization for Transsexuals)—was founded in January 2000 as a direct response to the closure of the health service. "The first office was in my home, but the second was in the smoking room in the offices of the LLH," says Tone Maria Hansen, one of the founders of the LFTS, referring to the main lesbian and gay organization, Landsforeningen for lesbisk og homofil frigjøring. "I made a phone call to the LLH and asked if they had room for us." The LLH welcomed the foundation of the LFTS: "We see the closure of the treatment center as a sign that society is punishing those who do not conform to traditional gender roles," it stated at the annual meeting in 2000. Still, trans rights were not on the agenda of lesbian and gay activists. "We cooperated very well but they were not interested at all," says Hansen. "They fought for the rights of gay and lesbian people, but we talked a lot, and it was very nice."

With the foundation of the LFTS, trans people came out as a community in public. In previous decades, there had been occasional stories about trans people in the press. These stories often highlighted the medical aspects of transitioning, usually with "before" and "after" photos, as when the tabloid *VG* in 1974 illustrated parts of Jan Morris's autobiography *Conundrum* with images of James and Jan.[44] Many of these stories focused on the surgical aspects of transitioning (for example, "Jon Becomes Linda," and "Eva (23) Was a Man" with the subtitle "But on November 26, the doctors transformed her into a woman on the operating table"[45]). A common angle was whether the "sex change" had made life better or worse, such as the stories about Eddie Espelid ("Norwegian Woman Became a Man and Is Happily Married") or Eva ("I have become who I am. Finally, I can begin my life").[46] Or take the story about "Bjørn" in the article "New Sex, Still as Lonely," who was happy about his surgical transition but had trouble finding work and making friends, or the 1975 article "—Don't Change Sex!" about a fifty-year-old Danish woman named Iris, who warned others against "changing sex." Though a Danish surgeon had refused to operate on her, she traveled to London and paid for the surgery out of her own pocket, but was struggling to adjust to her new life.[47]

With the founding of the LFTS, journalists had somewhere to direct their questions, and this entailed a shift in the press coverage. At the turn of the century, newspapers continued to publish life stories with headings such as "Soon a Woman," but at the same time, activists increasingly managed to set the agenda and give trans rights and those wanting medical transition a new public face.[48] Newspaper stories were no longer only about happy or unhappy individuals; trans people were increasingly portrayed as a community and as a patient group. The media now also reported on the deadlock at the Rikshospitalet. Up to ten patients committed suicide while waiting for assessment or treatment, a staggering number for such a small patient group, and newspapers and the main broadcasting channel brought these stories to the public with headings such as "Committing Suicide While in Queue Waiting for Surgery" or "Transsexuals Are Denied Treatment."[49] Following the suspension of the hospital's health service, the working group now began to cooperate directly with LFTS to stir up attention from the media, and the deadlock created momentum for activism to mobilize media and public attention.[50] Sex reassignment, which had been reported in previous decades through the lens of individual decisions and stories, was increasingly portrayed as a legitimate issue of patient rights. The press, playing into what biomediatization theorists Charles L. Briggs and Daniel C. Hallin defined as a "biocommunicable cartography," helped foster a new patient identity centered on access to health care services, reshaping "transsexuality" into a communal patient identity, both in public and in politics.[51]

The deadlock led physicians in the working group to lobby members of parliament: This transsexual patient group was "utterly deprived socially and economically," and they demanded that the Storting intervene. When a parliamentarian intervened to ask whether the minister of health could guarantee that patients who had begun the process of medical transition would be allowed to complete it, the question of the future of sex reassignment was now on the agenda at the highest political level.[52] The minister of health, however, made no promises to provide extraordinary state funding for the service.[53] As the sole hospital in Norway offering this type of treatment, the Rikshospitalet had the responsibility, he insisted, to ensure that necessary health care services were available: "The Rikshospitalet is a net budget cooperation where it is expected that various tasks are solved within the net grant and other general financial arrangements."[54] While the minister forced the Rikshospitalet to develop a plan to secure a future for sex reassignment surgery, the LFTS's lawyer wrote that "there was now a complete lack of confidence between the LFTS and the Rikshospitalet"—referring not to the clinicians involved in sex reassignment but to the head of the plastic surgery department.[55] Finally, in May 2001, the health authorities summoned representatives from the Rikshospitalet and the Norwegian Board of Health Supervision, along with experts and activists, to discuss the future structure of the health services.[56] This meeting and the resulting plan for the future organization of health care services for patients seeking sex reassignment initiated a process that ultimately led to the establishment of the Gender Identity Disorder Clinic in 2001/2002.

Norwegian activists at the time did not mobilize *trans* as an identity; rather, they positioned themselves as patients denied medical care. This made it much easier for them to gain support from across the political spectrum; at the end of the century, proper health care services for these patients had become a topic that united politicians across party lines. Even the minister of health from the Christian Democratic Party weighed in. Looking back on this episode, he wrote in his memoirs, "I have found much meaning in my political work meeting people who use their own experiences to help themselves or others in spite of the constraints posed by their illnesses or disabilities."[57] Tore Tønne, the minister of health from the Labor Party, which came to power in 2000, also supported the cause of patient advocates, and the first-anniversary celebration of the LFTS was opened with a greeting from the minister, leading to newspaper headlines such as "Trannies Celebrated by Tønne."[58] For Christian and conservative politicians, who otherwise opposed gay rights such as marriage and adoption, it was unproblematic to defend the rights of a patient group seeking medical treatment as it would take another couple of years before Norwegian lesbian, gay, and bisexual organizations adopted trans rights as part of their agenda. This was not yet a question of rights for a minority group threatening the ideal of the nuclear family, the heterosexual order, or even a binary understanding of sex, but a matter of patient rights and the provision of medical treatment for a group of psychiatric patients.

By adopting a "patient identity," it became increasingly important for the LFTS to distance themselves from queer activism. Tellingly, while the LLH made efforts to include trans rights in their advocacy, by considered, for example, changing their name to Skeive Folk (Queer Folk) to signal that it worked for "transsexuals, sado-masochists, fetishists, and transvestites," the LFTS distanced itself from the organization.[59] When the LLH and the youth organization Skeiv ungdom (Queer Youth) included the rights of trans people in their working program in 2007, for example, the LFTS disagreed and dissociated itself from the program because they did not feel comfortable with the use of the terms *trans* or *transgender* due to their "queer" connotations, nor with being included in a "working program for homosexuals."[60] Stein Wolff Frydenlund, who was one of the most active trans and queer activists in Bergen at the time and a member of the LFTS, played a fundamental role in including trans people in the new gay working program of the city. But when the leadership of the LFTS pulled out of the plans, he left the organization. He offers the following recollection in an interview: "Although gender identity is not the same as sexual orientation, it was difficult for me to distance myself from lesbian and gay politics."[61]

The Need for Objectivity

As sex reassignment raised numerous political, medical, and ethical questions, physicians throughout the second half of the twentieth century repeatedly sought to anchor the procedures in a public institution; but for the exact same reasons—the issue's delicate and complicated nature—the health authorities opposed such formalization or institutionalization. At the turn of the new century, the issue of formalizing the health services not only mobilized established institutions but engaged new actors as well: Activists, patients, bureaucrats, politicians, physicians, and journalists as well as documents, hormones, scalpels, and hospital buildings turned the organization of trans medicine into matters of public concern.[62] Ultimately, the relocation of the hospital created an acute situation that generated momentum for action. In this new, more complex and confusing landscape, physicians were no longer the only experts with authoritative knowledge on the topic, exemplifying what Sheila Jasanoff argued, that democratic processes and public involvement in modern knowledge societies have become inseparable from scientific and technological questions.[63] The entry of patient autonomy and patients' rights narrowed the doctors' room for maneuver. Activists now demanded that their voices be heard when bureaucrats and politicians weighed in on the future organization of health services. The neoliberalization and reform of the public health care system and the welfare state, including the introduction of DRGs and the centralization of expensive medical procedures, further challenged physicians' latitude in defining the frame for medical practice.

When the Gender Identity Disorder Clinic was finally institutionalized in 2001/2002, it signaled more than the simple formalization of a practice that had become routine in the previous decades. In the process of creating a clinic for sex reassignment, the whole diagnostic and therapeutic structure and authority for making treatment decisions shifted from a multidisciplinary group of doctors to expert psychiatrists. In this sense, the Norwegian history differed from that in neighboring countries such as Sweden, Denmark, and Finland, where psychiatrists were the primary conductors of these assessments all along.[64] At the beginning of the 1990s, the Rikshospitalet plastic surgery department still emphasized that it was the team of experts that together made communal decisions about who would have access to treatment.[65] In the annual report from the turn of the millennium, however, the role of the team was all but absent. "Psychiatrists observe, diagnose, and decide treatment indication," it stated. "Our surgical services depend on psychiatric assessment, pre-treatment and post-surgical follow-up."[66] Compared to the situation only a few years earlier, the different roles of the specialists—their expertise and responsibilities—were now compartmentalized in completely new ways. The assessment of patients became the task of the psychiatrist, who now took center stage while the other medical professionals were merely consulted for medical services. The role of the surgeon, for instance, became simply to carry out surgical-technical procedures at the recommendation of the psychiatrists. "The patients have been regarded as surgical patients," the hospital administration stated, but at the turn of the century the plastic surgeons had come to a different conclusion.[67] "These are psychiatric patients," the hospital administration wrote, adding that surgery was only "part of the total treatment." The hospital "had not taken sufficient account of the fact that it is a psychiatric clinical picture," but this was about to change.[68]

In the plan for the organization of sex reassignment that the Rikshospitalet provided to the Ministry of Health and Social Affairs in June 2001, psychiatrists played the role of administrators. Although the plan suggested that various medical specialists would meet every five weeks to discuss the patients, the psychiatrist and the psychiatric nurse made up the team's "permanent base." This was no metaphor but was a matter of fact. In the organization envisioned for the clinic in the future, the hospital psychiatrist was placed at the center of the organizational map, and all other specialists—endocrinologists, plastic surgeons, and gynecologists—were boxed off with unidirectional arrows pointing to the center: "These three—a leading psychiatrist, a psychiatric nurse, and a secretary—will serve as quality assurance for all other health services involved in the treatment and follow-up of this patient group."[69] Only a few years earlier, a new edition of the international guidelines on medical care for trans people had been published, *The Standards of Care for Gender Identity Disorders*, underscoring the role of mental health professionals in diagnosing and selecting patients for medical transition and providing support through psychotherapy.[70] So when the hospital requested more funding to formalize a health service for sex reassignment, and specifically for a full-time psychiatrist, it invoked these guidelines. Both the plastic surgeons' annual report and the plan for the organization of health services made arguments for funding the position of a psychiatrist as head of the new clinic: The hospital representatives wrote to the Ministry of Health and Social Affairs that the Rikshospitalet was "without a psychiatric department and therefore in need of a position for a psychiatrist."[71]

At the time, new positions for physicians were allocated by a national council on behalf of the ministry, and the Norwegian Medical Association exerted significant influence over these decisions.[72] The Rikshospitalet had neither the authority nor the funding to create a new position for a psychiatrist and it had no separate psychiatric department.[73] So to circumvent this obstacle, the administration appointed the psychiatrist who would become the head physician of the Gender Identity Disorder Clinic to a position on the hospital's administrative staff.[74] Because the DRG system did not allow for full reimbursement of surgical interventions, the hospital—as a last resort—created a new clinic under the leadership of a psychiatrist and hoped that the government would supplement the yearly block grant. In the end, the regulatory regime of DRGs and the funding framework, combined with the fact that the Rikshospitalet had neither a psychiatric department nor the authority to establish a position for a specialist, all ultimately contributed to the creation of a new therapeutic structure for sex reassignment with psychiatry in the leading role. The working group had dissolved, leaving a vacuum for the expert psychiatrists to fill. All these processes contributed to conceptualizing transsexualism as a psychiatric disorder with psychiatrists as the leading experts.

Activists also played a crucial role in this shift. Transsexualism was a "rare gender identity disorder," the LFTS wrote in information material for members, reproducing a pathologizing conceptualization of trans life.[75] As all treatment was shut down, the LFTS wrote to the minister of health that "it is not defensible that an increasing number of people with a psychiatric diagnosis walk around without a treatment option."[76] Activists pointed to the lack of treatment as a potential risk to society: "Transsexualism is a psychiatric diagnosis with a surgical solution," they wrote, arguing that psychiatric assessment was "a crucial and central part of the treatment of transsexuals."[77] Today, the advocacy organization uses different terms to describe themselves, but in an interview, Tone Maria Hansen recalls that physicians in the 1980s used the term "transsexual" and that she therefore also referred to herself as a transsexual woman.[78] It remained the case at the time, however, that the adoption of pathologizing terms and mobilizing an identity as psychiatric patients was the only way to access medical treatment.

How can this major displacement of expertise in relation to the question of medical transition in the early 2000s be understood? Several political, economic, and bureaucratic processes and historical actors were involved; thus, simply describing the complex changes in all-encompassing and homogenizing sociological terms such as "psychiatrization" offers little to a refined historical analysis. Rather, the shift must be considered within a broader medical and scientific crisis of legitimacy. From the early 1980s onward, the decisions regarding which patients should have access to hormonal and surgical treatment were made by the working group who resolved these clinical questions in an informal manner—often meeting privately after working hours. "It was like bingo," says the psychiatrist Ira Haraldsen in an interview, meaning that decisions were made randomly rather than systematically.[79] She joined the group in the late 1990s and later became the first director of the new gender identity clinic. The ways in which the medical experts made decisions, which in retrospect could be characterized—understandably but somewhat nonchalantly—as "bingo," nevertheless reflected its own medical logic, a logic rooted in professional clinical judgment regarding care and good medicine.

Alv A. Dahl explained, indeed, that the working group did not use diagnostic instruments, even though, in the 1980s, they were well aware of the DSM-III and the accompanying SCID interviews, as well as the diagnostic criteria set forth in ICD-9.[80] "We used clinical interviews and had a solid, professional gestalt as to what transsexualism was," he says.[81] By the turn of the century, however, with the increased neoliberalization, centralization, and bureaucratization of medicine, this logic had lost its credibility. The new psychiatric diagnostic regime, which replaced the former "nondisciplinary" practice based on clinical judgment, involved an enormous number of questionnaires and diagnostic instruments. People who went through the new diagnostic procedures explained in interviews that they had to fill out some seventy to eighty pages consisting of various diagnostic instruments and tests.

This shift in the psychiatric assessment of trans patients must be viewed against the backdrop of more general developments within psychiatry at the time. Historians of science and medicine have carefully documented how practices of standardization and classification have come to play a key role in modern medicine and psychiatry.[82] For psychiatry, this gradual change has been intimately tied to the increased importance of the DSM and the epistemological shift from the psychoanalytic-inspired second edition to the "atheoretical," operationalist third edition.[83] These changes, however, account only for shifting epistemologies within psychiatry and do not explain why a new psychiatric regime based on diagnostic instruments, questionnaires, and quantified knowledge replaced the cross-disciplinary, clinical judgment–based system for sex reassignment in the first place. Yaron Ezrahi has proposed that the modern liberal-democratic state has appropriated science and technology as a resource for shaping and legitimating political authority and that scientific knowledge has provided modern Western democracies with the tools to uphold a "neutral," objective, and depersonalized kind of power.[84] Similarly, in modern societies where democratic political culture is based above all on interests, the historian of science Theodore M. Porter has argued that *objectivity* or "strategies of impersonality"—that is, "knowledge that does not depend too much on the particular individuals who author it"—has become *the* most important defense against attempts to undermine a field.[85] This explains modern societies' "trust in numbers" or—in this particular story—the new diagnostic regime of trans medicine based on psychiatric questionnaires, tests, and instruments.

The shift in expertise in trans medicine must be understood along these lines. In the late twentieth century, the issue of medical transition entailed, on the one hand, political interests, bureaucratic autonomy, activists' demands, and media attention; on the other hand, medical professionals' latitude was increasingly challenged by patients' rights, activists, and liberalized financial systems. In this increasingly complex but restricted landscape of medical transition, the clinical judgment–based authority no longer enjoyed sufficient credibility. The working group had become an archaic entity in an increasingly specialized and standardized version of medicine. Neither endocrinology nor plastic surgery had the diagnostic competence or legitimacy to serve as a future home for trans medicine. Social medicine was a dying profession, and there were no experts in social medicine ready to pick up the baton after Grünfeld retired. The only way to protect the legitimacy of medical decisions regarding sex reassignment and to safeguard a future for trans medicine was to anchor it in the new "objective," quantified diagnostic regime of psychiatric expertise.

Medicine as Bureaucracy

The Norwegian story represents a distinct example of how nosological negotiations unfolded in the context of a welfare state and a public health care system. Since the early 1980s, people requesting sex reassignment were attended to by a working group consisting of physicians. They had a clear understanding of what transsexualism was and defended their decisions based on clinical judgment. Nevertheless, as this was a marginal field of medicine, bolstered by little clinical research that would serve to justify the soundness of their decisions, the physicians sought to embed their treatment practices in a formalized public health care structure. The health authorities, however, feared that such a formalization would lead more people to seek treatment to transition and preferred that the medical, ethical, and legal questions raised by sex reassignment be solved on an ad hoc basis at the discretion of the medical profession.

At the close of the century, hormonal and surgical transition was negotiated by physicians, patients, bureaucrats, politicians, journalists, and activists in the context of a public health care system under pressure from different angles, including from economic framings, institutional frictions, public debates, activism, and the positioning of medical experts. In an increasingly liberalized, centralized, and standardized health care system, physicians' room for maneuver was restricted by numerous factors: economics (reimbursement based on diagnosis-related groups), the legal system (activists and the new era of patients' rights that they helped to usher in), medicine (standardization of diagnostic practices and treatment courses), and politics (governmental regulation of specialized medical services). These discourses framed sex reassignment as a highly specialized type of care. The working group enacted sex reassignment as "a whole." The process of reforming the welfare state and public health care system and the hormonal and surgical practice of medical transition shaped each other in co-productive ways; since sex change was enacted as an either/or, one-size-fits-all regime, the binary model of sex was stabilized.

The new era of patient autonomy and emerging activism put pressure on the discretionary power of medical experts and the tradition of justifying decisions regarding who should gain access to medical transition through the logic of professional judgment. Restricted room for maneuver and an increased demand for unbiased, "objective," and dispassionate medical practice created a vacuum filled by expert psychiatrists who came armed with a battery of "objective" diagnostic instruments. By the end of the century, as physicians and psychiatrists increasingly organized clinical practice around diagnostic instruments and classifications, medical transition had to find a home in a specialized field of medicine. Ultimately, medical authority over trans medicine shifted in favor of psychiatric expertise. A strict pathologizing model rooted in psychiatric reasoning and practice became the only way to provide sex reassignment that conformed to the system of the reformed welfare state.

What do bureaucratic practices and the question of centralizing or decentralizing trans medicine have to do with a scientific and medical understanding of sex? As emphasized by Sheila Jasanoff, "Knowledge and its material embodiments are at once products of social work and constitutive of forms of social life."[86] Jasanoff proposed the concept of "co-production" to describe the facts, first, that knowledge about the world cannot be separated from how we choose to live in it, and second, that knowledge about the natural and the societal world is mutually constitutive. In the Norwegian story, practices of sex reassignment and politico-bureaucratic discussions about the future organization of the public health care system worked in co-productive ways: they constituted each other. While the rarity of the procedures and the highly specialized form of surgery was an argument for formalizing a national service, the debate about centralizing hospital services posited medical transition as a highly specialized form of medicine. Georges Canguilhem reminded us that "a norm offers itself as a possible mode of unifying diversity, resolving a difference, settling a disagreement."[87] The opening of the gender identity clinic in Norway demonstrates that, in the modern welfare state, the norms of medical practice and knowledge production are not only shaped within the walls of the clinic but are additionally shaped increasingly through the machinery of bureaucracy.

CONCLUSION

Social Medicine and the Norms of Health

In 2016, when Maria received a referral to the National Treatment Service for Transsexualism (NBTS) at the Rikshospitalet, the former Gender Identity Disorder Clinic, she believed that she had cleared the main hurdle to accessing gender-affirming hormonal and surgical treatment. However, it soon became clear that a referral from Maria's physician would not suffice; the NBTS refused to schedule an appointment until an outpatient psychiatric unit had conducted a comprehensive psychiatric evaluation. In the rejection letter, the chief psychiatrist explained that the Norwegian health care system consisted of three levels—primary, secondary, and tertiary services—and that the NBTS belonged to the last of these. Although Maria did not have any mental health issues, she must be examined by a psychiatrist and the referral had to come from the secondary level. "This is a medically sound, well-functioning procedure that makes the Norwegian health care system one of the three best in the world, according to a 2015 OECD assessment," the psychiatrist wrote.[1]

Around this time, trans medicine was undergoing major changes with profound consequences for trans people around the world. In 2019, the World Health Organization (WHO) removed the diagnosis of transsexualism from the chapter on mental disorders, and in the eleventh edition of the *International Classification of Diseases* the diagnosis was substituted by a diagnosis of gender incongruence in a new chapter on sexual health.[2] As trans and gender nonconforming people were no longer pathologized with psychiatric labels, this shift implied an uncertain future for mental health professionals in trans medicine: If transsexuality was no longer a psychiatric diagnosis, why should psychiatrists play the role of gatekeeper to gender-affirming treatment? Following the WHO decision, the Norwegian Directorate of Health published new national guidelines on gender incongruence, calling for the decentralization of these health services. With the establishment of the GID clinic in 2002, publicly funded gender-affirming treatment was effectively monopolized, but now the monopoly would be dismantled, and psychiatric assessment would no longer be obligatory in order to get treatment.

Undeterred by the WHO's change and the directorate's new guidelines, the NBTS refused to change its routines. In 2021, when Felix approached the gender clinic after getting the diagnosis of transsexuality from a psychiatrist two years earlier, the clinic required a thorough psychiatric evaluation at an outpatient psychiatric service before referral. Yet the outpatient service replied that they could not prioritize this examination: It was neither the case that they had any specific expertise on gender identity, nor that transsexuality was considered a psychiatric condition.[3] The NBTS refused to change their position: As a specialized tertiary health care service, they did not offer elementary psychiatric assessment, they stated. The clinic leadership referred to the LEON principle (lowest efficient level of care), which was introduced in 1974 to reduce costs and which stated that health care services should be provided at the lowest possible level of the health care system.[4] Felix was caught in a catch-22. He did not suffer from mental illness, but to get gender-affirming treatment, he first had to be examined by a psychiatrist; and to get this examination, a mental disorder had to be suspected.[5]

Around the same time, health services for trans patients were reorganized in Sweden and Denmark. In 2015, following increasing criticism from patients, trans advocacy groups, and human rights organizations, the Danish National Board of Health revised the national guidelines to shorten the evaluation time and to widen access to gender-affirming treatment. The monopoly at the Sexologisk Klinik at the Rigshospitalet was dismantled and new services were opened in Odense and Aalborg. What is more, in 2019, the service for adults was moved out of Sexologisk Klinik—where it had operated under the psychiatric department—and into a new Center for Gender Identity under the gynecological department. In Sweden, where practice was more decentralized, the National Board of Health and Welfare decided in December 2020 to centralize health services for trans patients in three national specialized centers.[6]

One hundred years after Martha Pedersen's bid at the Kommunehospitalet in Copenhagen for treatment to be recognized as a woman, trans medicine had become a highly formalized, specialized, and centralized form of care in all three Scandinavian countries. Although Martha, Maria, and Felix were requesting the same thing—medical help to transition and change their legal sex—Maria and Felix encountered a very different approach to trans medicine than Martha had found a century earlier. This book has analyzed the development of trans medicine from its emergence in the 1920s to gradual standardization in the 2000s with specialized gender clinics based on a specific diagnostic and therapeutic logic and routines for changing legal sex.

In a seminal book, Geoffrey C. Bowker and Susan Leigh Star explored how standards and classification systems organize modern societies, arguing that they profoundly shape moral and scientific possibilities and practices.[7] Diagnoses are one such standard: The classification of disease is constitutive not only of medical practice but is also a "social tool" for regulating relations between physicians, patients, and society.[8] The depathologization of trans and gender nonconforming people and the subsequent reorganization of trans medicine are striking examples of the role of diagnosis in modern medicine, what Charles E. Rosenberg characterized as the "tyranny of diagnosis."[9] At the same time, the Scandinavian case complicates the picture; it shows that trans medicine did not evolve in a uniform way but that practices were adjusted to local and regional contexts in interplay with international discourses and routines. Much more than a global standard, trans medicine was like a conglomerate of practices, an "amalgam of thoughts, a mixture of habits, an assemblage of techniques."[10] This realization calls for careful historicization of nosological categories, expertise, and professional authority in national and local contexts.[11]

In Scandinavia, diagnostic and therapeutic considerations were confined by the frames of the welfare state and a public health care system shaping trans medicine in a very different way from parallel developments in the United States. On the other side of the Atlantic, the medical market, physicians' fear of litigation, and the prestige project of the gender clinic at the Johns Hopkins Hospital in the 1960s and 1970s turned trans medicine into a highly specialized practice.[12] This development reflected broader trends in US medicine, described by sociologists of medicine as "biomedicalization" and characterized by corporatization and commodification, economic privatization and techno-scientization.[13] In Scandinavian countries, by contrast, trans medicine did not develop as part of a prestige project like the Hopkins clinic nor as specialized forms of medicine as in the many US gender identity clinics. Welfare state trans medicine was not associated with big money, large private companies, or elite universities, but developed in preestablished clinical institutions as part of common clinical practice.

For a long time, this type of care remained a dispersed practice and part of everyday medicine. For example, endocrinologists repeatedly argued that hormone replacement therapy was straightforward medicine that did not require specific expertise; primary care physicians, psychiatrists, and endocrinologists already prescribed hormones to their patients. In encounters with trans patients, health professionals—psychiatrists, endocrinologists, psychologists, sexologists, nurses, social workers, primary care physicians, and plastic surgeons—used what they already knew, tinkered with existing tools and technologies, and adapted practices to local contexts. In Sweden, psychiatrists played the dominant role, whereas in Denmark, trans medicine began as a cooperation between endocrinology, psychiatry, and plastic surgery but shifted to psychiatric sexology in the 1970s and 1980s. In Norway, trans medicine remained interdisciplinary until the turn of the century, and only then was it taken over by psychiatrists. Still, the diagnosis and treatment of trans patients in all countries was ancillary to regular tasks of treating patients with schizophrenia, diabetes, burn injuries, urinary tract infections, hypertension, and depression. Patients were received in regular medical and surgical departments or in the offices of private practitioners: Trans medicine was like any other kind of medicine.

At the same time, trans medicine would not have come into being without patients, communities, and activists who modified nosological categories and clinical practice by requesting access to treatment, sharing information, and making their bodies available to researchers. These exchanges opened new fields of knowledge, generated ethical questions, and provoked cultural and political change, but they also acted as catalysts for the professionalization and specialization of endocrinology, plastic surgery, clinical psychology, and sexology, legitimizing and bolstering expertise and authority. Personal relations and circles of knowledge between communities and medical professionals shaped trans medicine on both sides of the Atlantic. This played out differently in the Scandinavian countries, however, where the physicians involved in trans medicine often held key positions in academic medicine or in the state organs regulating sex—whether on the councils deciding on applications for castration or on the committees proposing state policy and legislation of sex reassignment—all the while cultivating a close relationship with trans communities. The distances between the state and the subcultures in Scandinavian countries were much shorter than in the United States. This, however, did not ease access to transition-related health services in the welfare state.

Trans medicine in Scandinavia grew out of the eugenic objective of controlling and regulating "sexual abnormality" in the interwar period and gradually developed into pragmatic responses to individual needs in the postwar decades. As the goal of trans medicine shifted from social control to the optimization of the individual, however, the alleged interests of society never disappeared from diagnostic and therapeutic considerations and decisions. To the contrary, the approach to people requesting medical transition was deeply influenced by the logic of social medicine, including skepticism toward somatic, pharmaceutical, and surgical solutions to health issues and the promotion instead of social solutions.

On a more general level, the example of trans medicine highlights the mediating capacity of medicine in the modern welfare state: the role of securing public trust and safeguarding bureaucratic intervention *beyond* the therapeutic qualities of the physician. Physicians cared for the patient and for society—and for the patient *in* society. At the interface between society and state bureaucracy, between medicine and public opinion, this history highlights the role of physicians in the welfare state (and increasingly too, the roles of other health professionals) in negotiating practical problems as a precondition for—and an enabler of—social and political change. The history of trans medicine in Scandinavia is a story of the co-constitutive relationship between the state and the clinic, and the role of the physician not only as a healer but as an evaluator. In Denmark, for example, Knud Sand in his role as chair of the Medico-Legal Council was a key figure in making the castration laws and in the medicolegal regulation of sex change. In the Norwegian Directorate of Health, the physicians acted as both bureaucrats and clinicians; Karl Evang, the director general of health, took time to correspond with individual patients, and Christofer Lohne Knudsen, the chief psychiatrist, saw trans patients in the clinic while handling the public regulation of sex reassignment in the health bureaucracy. Clinicians caring for trans patients, such as Per Anchersen in Norway or Jan Wålinder in Sweden, were also involved in the regulation and creation of a legal framework for sex reassignment and the change of sex status. They made policy recommendations based not only on what they believed to be in the best interests of their patients but also what they believed to be in the best interests of society as a whole.

Sexologists, such as Berthold Grünfeld in Norway or Preben Hertoft in Denmark, used their expertise on the role of society in health and illness to shape the public debate on transsexuality, transvestism, and sex reassignment, but also to formalize treatment and anchor it in the public health care system by creating specialized working groups and clinics. The expertise of social medicine and sexology was deployed on a grand scale in policymaking, but in turn also pragmatically, as when Grünfeld gave his patients testosterone injections in his office. This dual role of physician and bureaucrat, clinician and policymaker, sought to ensure the interests of patients and society in the context of an extensive public health system and free health care. Trans medicine was not a monolithic regime of regulation and control but a series of pragmatic responses to complex medical, ethical, and societal issues.

By the turn of the century, however, "sex change" was enacted as a totality, as an all-or-nothing medical and bureaucratic routine tightly tied to the official practices of name change, marriage, and the change of a person's legal sex. In Norway, this is why it ended up in a hyperspecialized and centralized clinic under the control of psychiatrists, and this also explains why this treatment has not been decentralized, democratized, and "despecialized," unlike other endocrinologic practices under the control of primary care physicians, such as hormone replacement for hypothyroidism or diabetic care.

On a fundamental level, this is the story of the production of a sex standard and binary state. Historians have demonstrated how trans medicine contributed to a reshuffling of the boundaries of sex in various contexts from the United States to Iran.[14] In Iran too, the state played an active role in regulating who should be allowed to transition, but a characteristic of the Scandinavian story is the logic of social medicine in shaping trans care and regulating sex: The boundaries of sex as a medical, legal, and bureaucratic category were reestablished as part of the political project of forming and reforming the welfare state. On the one hand, medical expertise was deployed toward the state's objective of administering sex, from the eugenic goal of shielding society from sex offenders and the control of "sexual abnormality" to the change of name, the regulation of marriage, and the management of reproductive rights; on the other hand, the neoliberal push to reform the health care system, including a new hospital structure and economic systems for reimbursement, indirectly reworked the category of sex, ultimately cementing a sex binary. In short, the sex standard was a bricolage of bureaucratic processes shaped by the logic of social medicine. Sex was not defined by an organ in the body but a tool for medical experts and state organs to handle moral and political issues. As Paisley Currah phrased it, sex is as sex does.[15]

The paternalism of social medicine provided patients with little direct influence on decision-making. Most people were denied treatment; the archives are silent about their stories. At the Rikshospitalet in Oslo in the late 1990s alone, up to ten people who were approved for treatment committed suicide while waiting for surgery. These numbers are staggering and point to the darkest side of trans care in the welfare state: the ephemerality of its practice, the rigidity of its outcomes, and the lack of follow-up. People who got treatment were also left to fend for themselves or were left to their primary care physicians who had little experience in adjusting hormone doses or knowledge about side effects and complications. In a tightly regulated drug market and health care system, there were few other ways for people to get hormones and surgery outside of the public health care system. Some managed to obtain hormones through acquaintances, but only the most affluent could afford to go to Casablanca for surgery. This remains an unexplored part of the history explored in the present book, but future research, hopefully, will shed more light on unofficial DIY (do-it-yourself) routes of medical transition in Scandinavia and beyond.

In the second half of the twentieth century, psychiatrists and psychologists repeatedly implemented testing regimes, diagnostic practices, and therapeutic routines to support sex reassignment, ranging from projective psychological testing and real-life tests to psychiatric diagnostic questionnaires and protracted waiting times. The question remains: Why could physicians and psychologists not simply trust their patients and leave the final decision to them? This question is still at the heart of current discussions about trans medicine between activists who advocate models of care based on informed consent and providers who defend a gatekeeping monopoly.[16] In an essay titled "My New Vagina Won't Make Me Happy," published in the *New York Times* in November 2018, Andrea Long Chu criticized the goal of gender-affirming treatment for only seeking to reduce gender dysphoria and alleviate pain. This has become a way to gatekeep treatment, she argued, but the principle of "first, do no harm" is not to protect patients from harm, she wrote, but "to install the medical professional as a little king of someone else's body."[17] For Chu, the discussion about the objectives of trans medicine should shift from alleviating dysphoria to fulfilling desire. "Desire implies deficiency; want implies want," wrote Chu. "To admit that what makes women like me transsexual is not identity, but desire is to admit just how much of transition takes place in the waiting rooms of wanting things."[18]

Chu writes from a US context, but the Scandinavian case demonstrates how challenging it has been for physicians to operationalize desire in welfare state medicine, both as a diagnostic category and as a therapeutic goal. These decisions are inseparable from the social goals of medicine in the welfare state and the double role of physicians and other health professionals in caring for the individual and for the society. The medical task of making "good politics," including the fair distribution of resources and the role of the physician in making decisions on behalf of society, is a very different logic from the free-market medicine in which patients pay and physicians deliver. In the public health system, alleviating suffering and curing disease have been the primary modes of intervention; state-employed physicians never represented only the individual patient—and their desires—but also acted as guardians of the social contract to distribute shared resources in a fair and equal way. As physicians and psychiatrists (and increasingly, psychologists also) were continually striving to translate desire into something workable for medical reasoning and practice, trans medicine became a question of medical epistemology and good practice: What is the condition, what should be the purpose of the treatment, and how can it be resolved without causing harm?

The question of what it means to heal and to harm has been a recurrent theme in the history of trans care. Trans medicine in the welfare state was restrictive, paternalistic, conformist, and normative, but it was also pragmatic, cautious, and concerned with a specific communal idea of society. It was a kind of care that sought to mediate between the social subject and the self. This paints a conflicted and ambiguous picture of social medicine in which the boundaries between care and harm are not always easy to distinguish. As this book has argued, the answer depends on whom you ask. In the 1984 book *Paradiset er ikke til salg* (*Paradise Is Not for Sale*), coauthored with the physician Teit Ritzau, Preben Hertoft wrote that "there is a difference between seeking the light and burning one's wings, let alone perishing in the flames." The boundaries of sex could not be transgressed—paradise was not for sale—"but having to give up essential aspects of your being to reach yourself can never become an ideal. Such a renunciation cannot be called a gain; it is and remains a loss."[19] For the Danish physicians, sex was given by nature; paradise could not be bought, and boundaries could not be transgressed without a fundamental loss. As much as Hertoft engaged with the trans communities, as much as he wanted to understand the people he studied, he seemed incapable of taking a step back and reflecting on how this concept of sex shaped norms of health and ideals of a fulfilling life.

Who can define health? What counts as harm? For experts like Hertoft, this was a question for the physician, even if, as he argued, it would be much more comfortable for physicians to give in to the desires and requests of their patients and let them live with the consequences. Put differently, if a physician is called upon to use their expertise in a clinical case, it is part of the contract that the physician is free to decide which measures to recommend. Annemarie Mol argued, on a similar note, that care is more than the promotion of patient autonomy and choice; the antipode of the logic of care is neglect, she wrote.[20] Ingrained in the logic of care is an ethical framework grounded in practice, an ethics different from the argumentational and analytical approaches of bioethics, with informed consent and patient autonomy as core values of good practice that came to dominate medicine in the 1960s and 1970s.[21] In the new regime of bioethics, "the self" replaced the "social subject," argued medical historian Roger Cooter.[22] This might be true for bioethics, but in the ethics of social medicine and the welfare model of trans care, the social subject—as I have argued in this book—remained the principal operator.

Yet that explanation remains unsatisfactory: The logic of care in trans medicine also grew out of a particular historical context, was enabled by a set of scientific concepts, and enforced norms of health. We have seen that the emergence and formulation of trans medicine was inextricably linked to the concomitant process of sex standardization. What would a trans medicine not based in a binary norm of sex, a teleological-developmental model of gender identity, or an etiological notion of transness look like? Historian of medicine François Delaporte has reflected, "The question of biological norms might very well exceed the competence of those who present themselves as experts on norms. The value of a quality of life lies, above all, *in the living*."[23] This brings us back to Chu's essay: An operation cannot be equated with an emotional state; it would be illusory to believe that an operation will alleviate all pain.[24] The stories in this book show that transitioning did not solve all problems; even with medical treatment, life as a trans person was not necessarily easy. But for many, it was the only true choice. "I always told the doctors that I wanted to live *one* life," Hanna says. "A whole life."[25] Sophia remembers standing in front of the mirror after the surgery to see and feel her vulva for the first time. "It was completely unfathomable, fantastic, wonderful. The tears came in streams," she says. "I felt so complete."[26]

The depathologization of trans and gender nonconforming people raises questions about the future role of mental health professionals in gender-affirming care. In an essay published in 2020, the psychoanalyst Avgi Saketopoulou made a plea for critical hospitality in discussions of medical transition, especially for young people, acknowledging the "uncharted territory."[27] In the present moment, it seems impossible to embrace the critical hospitality that is needed without acknowledging the wounds of the past.

At a deeper lever, beneath the shift from speaking about trans people to letting trans people speak is an epistemic difference between a theoretical, abstract norm of health and the knowledge of life. As Georges Canguilhem puts it, "the thought of the living must take from the living the idea of the living." Isn't that the most important lesson at the end of this story: how difficult, if not impossible, it can be to translate life into disembodied medical concepts? "To do biology, even with the aid of intelligence," wrote Canguilhelm, "we sometimes need to feel like beasts ourselves."[28] Taking this lesson seriously would imply a shift from what medicine can say about trans life to what trans lives can say about medicine.

Acknowledgments

This book would not have been possible without the many people who opened their homes and hearts to share their personal stories and encounters as patients within the health care system. I thank you for your trust in this research project. I also thank the physicians and psychologists who openly shared their professional experiences of working in trans health since the 1970s. Much has changed in the last fifty years, and I appreciate that you dared to openly share your doubts, beliefs, and regrets.

Many people have contributed to shaping this book since I started doing preliminary research ten years ago. I remain grateful for all discussions with the faculty and graduate students at the University of Oslo, especially my intellectual mentor Anne Kveim Lie, who was tremendously supportive and whom I am happy to call a friend. I thank Anne-Lise Orvin Middelthon for all her time and thoughtful comments, and Heidi Fjeld and Christoph Gradmann for stimulating discussions. The Norwegian Research Council funded the initial research for the book, and I thank all members of the Biomedicalization from the Inside Out project, especially Isa Dussauge for all their suggestions and Per Haave, who provided invaluable help and who unfortunately passed away much too soon.

A Fulbright fellowship made it possible to spend a semester at the Department of the History of Medicine at Johns Hopkins University. I thank Jeremy A. Greene for welcoming me so generously at the department and for commenting on various parts of my research. I am grateful to the graduate students and faculty for their contributions, particularly Jacob Moses with whom I organized the "Trans/Medicine" workshop in November 2021. I thank Beans Velocci, Eric Plemons, and Jules Gill-Peterson for the lively and thought-provoking discussion. Gill-Peterson's generous comments helped me draw comparisons between the Scandinavian and US stories. I particularly want to thank Susan Stryker for insightful comments and her thoughtful engagement with my work.

At the Berlin Institute for the History of Medicine and Ethics in Medicine I found a new academic home. Volker Hess has been immensely generous with his time and has been an inspiring discussion partner. I want to thank all colleagues for engaging discussions and help, particularly Birgit Nemec, Alexa Geisthövel, Susanne Michl, Mirjam Faissner, Henriette Voelker, Axel Hüntelmann, Oliver Falk, Amelie Kolandt, Tillmann Taape, Thomas Beddies, Annett Wienmeister, Carolin Pommert, Vera Seehausen, Susanne Doetz, and Anja Suter. A warm thanks to Stefanie Voth for helping me to navigate the German academic bureaucracy. I owe a great deal of gratitude to all participants at the "Geschlecht, Kjønn, Sex, Gender" workshop in Berlin in November 2023 at which I presented one chapter of the book.

There would have been no book without archives and libraries—and archivists and librarians. I want to thank Ola Devold (Norwegian Board of Health Supervision), Gunn Løwe (National Archives of Norway), Ragnhild Bjelland and Runar Jordåen (Skeivt arkiv), Solveig Schnitler (Norwegian Ministry of Justice and Public Security), Maria Storhaug-Meyer (Oslo City Archives), Knut Marstrander and Anne Marie Eng (Oslo University Hospital), Phoebe Evans Letocha (Alan Mason Chesney Medical Archives), Sean McGill (Kinsey Institute), Laura Søvsø Thomasen (Royal Danish Library), and the Danish National Archives. I also want to thank the university administration staff at the University of Oslo, Charité—Universitätsmedizin Berlin, and Johns Hopkins University who have facilitated the project by booking rooms, reimbursing expenses, and answering emails. From the beginning, the support this project received from activist communities and organizations has been extremely important to me. I am grateful for the support and help from FRI—the Norwegian Organization for Sexual and Gender Diversity—and the Harry Benjamin Ressurssenter who used their networks to recruit people for oral history interviews.

Over the years, I have met outstanding historians, scholars, and clinicians whose advice, wisdom, friendship, and knowledge have shaped this book: Richard A. McKay, Sandra Eder, Seth Holmes, Sahar Sadjadi, Lukas Engelmann, Dóra Vargha, Jacob Steer-Williams, Todd Sekuler, Tim Chandler, Reidar Schei Jessen, Helge Jordheim, Kristin Asdal, Susanne Bauer, Martine Schlünder, Solveig Jülich, Guillaume Lachenal, Espen

Ytreberg, Elisabeth Lund Engebretsen, Hannah J. Elizabeth, Asle Offerdal, Antoine Vialle, Claudia Stein, Hanna Worliczek, Mirjam Janet, Kate Davison, Bettina Wahrig, Heiko Stoff, Florence Vienne, Nick Hopwood, Staffan Müller-Wille, Janet Browne, Keith Wailoo, and Joanna Radin.

I have presented papers from this project at numerous seminars and conferences, and I would like to express my gratitude to everybody who attended and gave feedback: "Times in Plural" in Oslo at the Centre for Advanced Study in October 2018; the 16th Ischia Summer School on the History of Life Sciences in June 2019; the Science and Technology Studies seminar at the University of Oslo in December 2019; the Medical Anthropology and History seminar at the Oslo Institute of Health and Society in September 2020; the Stuttgarter Fortbildungsseminar für Medizingeschichte in October 2020; the Queer Research Group in Norway in November 2020; the Material Life of Time conference in March 2021; the annual conference of the American Association for the History of Medicine in May 2021; the colloquium of the European Association for the History of Medicine and Health in May 2021; Arbeitsgruppe für Psychiatriegeschichte at Institut für Geschichte der Medizin, Berlin, in June 2021; the 4S conference in October 2021; the book workshop on psychiatric practices in European psychiatry throughout 2020 and 2021, particularly the helpful suggestions by Jean-Christophe Coffin and Cornelius Borck; the Centre for History in Public Health at the London School of Hygiene and Tropical Medicine in February 2022; Berlin-Brandeburgische Gesellschaft für Geschichte der Medizin in March 2023; Fachverband Medizingeschichte in June 2023; the Oslo Science Studies Colloquium in March 2024.

Karen Darling, my editor at the University of Chicago Press, believed in the project and masterfully shepherded the manuscript through the review and publication process. The comments from the anonymous reviewers greatly improved the manuscript. Brian Clarke skillfully copyedited and proofread the book.

Finally, I want to thank my mum and dad, my family and friends for supporting me and cheering me along the way.

Abbreviations

BSA	Bodil Solberg Private Archive
DNA	The Danish National Archives, Copenhagen
DRGs	Diagnosis-related groups
DSM-III	*Diagnostic and Statistical Manual of Mental Disorders*, 3rd edition
FPE	Full Personality Expression (Phi Pi Epsilon; US trans association)
FPE-NE	Full Personality Expression, Northern Europe
GID	Gender Identity Disorder
HBC	Harry Benjamin Collection, Kinsey Institute, Bloomington, Indiana
HHA	Haeberle-Hirschfeld-Archiv für Sexualwissenschaft, Universitätsbibliothek, Humboldt Universität, Berlin
ICD-9	*International Classification of Diseases*, 9th edition
IHA	Ira Haraldsen Private Archive
JMC	John Money Collection, Alan Mason Chesney Archive, the Johns Hopkins Hospital, Baltimore
LEON	Lowest efficient level of care
LFTS	Landsforeningen for Transseksuelle (National Organization for Transsexuals)
LLH	Landsforeningen for lesbisk og homofil frigjøring
MJPS	The Ministry of Justice and Public Security, Oslo
NAN	The National Archives of Norway, Oslo
NBHSA	The Norwegian Board of Health Supervision Archive, Oslo
NBTS	The National Treatment Service for Transsexualism, Oslo
NEM	The Norwegian National Research Ethics Committee
NLN	The National Library of Norway, Oslo

NMA The Norwegian Medicines Agency, Oslo

NSD The Norwegian Center for Research Data

OCA Oslo City Archives, Oslo

OUH Oslo University Hospital, Oslo

OUHA Oslo University Hospital Archive

RDL The Royal Danish Library, Copenhagen

REK Regional Committees for Medical and Health Research Ethics in Norway

RSAO The Regional State Archives in Oslo

SA Skeivt arkiv, Bergen

SCID Structured Clinical Interviews

WHO World Health Organization

WPATH World Professional Association for Transgender Health

Notes

Introduction

1. As with all the other patient names in this book—unless they were already publicly known—the name Martha Pedersen is a pseudonym. Pedersen's case was first described in Christian Graugaard, "Professor Sands høns—om sexualbiologi i mellemkrigstidens Danmark" (PhD thesis, University of Copenhagen, 1997), 60–62.

2. Katie Sutton, *Sex Between Body and Mind: Psychoanalysis and Sexology in the German-Speaking World, 1890s–1930s* (Ann Arbor: University of Michigan Press, 2019), 118–44.

3. Richard von Krafft-Ebing, *Psychopathia Sexualis: Eine klinisch-forensiche Studie* (Stuttgart: Verlag von Ferdinand Enke, 1886); Carl Westphal, "Die conträre Sexualempfindung, Symptom eines neuropathischen (psychopathischen) Zustandes," *Archiv für Psychiatrie und Nervenkrankheiten* 2, no. 1 (1870), https://doi.org/10.1007/BF01796143. See also Rainer Herrn, *Schnittmuster des Geschlechts: Transvestismus und Transsexualität in der frühen Sexualwissenschaft* (Gießen: Psychosozial-Verlag, 2005). For the changing concepts and organic development of Krafft-Ebing's taxonomic system, see Harry Oosterhuis, *Stepchildren of Nature: Krafft-Ebing, Psychiatry, and the Making of Sexual Identity* (Chicago: University of Chicago Press, 2000).

4. Karl Heinrich Ulrichs, *Forschungen über das Räthsel der mannmännlichen Liebe* (Leipzig: Selbstverlag des Verfassers, 1864).

5. Westphal, "Conträre Sexualempfindung."

6. Chandak Sengoopta, *The Most Secret Quintessence of Life: Sex, Glands, and Hormones, 1850–1950* (Chicago: University of Chicago Press, 2006), 70–75.

7. Magnus Hirschfeld, *Was soll das Volk vom dritten Geschlecht wissen! Eine Aufklärungsschrift* (Leipzig: Verlag von Max Spohr, 1901). As Marhoefer has argued, however, the model of the modern homosexual was based on a eugenic and racist framework: see Laurie Marhoefer, *Racism and the Making of Gay Rights: A Sexologist, His Student, and the Empire of Queer Love* (Toronto: University of Toronto Press, 2022).

8. Magnus Hirschfeld, *Die Transvestiten: Eine Untersuchung über den erotischen Verkleidungstrieb* (Berlin: Alfred Pulvermacher, 1910), 281–89.

9. Magnus Hirschfeld, "Die intersexuelle Konstitution," in *Jahrbuch für sexuelle Zwischenstufen* (Stuttgart: Julius Püttmann, 1923), 24 (emphasis added).

10. See Rainer Herrn, *Der Liebe und dem Leid: Das Institut für Sexualwissenschaft 1919–1933* (Berlin: Suhrkamp, 2022), 31–48.

11. Hirschfeld, *Transvestiten*; Magnus Hirschfeld, "Der Transvestitismus," in *Sexualpathologie* (Bonn: A. Marcus & E. Webers Verlag, 1918). For the birth of modern identity categories, see Emma Heaney, *The New Woman: Literary Modernism, Queer Theory, and the Trans Feminine Allegory* (Evanston: Northwestern University Press, 2017); Marhoefer, *Racism and the Making of Gay Rights*.

12. Chandak Sengoopta, "Glandular Politics: Experimental Biology, Clinical Medicine, and Homosexual Emancipation in Fin-de-Siècle Central Europe," *Isis* 89, no. 3 (1998), https://doi.org/10.1086/384073; Sengoopta, *Secret Quintessence*, 55–67; Heiko Stoff, *Ewige Jugend: Konzepte der Verjüngung vom späten 19. Jahrhundert bis ins Dritte Reich* (Cologne: Böhlau Verlag, 2004), 435–69; Heiko Stoff, "Identität und Differenz: Zur Diskursgeschichte der Sexualität zu Beginn des 21. Jahrhunderts," in *Sexualität als Experiment: Identität, Lust und Reproduktion zwischen Science und Fiction*, ed. Nicolas Pethes and Silke Schicktanz (Frankfurt: Campus Verlag, 2008).

13. Graugaard, "Professor Sands høns," 38–39; Knud Sand, "Experiments on the Endocrinology of the Sexual Glands," *Endocrinology* 7, no. 2 (1923): 273–301. https://doi.org/10.1210/endo-7-2-273.

14. Sand, "Experiments on the Endocrinology of the Sexual Glands," 299.

15. Graugaard, "Professor Sands høns," 201–13; Marie-Louise Holm, "Fleshing Out the Self: Reimagining Intersexed and Trans Embodied Lives Through (Auto) Biographical Accounts of the Past" (PhD thesis, Linköping University, 2017), 196.

16. Knud Sand, *Experimentelle Studier over Kønskarakterer hos Pattedyr* (Copenhagen: Steen Hasselbalchs Forlag, 1918), 188.

17. Sand, *Experimentelle Studier over Kønskarakterer hos Pattedyr*, 153–54.

18. Holm, "Fleshing Out the Self," 310–14.

19. Lili Elbe, *Fra Mand til Kvinde: Lili Elbes Bekendelser*, a compilation of autobiographic material by Niels Hoyer (Copenhagen: Hage & Clausens Forlag, 1931), 36. The book was written by three persons—Lili Elbe, Ernst Ludwig Harthern Jakobson (pseudonym Niels Hoyer), and Kurt Warnekros—which makes it difficult to say which events are true and which are not. See also, Lili Elbe, *Man into Woman: The First Sex Change: A Portrait of Lili Elbe*, ed. Niels Hoyer (London: Blue Boat Books, 2004 [1933]).

20. Herrn, *Der Liebe und dem Leid*, 428–31.

21. Elbe, *Fra Mand til Kvinde*, 17–18.

22. Kurt Warnekros to Einar Wegner, July 18, 1930, Justitsministeriet, 1. Kontor, 1848–1967 journalsager, 1930 E, 1931–1985, file no. E1953, Danish National Archives, Copenhagen (hereafter cited as DNA). For the sake of clarity, many details have been omitted. For the complete story, see Sabine Meyer, *"Wie Lili zu*

einem richtingen Mädchen wurde." Lili Elbe: Zur Konstruktion von Geschlecht und Identität zwischen Medialisierung, Regulierung und Subjektivierung (Bielefeld: Transcript Verlag, 2015), 301–18.

23. Medico-Legal Council to the Ministry of Justice, August 9, 1930, Justitsministeriet, 1. Kontor, 1848–1967 journalsager, 1930 E, 1931–1985, file no. E1953, DNA.

24. Herrn, *Der Liebe und dem Leid*, 425, 33.

25. J. Gill-Peterson, *Histories of the Transgender Child* (Minneapolis: University of Minnesota Press, 2018), 16.

26. Joanne Meyerowitz, *How Sex Changed: A History of Transsexuality in the United States* (Cambridge, MA: Harvard University Press, 2002); Susan Stryker, *Transgender History: The Roots of Today's Revolution*, 2nd ed. (Berkeley: Seal Press, 2017); Herrn, *Schnittmuster*; Herrn, *Der Liebe und dem Leid*.

27. There are still very few monographs on European trans history beyond the German case. A notable exception is Alex Bakker's book on the trans care in the Netherlands with the opening of a specialized clinic in Amsterdam in 1972. See Alex Bakker, *The Dutch Approach: Fifty Years of Transgender Health Care at the CVU Amsterdam Gender Clinic* (Amsterdam: Boom, 2021).

28. Gill-Peterson, *Histories of the Transgender Child*.

29. Graugaard, "Professor Sands høns."

30. Holm, "Fleshing Out the Self."

31. Knud Sand to the Copenhagen police, March 7, 1922, Retslægerådet, Box Manglende sager i tidligere afleveringer (1922–1964), D996: 1922–1956, Folder unnumbered Transvestit, DNA.

32. Holm, "Fleshing Out the Self," 308–9.

33. Emma Heaney argued that "sex became cis" in the late nineteenth and early twentieth centuries, in the sense that sexological and psychoanalytic discourse reduced trans femininity to an allegory that confirmed a general cis model of sex. This model was incapable of incorporating how some trans women experienced sex on the level of organ and sensation, and in social interactions; see Heaney, *New Woman*, 4, 14, 160.

34. Heaney, *New Woman*, 309.

35. Mary Hilson, *The Nordic Model: Scandinavia Since 1945* (London: Reaktion Books, 2008), 88.

36. Niels Finn Christiansen and Pirjo Markkola, "Introduction," in *The Nordic Model of Welfare—a Historical Reappraisal*, ed. Niels Finn Christiansen et al. (Copenhagen: Museum Tusculanum Press, 2006); Jenny Andersson, "Mellan tillväxt och trygghet: Idéer om produktiv socialpolitik i socialdemokratisk socialpolitisk ideologi under efterkrigstiden" (PhD thesis, Uppsala University, 2003).

37. Anne-Lise Seip, *Veiene til velferdsstaten: Norsk sosialpolitikk 1920–75* (Oslo: Gyldendal, 1994).

38. Synnøve Bendixsen, Mary Bente Bringslid, and Halvard Vike, eds., *Egalitarianism in Scandinavia: Historical and Contemporary Perspectives* (Cham: Palgrave Macmillan, 2018); Karen Fog Olwig and Karsten Paerregaard, eds., *The Question*

of Integration: Immigration, Exclusion and the Danish Welfare State (Newcastle: Cambridge Scholars Publishing, 2011); Øystein Sørensen and Bo Stråth, eds., *The Cultural Construction of Norden* (Oslo: Scandinavian University Press, 1997).

39. Francis Sejersted, *The Age of Social Democracy: Norway and Sweden in the Twentieth Century* (Princeton: Princeton University Press, 2011), 7.

40. Marianne Gullestad, "The Scandinavian Version of Egalitarian Individualism," *Ethnologia Scandinavica: A Journal for Nordic Ethnology* 21 (1991); Marianne Gullestad, *Det norske sett med nye øyne: Kristisk analyse av norsk innvandringsdebatt* (Oslo: Universitetsforlaget, 2002), 82–85.

41. For example, historians have argued that even if equivalence gained importance from the 1970s, which in theory could have led to more differentiated health services for minority groups such as immigrants and the Sámi, the measures were still harnessed to meet the needs of the majority: see Teemu Ryymin and Kari Ludvigsen, "From Equality to Equivalence? Norwegian Health Policies Towards Immigrants and the Sámi, 1970–2009," *Nordic Journal of Migration Research* 3, no. 1 (2013), https://doi.org/10.2478/v10202-012-0011-y.

42. Hilson, *Nordic Model: Scandinavia Since 1945*, 88.

43. Thomas Etzemüller, "Rationalizing the Individual—Engineering Society: The Case of Sweden," in *Engineering Society: The Role of the Human and Social Sciences in Modern Societies, 1880–1980*, ed. Kerstin Brückweh et al. (Houndmills: Palgrave Macmillan, 2012), 97.

44. Yvonne Hirdman, *Att lägga livet til rätta: Studier i svensk folkhemspolitik* (Stockholm: Carlssons, 2018 [1989]).

45. Helga Maria Hernes, *Staten—kvinner ingen adgang?* (Oslo: Universitetsforlaget, 1982); Helga Maria Hernes, *Welfare State and Woman Power: Essays in State Feminism* (Oslo: Norwegian University Press, 1987).

46. Birte Siim and Hege Skjeie, "Tracks, Intersections and Dead Ends: Multicultural challenges to state feminism in Denmark and Norway," *Ethnicities* 8, no. 3 (2008), https://doi.org/10.1177/1468796808092446.

47. Diana Mulinari et al., "Introduction: Postcolonialism and the Nordic Models of Welfare and Gender," in *Complying with Colonialism: Gender, Race, and Ethnicity in the Nordic Region*, ed. Suvi Keskinen et al. (Farnham: Ashgate, 2009).

48. Didier Fassin, "Introduction: Governing Precarity," in *At the Heart of the State: The Moral World of Institutions*, ed. Didier Fassin et al. (London: Pluto Press, 2015); Veena Das and Deborah Poole, "State and Its Margins: Comparative Ethnographies," in *Anthropology in the Margins of the State*, ed. Veena Das and Deborah Poole (New Delhi: Oxford University Press, 2004).

49. Vilhelm Aubert, *The Hidden Society* (Totowa, NJ: Bedminster Press, 1965).

50. Alva Myrdal and Gunnar Myrdal, *Kris i befolkningsfrågan* (Stockholm: Albert Bonniers forlag, 1934).

51. Etzemüller, "Rationalizing the Individual—Engineering Society"; Aina Schiøtz, *Det offentlige helsevesen i Norge 1603–2003. Bind 2. Folkets helse—landets*

styrke, 1850–2003 (Oslo: Universitetsforlaget, 2003); Sejersted, *Age of Social Democracy*; Rune Slagstad, *De nasjonale strateger* (Oslo: Pax forlag, 1998).

52. Per Haave, *Sterilisering av tatere 1934–1977: En historisk undersøkelse av lov og praksis* (Oslo: Norges forskningsråd, 2000); Mattias Tydén, *Från politik till praktik. De svenska steriliseringslagarna 1935–1975*, 2nd ed. (Stockholm: Almqvist & Wiksell International, 2002); Nils Roll-Hansen, "Norwegian Eugenics: Sterilization as Social Reform," in *Eugenics and the Welfare State: Sterilization Policy in Denmark, Sweden, Norway, and Finland*, ed. Gunnar Broberg and Nils Roll-Hansen (East Lansing: Michigan State University Press, 2005).

53. Ida Ohlsson Al Fakir, "Nya rum för socialt medborgarskap. Om vetenskap och politik i 'Zigenarundersökningen'—en socialmedicinsk studie av svenska romer 1962–1965" (PhD thesis, Linnaeus University, 2015).

54. Annika Berg, *De samhällsbesvärliga. Förhandlingar om psykopati och kverulans i 1930- och 1940-talens Sverige* (Gothenburg: Makadam förlag, 2018).

55. Adele E. Clarke et al., eds., *Biomedicalization: Technoscience, Health, and Illness in the US* (Durham, NC: Duke University Press, 2010).

56. J. Andrew Mendelsohn developed this argument about the role of physicians in early modern Europe: see J. Andrew Mendelsohn, Annemarie Kinzelbach, and Ruth Schilling, eds., *Civic Medicine: Physician, Polity, and Pen in Early Modern Europe* (London: Routledge, 2020).

57. Ole Berg, *Spesialisering og profesjonalisering: En beretning om den sivile norske helseforvaltnings utvikling fra 1809 til 2009. Del 1: 1809–1983—Den gamle helseforvaltning* (Oslo: Statens helsetilsyn, 2009); Trond Nordby, *Karl Evang: En biografi* (Oslo: Aschehoug, 1989).

58. Susanne Bauer, "From Administrative Infrastructure to Biomedical Resource: Danish Population Registries, the 'Scandinavian Laboratory,' and the 'Epidemiologist's Dream,'" *Science in Context* 27, no. 2 (2014), https://doi.org/10.1017/S0269889714000040; Sheila Jasanoff, *Designs on Nature: Science and Democracy in Europe and the United States* (New Jersey: Princeton University Press, 2005); Anne Kveim Lie, "Producing Standards, Producing the Nordic Region: Antibiotic Susceptibility Testing, from 1950–1970," *Science in Context* 27, no. 2 (2014), https://doi.org/10.1017/S0269889714000052.

59. Kristin Asdal and Christoph Gradmann, "Introduction: Science, Technology, Medicine—and the State: The Science-State Nexus in Scandinavia, 1850–1980," *Science in Context* 27, no. 2 (2014), https://doi.org/10.1017/S0269889714000039.

60. Dorothy Porter, ed., *Social Medicine and Medical Sociology in the Twentieth Century*, vol. 43, Clio Medica (Amsterdam: Rodopi, 1997); Dorothy Porter, "How Did Social Medicine Evolve, and Where Is It Heading?," *PLOS Medicine* 3, no. 10 (2006), https://doi.org/10.1371/journal.pmed.0030399. There are a few exceptions, highlighting the roots of social medicine in the Global South: see, e.g., Michelle Pentecost et al., "Revitalising Global Social Medicine," *The Lancet* 398, no. 10300 (2021), https://doi.org/10.1016/S0140-6736(21)01003-5; Howard Waitzkin

et al., "Social Medicine Then and Now: Lessons from Latin America," *American Journal of Public Health* 91, no. 10 (2001), https://doi.org/10.2105/ajph.91.10.1592.

61. Dorothy Porter and Roy Porter, "What Was Social Medicine? An Historiographical Essay," *Journal of Historical Sociology* 1, no. 1 (1988).

62. Marc Berg and Annemarie Mol, eds., *Differences in Medicine: Unraveling Practices, Techniques, and Bodies* (Durham, NC: Duke University Press, 1998).

63. Bruno Latour and Steve Woolgar, *Laboratory Life: The Construction of Scientific Facts* (New Jersey: Princeton, 1986); Ilana Löwy, *Between Bench and Bedside: Science, Healing and Interleukine-2 in a Cancer Ward* (Cambridge, MA: Harvard University Press, 1996); Hans-Jörg Rheinberger, *Toward a History of Epistemic Things: Synthetizing Proteins in the Test Tube* (Stanford, CA: Stanford University Press, 1997).

64. Annemarie Mol, *The Body Multiple: Ontology in Medical Practice* (Durham: Duke University Press, 2002); Stefan Hirschauer, *Die soziale Konstruktion der Transsexualität: Über die Medizin und den Geschlechtswechsel* (Frankfurt: Suhrkamp, 1993).

65. Geertje Mak, *Doubting Sex: Inscriptions, Bodies and Selves in Nineteenth-Century Hermaphrodite Case Histories* (Manchester: Manchester University Press, 2012).

66. Das and Poole, "State and Its Margins: Comparative Ethnographies," 19.

67. Fassin, "Introduction: Governing Precarity," 3–4.

68. Quentin Skinner, *Die drei Körper des Staates*, trans. Karin Wördemann (Göttingen: Wallstein Verlag, 2012), 85–88.

69. Erwin H. Ackerknecht, "A Plea for a 'Behaviorist' Approach in Writing the History of Medicine," *Journal of the History of Medicine and Allied Sciences* XXII, no. 3 (1967), https://doi.org/10.1093/jhmas/XXII.3.211; Guenter B. Risse and John Harley Warner, "Reconstructing Clinical Activities: Patient Records in Medical History," *Social History of Medicine* 5, no. 2 (1992), https://doi.org/10.1093/shm/5.2.183.

70. For Denmark, cases from the Medico-Legal Council from the early 1920s to the 1960s provide the main material. Medical journals from the 1950s have been destroyed or digitized as part of patients' current medical records and are not available for historical research. In Norway, I have analyzed the applications to the Expert Council on Cases of Sex Operations from the 1950s and 1960s and medical records from the 1950s to the early 2000s. Unfortunately, it has not been possible to locate or access relevant archival material or medical records in Sweden. Moreover, no material from Jan Wålinder's research project has been preserved. To access the medical records, either at Karolinska Hospital or at St. Jörgen Hospital, one would need the full name and personal ID number of every individual documented in the records. Because the names are not public, it is impossible to locate the files unless one gains access to all the medical records from the two hospitals in which Wålinder practiced. Not only would this be an insurmountable task,

but it would most likely be ethically impossible; the analysis of the developments in Sweden is therefore based on published sources.

71. Bernice L. Hausman, *Changing Sex: Transsexualism, Technology, and the Idea of Gender* (Durham, NC: Duke University Press, 1995); Hirschauer, *Soziale Konstruktion der Transsexualität*.

72. Jay Prosser, *Second Skins: The Body Narrative of Transsexuality* (New York: Columbia University Press, 1998), 7–8.

73. Emmett H. Drager and Lucas Platero, "At the Margins of Time and Place: Transsexuals and the Transvestites in Trans Studies," *TSQ: Transgender Studies Quarterly* 8, no. 4 (2021), https://doi.org/10.1215/23289252-9311018.

74. Ludwik Fleck, *Entstehung und Entwicklung einer wissenschaftlichen Tatsache* (Frankfurt am Main: Suhrkamp, 1980 [1935]).

75. The interviews were semistructured, between one and three hours long, and mostly conducted in person. Most people were interviewed once, but I had follow-up interviews with a handful of people. The interviews were recorded using the TSD (Services for Sensitive Data) audio app developed by the University of Oslo. All interviews were recorded and transcribed by the author, and selected quotes were translated into English. All interviewees had the option of appearing under their full name, and transcripts were made available to those who requested them.

76. Paul Thompson, *The Voice of the Past: Oral History*, 3rd ed. (New York: Oxford University Press, 2000), 9.

77. Richard Andrew McKay, *Patient Zero and the Making of the AIDS Epidemic* (Chicago: University of Chicago Press, 2017); Chloe Silverman, *Understanding Autism: Parents, Doctors, and the History of a Disorder* (Princeton: Princeton University Press, 2012); Matthew Smith, *An Alternative History of Hyperactivity: Food Additives and the Feingold Diet* (New Brunswick: Rutgers University Press, 2011); Dóra Vargha, *Polio Across the Iron Curtain: Hungary's Cold War with an Epidemic* (Cambridge: Cambridge University Press, 2018).

78. I actively searched for empirical material that could alleviate the "whiteness" of Scandinavian trans history, but in this I was unsuccessful. All my interviewees are from Norway, none have a migrant background, and none are Sámi, Kven, or Roma. The youngest person is aged between 30 and 40, and the oldest between 70 and 80. Even though the Norwegian history of trans medicine is very much an Oslo history, since that is where the main medical and bureaucratic institutions were located, people from all class backgrounds came to the city from all over the country. This is reflected in the background of my interviewees; some people grew up in the northernmost parts of Norway, others in small communities along the fjords on the west coast, and some grew up in the capital. Some people were raised in working-class families, others were the children of academics. Some worked as strippers, others as high-ranked military officers. This is also reflected in a handful of published autobiographies in Norway. One was by a manual laborer who worked in a warehouse in the 1950s; another by a woman selling sex on the

streets of Oslo in the 1960s and 1970s; and yet another by a businesswoman rising through a populist political party in the 1990s and 2000s: see, respectively, Bernt Eggen, *Bastard! En sannferdig roman om kjønnsskifte i Norge* (Oslo: Aschehoug, 1979); Marianne Nordli, *Det sterke kjønn—min kamp for å bli kvinne* (Oslo: Juritzen forlag, 2017); Janni Christin Wintherbauer and Ane Rostad Stokholm, *Janni—slik ble mitt liv* (Oslo: Pegasus forlag, 2010).

79. Nancy Tomes, "Oral History in the History of Medicine," *Journal of American History* 78, no. 2 (1991), https://doi.org/10.2307/2079538.

80. University of Minnesota, "The Tretter Transgender Oral History Project," February 22, 2022, https://www.lib.umn.edu/collections/special/tretter/transgender-oral-history-project.

81. Emily R. Novak Gustainis and Phoebe Evans Letocha, "The Practice of Privacy," in *Innovation, Collaboration, and Models: Proceedings of the CLIR Cataloging Hidden Special Collections and Archives*, ed. Cheryl Oestreicher (Washington, DC: Council on Library and Information Resources, 2015); Susan C. Lawrence, "Access Anxiety: HIPAA and Historical Research," *Journal of the History of Medicine and Allied Sciences* 62, no. 4 (2007).

82. A large part of the research on which this book is based was conducted within the Biomedicalization From the Inside Out project at the University of Oslo (project leader: Anne Kveim Lie). Data handling and storage was conducted according to Norwegian research ethics guidelines. The project was approved by the Norwegian Center for Research Data (NSD, 395915, August 15, 2019, April 27, 2020, and December 2, 2021). Sensitive material was photographed with the TSD app. Data processing was encrypted, and it is stored on a server that can only be accessed through two-step authentication. All medical files were given a case file number. The key is stored on the University of Oslo's secure database (TSD) along with the list of pseudonyms of the people interviewed by me. In the present book, all names of former patients and care providers have been anonymized, except for those care providers with a publication history. All dates of contacts with the health care system have been modified to further protect the anonymity of individuals.

83. NEM to Anne Kveim Lie, NEM 2019/260, March 20, 2020.

84. REK to Anne Kveim Lie, 29326, November 15, 2019.

85. Charles E. Rosenberg, "Meanings, Policies, and Medicine: On the Bioethical Enterprise and History," *Daedalus* 128, no. 4 (1999).

86. Roger Cooter, "Inside the Whale: Bioethics in History and Discourse," *Social History of Medicine* 23, no. 3 (2010), https://doi.org/10.1093/shm/hkq058.

87. Parts of chapter 8 were first published as Ketil Slagstad, "Society as Cause and Cure: The Norms of Transgender Social Medicine," *Culture, Medicine, and Psychiatry* 45, no. 3 (2021), https://doi.org/10.1007/s11013-021-09727-4, and Ketil Slagstad, "Psychiatric Practices Beyond Psychiatry: The Sexological Administration of Transgender Life Around 1980," in *Doing Psychiatry in Postwar Europe: Practices,*

Routines and Experiences, ed. Gundula Gahlen et al. (Manchester: Manchester University Press, 2024), under CC BY 4.0 license (https://creativecommons.org/licenses/by/4.0/).

88. Chapter 11 is an adaptation of an essay first published as Ketil Slagstad, "Bureaucratizing Medicine: Creating a Gender Identity Clinic in the Welfare State," *Isis* 113, no. 3 (2022), https://doi.org/10.1086/721140. ©2022 History of Science Society.

Chapter One

1. Dr. NN to the Directorate of Health, 25 August 1953, S-3212 Justisdepartementet, Lovavdelingen, De, Box 58, Folder Straffeloven, NAN (hereafter cited as "Folder Straffeloven").

2. Lorraine Daston and Katharine Park, "The Hermaphrodite and the Orders of Nature," *GLQ* 1 (1995); Leah DeVun, *The Shape of Sex: Nonbinary Gender from Genesis to the Renaissance* (Columbia University Press, 2021), https://doi.org/10.7312/devu19550; Alice Domurat Dreger, *Hermaphrodites and the Medical Invention of Sex* (Cambridge, MA: Harvard University Press, 1998); Herrn, *Schnittmuster*; Ulrike Klöppel, *XXoXY ungelöst: Hermaphroditismus, Sex und Gender in der deutschen Medizin: Eine historische Studie zu Intersexualität* (Bielefeld: transcript Verlag, 2010); Thomas Laqueur, *Making Sex: Body and Gender from the Greeks to Freud* (Cambridge, MA: Harvard University Press, 1990); Elizabeth Reis, *Bodies in Doubt: An American History of Intersex* (Baltimore: Johns Hopkins University Press, 2009); Ketil Slagstad, "The Political Nature of Sex—Transgender in the History of Medicine," *New England Journal of Medicine* 384, no. 11 (2021), https://doi.org/10.1056/NEJMms2029814.

3. Denmark (1929; amended in 1935 to include castration and sterilization on the basis of social-humanitarian and eugenic indications), Sweden (1934), Norway (1934), Finland (1935), and Iceland (1938). See Haave, *Tatere*, 83–85. For the medico-legal discourse on homosexuality, see Runar Jordåen, "Inversjon og perversjon: Homoseksualitet i norsk psykiatri og psykologi frå slutten av 1800-talet til 1960" (PhD thesis, University of Bergen, 2010).

4. Alison Bashford and Philippa Levine, eds., *The Oxford Handbook of the History of Eugenics* (New York: OUP, 2010).

5. Lene Koch, "Eugenic Sterilisation in Scandinavia," *European Legacy* 11, no. 3 (2006), https://doi.org/10.1080/10848770600668340; Koch, *Racehygiejne*, 81–87.

6. Lene Koch, *Racehygiejne i Danmark 1920–56* (Copenhagen: Informations Forlag, 2010), 41–73; Tydén, *Från politik till praktik*, 73–143; Nils Roll-Hansen and Gunnar Broberg, eds., *Eugenics and the Welfare State: Norway, Sweden, Denmark, and Finland* (East Lansing: Michigan State University Press, 2005).

7. See Haave, *Tatere*; Koch, *Racehygiejne*; Tydén, *Från politik till praktik*.

8. Tydén, *Från politik till praktik*, 522.

9. Hilson, *Nordic Model: Scandinavia Since 1945*, 104.

10. Gunnar Broberg and Mattias Tydén, "Eugenics in Sweden: Efficient Care," in *Eugenics and the Welfare State: Sterilization Policy in Denmark, Sweden, Norway, and Finland*, ed. Gunnar Broberg and Nils Roll-Hansen (East Lansing: Michigan State University Press, 2005), 120.

11. Audvar Ragnar Os, memo, "Omskaping" av mann til kvinne ved operative inngrep, November 27, 1953, S-3212, Folder Straffeloven, NAN (hereafter cited as "Memo Omskaping").

12. Koch, *Racehygiejne*, 198–210.

13. Svein Atle Skålevåg, *Utilregnelighet: En historie om rett og medisin* (Oslo: Pax forlag, 2016), 124–27; Graugaard, "Professor Sands høns," 141–43.

14. Ingeborg Aas, *Hvordan kan samfundet beskytte sig mot åndssvake og sedelighetsforbrytere* (Oslo: Olaf Norlis forlag, 1932), 12. For the political context, see Haave, *Tatere*, 45–50.

15. Although the ministry made the final decision, it almost always requested a statement from the Medico-Legal Council; see Holm, "Fleshing Out the Self," 190–202; Retslægerådet, *Retslægerådet 1909–2009* (Copenhagen: Retslægerådet, 2009).

16. Under the Nazi occupation, however, the law was amended to include coercive measures and was thereby shifted in a direction based purely on eugenics.

17. Jordåen, "Inversjon og perversjon," 169–72.

18. We do not know how many men were castrated in Norway before the 1934 law, but at least some homosexual men were castrated in the 1920s at Ullevål Hospital in Oslo: see Haave, *Tatere*, 138; Aas, *Åndssvake og sedelighetsforbrytere*, 38.

19. Knud Sand to the Copenhagen police, December 12, 1928, Retslægerådet, Box Manglende sager i tidligere afleveringer (1922–1964), D996: 1922–1956, Folder 1569/1928, DNA. This case was first described by Christian Graugaard in his dissertation: see Graugaard, "Professor Sands høns," 165–66. All cases from the archive of the Medico-Legal Council in this chapter have been identified either through Graugaard's or Sølve Holm's theses; see Holm, "Fleshing Out the Self." I am indebted to Graugaard and Holm for their meticulous work in the archives.

20. Graugaard, "Professor Sands høns," 203–09.

21. Sand, *Experimentelle Studier over Kønskarakterer hos Pattedyr*, 166.

22. Holm, "Fleshing Out the Self," 311.

23. Knud Sand, *Die Physiologie des Hodens* (Leipzig: Verlag von Curt Kabitzsch, 1933), 216.

24. Knud Sand, undated, probably January 1935, Retslægerådet, Box Manglende sager i tidligere afleveringer (1922–1964), D996: 1922–1956, Folder 261/35, DNA.

25. Sølve Holm's PhD thesis provides an in-depth reconstruction and reading of her life story: see Holm, "Fleshing Out the Self," 275–360.

26. NN to Knud Sand, December 19, 1941, Retslægerådet, Ka-sager (1929–1968), Box L59, File 699, DNA.

27. Holm, "Fleshing Out the Self," 310–12.

28. Georg K. Stürup, undated 1953, Retslægerådet, Ka-sager (1929–1968), Box L59, File 699, DNA.

29. An application for castration was never submitted, however. As Sølve Holm argues, this probably had religious reasons. She would not agree to a simple castration, "where I am made sexless," unless she could have full surgery, including the transposition of the urethra and the implantation of an ovary: see Holm, "Fleshing Out the Self," 321.

30. Holm, "Fleshing Out the Self," 312.

31. Report by the county medical doctor in Aalborg, September 6, 1943, Retslægerådet, Box Manglende sager i tidligere afleveringer (1922–1964), D996: 1922–1956, Folder 1809/43, DNA.

32. Statements by Knud Sand, October 7, 1943, and Hjalmar Helweg, September 14, 1943, both in Retslægerådet, Box Manglende sager i tidligere afleveringer (1922–1964), D996: 1922–1956, Folder 1809/43, DNA.

33. Louis Le Maire, "Danish Experiences Regarding the Castration of Sexual Offenders," *Journal of Criminal Law, Criminology, and Police Science* 47, no. 3 (1956): 297, https://doi.org/10.2307/1140320.

34. George Jorgensen Jr.'s letter to Dr. Stürup, undated (1950 or 1951), Christine Jorgensens efterladte papirer, box 8, DRL. The statement was reproduced in the psychiatrist Georg K. Stürup's application for castration, see Georg K. Stürup to the Ministry of Justice, April 1, 1951, Retslægerådet, Ka-sager (1929–1968), Box L55, Folder 641 "Chris," DNA.

35. Chris Jorgensen to Mr. and Mrs. Jorgensen, May 1, 1950, Christine Jorgensens efterladte papirer, box 21, DRL.

36. Postcards to Mr. and Mrs. Jorgensen, June 20, June 23, July 1, 1950, Christine Jorgensens efterladte papirer, box 21, DRL.

37. Preben Hertoft and Teit Ritzau, *Paradiset er ikke til salg: Trangen til at være begge køn* (Viborg: Lindhardt og Ringhof, 1984), 34.

38. Christine Jorgensen to Knud Sand, March 20, 1952, Retslægerådet, Ka-sager (1929–1968), Box L55, Folder 641 "Chris," DNA.

39. George Jorgensen Jr.'s letter to Dr. Stürup, undated (1950 or 1951), Christine Jorgensens efterladte papirer, box 8, DRL.

40. Christine Jorgensen ("Brud") to Mr. and Mrs. Jorgensen, August 3, 1950, Christine Jorgensens efterladte papirer, box 21, DRL.

41. Christian Hamburger, "Intersexualität," in *Mensch, Geschlecht, Gesellschaft: Das Geschlechtsleben unserer Zeit gemeinverständlich dargestellt*, eds. Hans Giese and A. Willy (Paris: Guillaume Aldor; 1954).

42. Christine Jorgensen to Gen and Dr. Joe, July 20, 1950, Christine Jorgensens efterladte papirer, Box 28, DRL.

43. Hertoft and Ritzau, *Paradiset*, 111.

44. Georg K. Stürup, "Transvestisme i klinisk kriminologi," *Nordisk Medicin* 56, no. 35 (1956).

45. The patient's perseverance in continuing to push for surgery paid off almost

two decades after the first contact, however, when Dahl-Iversen finally agreed to recommend castration. In the request to the Medico-Legal Council, it was stated that the patient had "the feeling that life is passing him by as he sees people around him planning their lives, starting families, etc., in short, living, while he sees only darkness and death." Georg K. Stürup, endorsement of application for castration, January 17, 1955, Retslægerådet, Ka-sager (1929–1968), Box L66, Folder 790, DNA.

46. Stürup, endorsement of application for castration.

47. Hertoft and Ritzau, *Paradiset*, 116–17.

48. Georg K. Stürup to Knud Sand, April 1, 1951, Retslægerådet, Ka-sager (1929–1968), Box L55, Folder 641 "Chris," DNA.

49. Holm, "Fleshing Out the Self," 326.

50. Georg William Jorgensen to the Ministry of Justice's 3. Office, April 1, 1951, Retslægerådet, Ka-sager (1929–1968), Box L55, Folder 641 "Chris," DNA.

51. Holm, "Fleshing Out the Self," 326–27 n. 398; Thorkil Sørensen and Preben Hertoft, "Sexmodifying Operations on Transsexuals in Denmark in the Period 1950–1977," *Acta Psychiatrica Scandinavica* 61, no. 1 (1980), https://doi.org/10.1111/j.1600-0447.1980.tb00565.x.

52. Medical note, admission November 18–December 5, 1952, Rigshospitalet's Surgical Clinic C, Retslægerådet, Ka-sager (1929–1968), Box L55, Folder 641 "Chris," DNA.

53. Hertoft and Ritzau, *Paradiset*, 101.

54. Source: National Library of Norway.

55. Jens Rydström and David Tjeder, *Kvinnor, män och alla andra—En svensk genushistoria* (Lund: Studentlitteratur, 2009), 161.

56. Contract between the Turistforeningen for Danmark and Christine Jorgensen, Copenhagen, 18 November, 1952, Christine Jorgensens efterladte papirer, Box 8, DRL.

57. Harry R. Berglind to Christine Jorgensen, February 14, 1953, Christine Jorgensens efterladte papirer, Box 1, DRL.

58. Christine Jorgensen to Gen and Dr. Joe, September 17, 1951, Christine Jorgensens efterladte papirer, Box 28, DRL.

59. Chris Jorgensen to Mr. and Mrs. Jorgensen, undated, sometime in the autumn of 1952, Christine Jorgensens efterladte papirer, Box 21, RDL.

60. Christian Hamburger, Georg K. Stürup, and Erling Dahl-Iversen, "Transvestism: Hormonal, Psychiatric, and Surgical Treatment," *Journal of the American Medical Association* 152, no. 5 (1953), https://doi.org/10.1001/jama.1953.03690050015006; Christian Hamburger, Georg K. Stürup, and E. Dahl-Iversen, "Transvestisme. Hormonal, psykiatrisk og kirurgisk behandling af et tilfælde," *Nordisk Medicin* 12, no. VI (1953).

61. Hamburger, Stürup, and Dahl-Iversen, "Transvestism," 395. Sølve M. Holm interprets this primarily as a protective measure against homosexual men's perceived hypersexuality and promiscuity and the associated risk of venereal diseases: see Holm, "Fleshing Out the Self," 329–30, n. 404.

62. Holm, "Fleshing Out the Self," 200, 327.

63. Holm, "Fleshing Out the Self," 242.

64. Holm, "Fleshing Out the Self," 337.

65. For a detailed analysis of this case, see Holm, "Fleshing Out the Self," 337–43.

66. The Medico-Legal Council, July 20, 1953, Retslægerådet, Ka-sager (1929–1968), Box L59, Folder 699, DNA.

67. See the following cases in Retslægerådet, Ka-sager (1929–1968): Box 62, Folder 734; Box 64, Folder 774; Box 65, Folder 789; Box 66, Folder 790; Box 66, Folder 798; Box 73, Folder 884, all in DNA.

68. Hjalmar Helweg, April 30, 1953, Retslægerådet, Ka-sager (1929–1968), Box L59, Folder 699, DNA.

69. Christofer Lohne Knudsen, note, November 13, 1961, Folder Kjønnsskifte, NAN.

70. NN to the Norwegian Medical Association, September 11, 1953, PA-0280 Den norske legeforening, Db, Box 11, Folder Steriliseringssaken, NAN.

71. Admission note, 1954, case 1012, Aker Hospital, Oslo University Hospital Archive (hereafter cited as OUHA).

72. Christofer Lohne Knudsen, psychiatric examination, 1954, case 1012, Aker Hospital, OUHA.

73. The Legislation Department to the Directorate of Health, Hormonbehandling og tillatelse for en mann til å opptre som kvinne, December 7, 1953, S-3212 Justisdepartementet, Lovavdelingen, Kopibøker, Box 56, Folder 56, NAN (hereafter cited as "Hormonbehandling letter").

74. Audvar Ragnar Os, Memo Omskaping, S-3212, Folder Straffeloven, NAN.

75. Audvar Ragnar Os, Memo Omskaping, S-3212, handwritten remark by Hans Fredrik Marthinussen.

76. Hans Fredrik Marthinussen, handwritten remark, memo, Hormonbehandling og tillatelse for en mann til å opptre som en kvinne, November 28, 1953, S-3212, Folder Straffeloven, NAN.

77. The Legislation Department to the Directorate of Health, "Omskaping" av mann til kvinne ved operative inngrep, December 7, 1953, S-3212 Justisdepartementet, Lovavdelingen, Ba—Kopibøker, Box 56, NAN (hereafter cited as "Letter Omskaping").

78. Letter Omskaping, S-3212, Folder Straffeloven, NAN.

79. Heaney, *New Woman*.

80. Memo Omskaping, S-3212, Folder Straffeloven, NAN.

81. Memo Omskaping, S-3212.

82. For the concept of "moral economy," see Fassin, "Introduction: Governing Precarity," 9–10.

83. Ann Laura Stoler, *Along the Archival Grain: Epistemic Anxieties and Colonial Common Sense* (Princeton: Princeton University Press, 2009), 2.

84. Kristin Asdal and Tone Druglitrø, "Modifying the Biopolitical Collective: The Law as a Moral Technology," in *Humans, Animals and Biopolitics: The*

More-Than-Human Condition, ed. Kristin Asdal, Tone Druglitrø, and Steve Hinchliffe (London: Routledge, 2017).

85. Tal Golan, *Laws of Men and Laws of Nature* (Cambridge, MA: Harvard University Press, 2004).

Chapter Two

1. "De tiltalte i homosex-saken ville skifte kjønn," *Demokraten* (April 24, 1954), 1.

2. Forensic psychiatric report by Dr. NN, 1953, A-10055, Gj, Box 172, The Regional State Archives in Oslo (hereafter RSAO) (hereafter cited as "Folder Homosex").

3. Declaration by L. Nielsen, A-10055, Gj, Box 172, RSAO.

4. As already noted, after approving the application for castration of Christine Jorgensen, the Danish minister of justice decided that in the future, only Danish citizens could apply for the procedure. To help the trans people abroad who contacted him, Hamburger established a cooperation with the Dutch psychiatrist Frederik Hartsuiker. After a while, the psychiatrist Coen van Emde Boas took over the responsibility and cooperated with the endocrinologist Harry Benjamin in the United States. From 1954 to 1956, around ten individuals, primarily trans women from the United States, underwent sex reassignment surgery in the Netherlands. See Alex Bakker, *The Dutch Approach*, 20–21.

5. Declaration by L. Nielsen, forensic psychiatric report, S-4249 Den rettsmedisinske kommisjon, Db Rettspsykiatriske erklæringer, Box 95, NAN.

6. Police report, A-10055, Folder Homosex, RSAO.

7. Skålevåg, *Utilregnelighet. En historie om rett og medisin*, 133–36.

8. Per Anchersen, "Sammenfatning og utsyn," in *Nervøse lidelser og sinnets helse*, ed. Per Anchersen and Leo Eitinger (Oslo: H. Aschehough, 1955), 232.

9. "De tiltalte i homosex-saken ville skifte kjønn."

10. Per Anchersen and Jon Leikvam, forensic psychiatric report, April 9, 1954, S-4249 Den rettsmedisinske kommisjon, Dbb, Box 95, NAN; Police report, 1954, A-10055, Gj, Box 172, NAN.

11. NN Court of Appeal, 1954, A-10055, Gj, Box 172, NAN.

12. Sheila Jasanoff, "Science, Common Sense & Judicial Power in U.S. Courts," *Daedalus* 147, no. 4 (2018), https://doi.org/10.1162/daed_a_00517.

13. Per Anchersen and Nils H. Houge, forensic psychiatric report, 1954, S-4736, Box 118, jnr. 465/58, NAN.

14. Autobiographical statement by K. Karlsen, 1954, S-4736, Box 118, No. 465/58, NAN.

15. Anchersen and Houge, report, S-4736, Box 118, No. 465/58, NAN.

16. Per Anchersen to the Regional Public Persecution Office in Oslo, 1955, Box 118, No. 465/58, NAN.

17. Rikshospitalet, 1954, Case 1012, OUHA.

18. Per Anchersen, "Problems of Transvestism," *Acta Psychiatrica et Neurologica Scandinavica* 106 (1957): 254–55.

19. Nils Kinnerød to the Director General of Health, October 6, 1954, S-4736 Det sakkyndige råd i saker om seksualinngrep, D, no. 118, NAN.

20. NN to the Director General of Health, October 14, 1954, S-4736, Box 118, NAN.

21. Kinnerød to the Director General of Health, S-4736, Box 118, NAN.

22. NN to the Director General of Health, October 14, 1954, S-4736, Box 118, NAN.

23. Christofer Lohne Knudsen to Det sakkyndige utvalg for utredning av behandlingsspørsmål vedrørende transvestitisme, April 2, 1955, S-4736, series D, no. 57, Folder 2: Kjønnsskifte, NAN (hereafter cited as "Folder Kjønnsskifte").

24. Majority report, April 10, 1956, S-4736, Folder Kjønnsskifte, NAN, p. 16.

25. Application for castration, April 16, 1956, S-4736 Det sakkyndige råd i saker om seksualinngrep, D, Box 123, NAN.

26. Minority report, September 3, 1956, p. 24, S-4736, Folder Kjønnsskifte, NAN.

27. Johan Bremer, *Asexualization: A Follow-Up Study of 244 Cases* (Oslo: Oslo University Press, 1958).

28. Bremer, *Asexualization*, 318.

29. Bremer, *Asexualization*, 99.

30. Minority report, p. 23, S-4736, Folder Kjønnsskifte, NAN.

31. See, for instance, applications for sex change operations in 1959, S-4736, Box 140, NAN.

32. Johan Bremer, "Mutilerende behandling av transseksualisme?" *Tidsskrift for Den Norske Lægeforening* 68, no. 13–14 (1961): 923.

33. For the making of the "pharmaceutical revolution," see Jeremy A. Greene, Flurin Condrau, and Elizabeth Siegel Watkins, "Medicine Made Modern by Medicines," in *Therapeutic Revolutions: Pharmaceuticals and Social Change in the Twentieth Century*, ed. Jeremy A. Greene, Flurin Condrau, and Elizabeth Siegel Watkins (Chicago: University of Chicago Press, 2016). See particularly Nicolas Henckes's chapter on psychopharmaceuticals, "Magic Bullet in the Head? Psychiatric Revolutions and Their Aftermath."

34. Anchersen, "Problems of Transvestism," 255.

35. Karl Evang, memo, November 26, 1956, S-4736, Folder Kjønnsskifte, NAN.

36. Application for castration, 1955, S-4736, Box 118, NAN.

37. K. Karlsen to John Leikvam, 1958, S-4736, Box 118, NAN.

38. K. Karlsen to the Expert Council on Sex Operations, 1958, S-4736, Box 118, NAN.

39. Nic Waal, Johan Lofthus, and Ole Harbek, discussion paper, September 1958, S-4736, Box 118, NAN.

40. Johan Bremer, October 11, 1958, S-4736, Box 118, NAN.

41. Jon Leikvam to the Directorate of Health, July 16, 1959, S-4736, Box 118, NAN.

42. Karl Evang to Jørgen H. Vogt, October 6, 1959, S-4736, Folder Kjønnsskifte, NAN.

43. Memo to Director General Knut Munch-Søegaard, December 28, 1961, S-4736 Folder Kjønnsskifte, NAN.

44. Jørgen H. Vogt to Christofer Lohne Knudsen, December 6, 1961, S-4736 Folder Kjønnsskifte, NAN.

45. Johs. Andenæs, *Strafferettens alminnelige del* (Oslo: Universitetets studentkontor, 1952), 141–42; Johs. Andenæs, *Alminnelig strafferett* (Oslo: Akademisk forlag, 1956), 173.

46. Director General Carl L. Stabel, November 5, 1960, S-3212, Folder Straffeloven, NAN.

47. Stabel to the Directorate of Health, December 3, 1960, S-4736, Folder Kjønnsskifte, NAN.

48. Christofer Lohne Knudsen to Jørgen H. Vogt, December 18, 1961, medical record, case 1012, OUHA.

49. From 1967 to 1977, the Expert Council assessed fifteen applications for castration, three of which concerned trans people. From 1972, all cases regarding castration as part of medical transition were referred to "other instances," but it is not clear what this entailed; see "Kastrasjon," undated memo, S-4736 Folder Kjønnsskifte, NAN.

50. Jon Bjørnsson to Per Anchersen, November 14, 1961, S-4736 Folder Kjønnsskifte, NAN.

51. I have examined all correspondence in the Karl Evang archive: see PA-0386 Karl Evang, M, Seksualkorrespondanse, NAN.

52. Siv Frøydis Berg, *Den unge Karl Evang og utvidelsen av helsebegrepet* (Oslo: Solum Forlag, 2002); Kari H. Nordberg, "Frigjøring gjennom vitenskap—Karl Evang som seksualopplyser," *Tidsskrift for Den Norske Legeforening* 141, no. 9 (2021), https://doi.org/10.4045/tidsskr.20.0558.

53. This is a pseudonym.

54. Margrete Hovd to Karl Evang, April 22, 1959, S-4736, Folder Kjønnsskifte, NAN.

55. Margrete Hovd to the Minister of Justice, January 12, 1961, S-4736, Folder Kjønnsskifte, NAN.

56. Margrete Hovd's mother to Karl Evang, April 14, 1959, S-4736, Folder Kjønnsskifte, NAN.

57. Margrete Hovd's mother to Karl Evang, March 16, 1960, S-4736, Folder Kjønnsskifte, NAN.

58. Mattis Kvaal to Karl Evang, undated, received July 28, 1959, S-4736, Folder Kjønnsskifte, NAN.

59. Mattis Kvaal's mother to Karl Evang, undated, S-4736, Folder Kjønnsskifte, NAN.

60. Svein Atle Skålevåg, "A Culture of Consensus: Organising Expertise in Norwegian Forensic Psychiatry, Late Nineteenth to Early Twentieth Century," in *Forensic Cultures in Modern Europe*, ed. Willemijn Ruberg et al. (Manchester: Manchester University Press, 2023), 256.

61. Alexa Geisthövel and Volker Hess, "Handelndes Wissen: Die Praxis des Gutachtens," in *Medizinisches Gutachten: Geschichte einer neuzeitlichen Praxis*, ed. Alexa Geisthövel and Volker Hess (Göttingen: Wallstein Verlag, 2017); J. Andrew Mendelsohn, "Public Practice: The European *Longue Durée* of Knowing for Health and Polity," in *Civic Medicine: Physician, Polity, and Pen in Early Modern Europe*, ed. John Andrew Mendelsohn, Annemarie Kinzelbach, and Ruth Schilling (London: Routledge, 2020).

62. For the deep roots of trans misogyny and its relations to colonialist regimes, segregation, and capitalist exploitation, see Jules Gill-Peterson, *A Short History of Trans Misogyny* (London: Verso Books, 2024).

Chapter Three

1. The concept was coined by sexologists earlier in the century and was even used by some trans people; see Sutton, *Sex Between Body and Mind*.

2. Gill-Peterson, *Transgender Child*, 68.

3. Hirschfeld, *Transvestiten*, 159.

4. Hirschfeld, "Transvestitismus," 140.

5. Havelock Ellis, "Sexo-ästhetische Inversion," *Zeitschrift für Psychotherapie und medizinische Psychologie* V (1914); Havelock Ellis, "Eonism and Other Supplementary Studies," in *Studies in the Psychology of Sex* (Philadelphia: F. A. Davis, 1928), 100.

6. Rainer Herrn, "Geschlecht als Option: Selbstversuche und medizinische Experimente zur Geschlechtsumwandlung im frühen 20. Jahrhundert," in *Sexualität als Experiment: Identität, Lust und Reproduktion zwischen Science und Fiction*, ed. Nicolas Pethes and Silke Schicktanz (Frankfurt: Campus Verlag, 2008), 52.

7. Hirschfeld, "Intersexuelle Konstitution." In fact, Hirschfeld suggested an individual's sex was defined by the soul, not the body. "A person's sex is defined not by the body but by the soul." See Herrn, *Der Liebe und dem Leid*, 273–74.

8. Hirschfeld, "Transvestitismus," 177.

9. Emil Gutheil, "Analysis of a Case of Transvestism," in *Sexual Aberrations: The Phenomena of Fetishism in Relations to Sex*, ed. Wilhelm Stekel (New York: Horace Liveright, 1930).

10. Fritz Bättig, "Beitrag zur Frage des Transvestitismus" (Dissertation, Buchdrückerei Fluntern, 1952); Benno Dukor, "Probleme um den Transvestitismus," *Schweizerische Medizinische Wochenschrift* 81, no. 22 (1951).

11. Frank Töpfer, ed., *Verstümmelung oder Selbsverwirklichung? Die Boss-Mitscherlich-Kontroverse* (Stuttgart-Bad Cannstatt: frommann-holzboog, 2012).

12. Minority report, 1956, p. 3, S-4736, Folder Kjønnsskifte, NAN.

13. Johan Bremer to the Directorate of Health, July 22, 1974, S-1286 Sosialdepartementet, Helsedirektoratet, Kontoret for psykiatri H4, Box 611, Folder: "824 Transseksualitet," NAN (hereafter cited as "Folder Transseksualitet").

14. The case has followed medicine from its beginning. For example, books 1 and 3 of Hippocrates's *Epidemics* consist of case stories on crises and disease progress. See, for example: Sibylle Brändli, Barbara Lüthi, and Gregor Spuhler, eds., *Zum Fall machen, zum Fall werden: Wissensproduktion und Patientenerfahrung in Medizin und Psychiatrie des 19. und 20. Jahrhunderts* (Frankfurt am Main: Campus Verlag, 2009); Julia Epstein, "Historiography, Diagnosis, and Poetics," *Literature and Medicine* 11, no. 1 (1992); John Forrester, "If *p*, Then What? Thinking in Cases," *History of the Human Sciences* 9, no. 3 (1996), https://doi.org/10.1177/095269519600900301; Volker Hess and J. Andrew Mendelsohn, "Case and Series: Medical Knowledge and Paper Technology, 1600–1900," *History of Science* 48, no. 3–4 (2010), https://doi.org/10.1177/007327531004800302; Volker Hess and J. Andrew Mendelsohn, "Fallgeschichte, Historia, Klassifikation," N.T.M 21 (2013), https://doi.org/10.1007/s00048-013-0086-0; Volker Hess, "Observatio und Casus. Status und Funktion der medizinischen Fallgeschichte," in *Fall, Fallgeschichte, Fallstudie: Theorie und Geschichte einer Wissensform*, ed. Susanne Düwell and Nicolas Pethes (Frankfurt am Main: Campus Verlag, 2014); J. Andrew Mendelsohn, "Empiricism in the Library: Medicine's Case Histories," in *Science in the Archives: Pasts, Presents, Futures*, ed. Lorraine Daston (Chicago: University of Chicago Press, 2017); Michael Stolberg, "Formen und Funktionen medizinischer Fallberichte in der Frühen Neuzeit (1500–1800)," in *Fallstudien: Theorie—Geschichte—Methode*, ed. Johannes Süßmann, Susanne Scholz, and Gisela Engel (Berlin: trafo, 2007); Steve Sturdy, "Knowing Cases: Biomedicine in Edinburgh, 1887–1920," *Social Studies of Science* 37, no. 5 (2007), https://doi.org/10.1177/0306312707076597.

15. Mendelsohn, "Empiricism."

16. Minority report, p. 6, S-4736, Folder Kjønnsskifte, NAN.

17. For the use of text modules in case stories, see Hess, "Observatio und Casus."

18. Bättig, "Beitrag zur Frage des Transvestitismus." This paragraph is based on Bremer's minority report, 1956, pp. 13–14, S-4736, Folder Kjønnsskifte, NAN.

19. Bättig, "Beitrag zur Frage des Transvestitismus," 13.

20. Bättig, "Beitrag zur Frage des Transvestitismus," 14.

21. Bättig, "Beitrag zur Frage des Transvestitismus," 22.

22. Minority report, p. 24, S-4736, Folder Kjønnsskifte, NAN.

23. Bättig, "Beitrag zur Frage des Transvestitismus," 34.

24. Richard Mühsam, "Chirurgische Eingriffe bei Anomalien des Sexuallebens," in *Therapie der Gegenwart* (Berlin: Urban & Schwarzenberg, 1926), 452, HHA. This case has also been described in detail in Herrn, "Geschlecht als Option."

25. Mühsam, "Chirurgische Eingriffe," 453.

26. Minority report, p. 24, S-4736, Folder Kjønnsskifte, NAN.

27. Bruno Latour, *We Have Never Been Modern* (Cambridge, MA: Harvard University Press, 1993), 14.

28. Hess and Mendelsohn, "Fallgeschichte, Historia, Klassifikation."

29. Volker Hess, "Formalisierte Beobachtung. Die Genese der modernen Krankenakte am Beispiel der Berliner und Pariser Medizin (1725–1830)," *Medizinhistorisches Journal* 45, no. 3/4 (2010); Hess and Mendelsohn, "Case and Series: Medical Knowledge and Paper Technology, 1600–1900," *History of Science* 48, no. 3–4 (2010): 287–314, https://doi.org/10.1177/007327531004800302; Hess, "Observatio und Casus."; Volker Hess, "A Paper Machine of Clinical Research in the Early Twentieth Century," *Isis* 109, no. 3 (2018), https://doi.org/10.1086/699619; Anke te Heesen, "The Notebook: A Paper Technology," in *Making Things Public: Atmospheres of Democracy*, ed. Bruno Latour and Peter Weibel (Cambridge, MA: MIT Press, 2005).

30. Minority report, p. 24, S-4736, Folder Kjønnsskifte, NAN.

31. Minority report, p. 24, S-4736, Folder Kjønnsskifte, NAN.

32. Minority report, p. 24, S-4736, Folder Kjønnsskifte, NAN. The idea that people sought medical treatment for sensationalist purposes had deep roots in psychiatry and sexual pathology; see Hans Binder, "Das Verlangen nach Geschlechtsumwandlung," *Zeitschrift für die gesamte Neurologie und Psychiatrie* 110, no. 4/6 (1933).

33. Geisthövel and Hess, "Handelndes Wissen," 12–14.

34. Kristin Asdal and Bård Hobæk, "The Modified Issue: Turning Around Parliaments, Politics as Usual and How to Extend Issue-Politics with a Little Help from Max Weber," *Social Studies of Science* 50, no. 2 (2020), https://doi.org/10.1177/0306312720902847.

35. See, for example, Johan Bremer, 1958, S-4736, D, box 118, jnr. 465/58, NAN.

36. Hans Bürger-Prinz, Heinrich Albrecht, and Hans Giese, "Zur Phänomenologie des Transvestitismus bei Männern," *Beiträge zur Sexualforschung* 3 (1953): 41. In 1949, Hans Giese founded the German Society for Sex Research and was a dominant figure in sexology in postwar West Germany, along with Hans Bürger-Prinz, who supervised his doctorate on homosexuality and who acted as the president of the society. Giese developed a pathologizing model of homosexuality as he tried to come to terms with his own homosexual desires; see Moritz Liebeknecht, *Wissen über Sex. Die Deutsche Gesellschaft für Sexualforschung im Spannungsfeld westdeutscher Wandlungsprozesse* (Göttingen: Wallstein Verlag, 2020).

37. Harry Benjamin, "Transsexualism and Transvestism as Psycho-Somatic and Somato-Psychic Syndromes," *American Journal Psychotherapy* 8, no. 2 (1954), https://doi.org/10.1176/appi.psychotherapy.1954.8.2.219. He presented the distinction for the first time in 1953: see Harry Benjamin, "Transvestism and Transsexualism," *International Journal of Sexology* VII, no. 1 (1953).

38. Beans Velocci, "Standards of Care: Uncertainty and Risk in Harry Benjamin's Transsexual Classifications," *TSQ: Transgender Studies Quarterly* 8, no. 4 (2021), https://doi.org/10.1215/23289252-9311060.

39. NN to Karl Evang, January 3, 1955, PA-0386 Karl Evang papers, M, box 203 seksualkorrespondanse, NAN.

40. Per Anchersen to Karl Evang, January 10, 1955, PA-0386 Karl Evang papers, F, box 16 korrespondanse, NAN.

41. Christian Hamburger, "The Desire for Change of Sex as Shown by Personal Letters from 465 Men and Women," *Acta Endocrinologica* 14, no. 4 (1953), https://doi.org/10.1530/acta.0.0140361.

42. In the late nineteenth and early twentieth centuries, the process of differentiating sex and sexuality simultaneously took place in medicine and the subcultures. Hirschfeld's theory of sexual intermediates contributed to the separation—the "masculine" part of the German homosexual movement sought to distance itself from the effeminacy of cross-dressing, and many "heterosexual transvestites" wanted to avoid the stigma of homosexuality; see Herrn, *Schnittmuster*, 31–42.

43. The society was later renamed the European Society of Endocrinology.

44. Jens Rydström, "Into the Wild and Back Again: Pornographic Discourse and Sexual Liberation in Sweden, 1954–1986," in *Gender, Materiality, and Politics: Essays on the Making of Power*, ed. Anna Nilsson Hammar, Daniel Nyström, and Martin Almbjär (Lund: Nordic Academic Press, 2022).

45. Signe Bremer, "'Jag söker gemenskap': Svenska sextidningar som transmöjliggörande rum under 1960-talet," *Lamda Nordica* 28, no. 1 (2023), https://doi.org/10.34041/ln.v28.867.

46. "Hvorfor akkurat jeg?" *Norsk Dameblad*, no. 15 (1955), 38, NLN.

47. "Menn i kvinneklær," *Norsk Dameblad*, no. 3 (1953), 7, NLN.

48. Sigrid Sandal, "En særlig trang til å ville forandre sitt kjønn—Kjønnsskiftebehandling i Norge 1952–1982" (Master's thesis, University of Bergen, 2017), 36–37.

49. Kari Melby, "Husmortid. 1900–1950," in *Med kjønnsperspektiv på norsk historie: fra viktingtid til 2000-årsskiftet*, ed. Ida Blom and Søvi Sogner (Oslo: Cappelen akademisk forlag, 2005), 262–64.

50. Reidun Kvaale, *Kvinner i norsk presse gjennom 150 år* (Oslo: Gyldendal, 1986), 151–55, 70–71.

51. "Menn i kvinneklær," 7.

52. "Han føler seg som kvinne," *Norsk Dameblad*, no. 5 (1955), 8, NLN.

53. "Skjulte tragedier," 21.

54. Ian Hacking, "The Looping Effect of Human Kinds," in *Causal Cognition: A Multi-Disciplinary Debate*, ed. D. Sperber, D. Premack, and A. J. Premack (New York: Oxford University Press, 1995); Ian Hacking, *Historical Ontology* (Cambridge, MA: Harvard University Press, 2002), 99–114; Ian Hacking, "Kinds of People: Moving Targets: British Academy Lecture," *Proceedings of the British Academy* no. 151 (2007), https://doi.org/10.5871/bacad/9780197264249.003.0010.

55. Joanne Meyerowitz, "Sex Change and the Popular Press: Historical Notes on Transsexuality in the United States, 1930–1955," *GLQ: A Journal of Lesbian and Gay Studies* 4, no. 2 (1998), https://doi.org/10.1215/10642684-4-2-159.

56. Norwegian and Danish newspapers published several articles about her transition, her book, and her death. I have found 53 articles about Lili Elbe published by Norwegian newspapers just in 1931; see also Sabine Meyer, *"Wie Lili zu einem richtingen Mädchen wurde." Lili Elbe: Zur Konstruktion von Geschlecht und Identität zwischen Medialisierung, Regulierung und Subjektivierung* (Bielefeld: Transcript Verlag, 2015), 180–216.

57. Jeremy A. Greene, "Knowledge in Medias Res: Toward a Media History of Science, Medicine, and Technology," *History and Theory* 59, no. 4 (2020), https://doi.org/10.1111/hith.12181.

58. Rainer Herrn, "Die falsche Hofdame vor Gericht: Transvestitismus in Psychiatrie und Sexualwissenschaftoder die Regulierung der öffentlichen Kleiderordnung," *Medizinhistorisches Journal* 49, no. 3 (2014); Herrn, *Schnittmuster*; Oosterhuis, *Stepchildren*; Scott Spector, *Violent Sensations: Sex, Crime, and Utopia in Vienna and Berlin, 1860–1914* (Chicago: University of Chicago Press, 2016); Sutton, *Sex Between Body and Mind*.

59. Alex Bakker et al., eds., *Others of My Kind: Transatlantic Transgender Histories* (Calgary, AB: University of Calgary Press, 2020); Annette F. Timm, "'I am so grateful to all you men of medicine': Trans Circles of Knowledge and Intimacy," in *Others of My Kind: Transatlantic Transgender Histories*, ed. Alex Bakker et al. (Calgary, AB: University of Calgary Press, 2020).

60. Timm, "Trans Circles of Knowledge."

61. Julian Honkasalo, "Transfeminine Letter Clubs, Community Care and the Radical Politics of the Erotic," *European Journal of Women's Studies* 30, no. 2 (2023), https://doi.org/10.1177/13505068231164215.

62. "Ikke kvinne og ikke mann," *Norsk Dameblad*, no. 14 (1955), 24, NLN.

63. "Skjulte tragedier," *Norsk Dameblad*, no. 7 (1953), 21, NLN.

64. "Kan vi hjelpe?" *Norsk Dameblad*, no. 13 (1955), 3, NLN.

65. Majority report, p. 2–3, S-4736, Folder Kjønnsskifte, NAN.

66. Ib Ostenfeld, "Paradoks kønsindstilling—Genuin transvestisme," *Ugeskrift for læger* 121, no. 13 (1959): 488–89.

67. Jan Wålinder, *Transsexualism: A Study of Forty-Three Cases* (Gothenburg: Akademiförlaget, 1967), VI.

68. Theodore M. Porter, *Trust in Numbers: The Pursuit of Objectivity in Science and Public Life* (New Jersey: Princeton University Press, 1995). There was a much longer history in medicine and psychiatry, however, of creating censuses and organizing large sets of data in tables: see Theodore M. Porter, *Genetics in the Madhouse: The Unknown History of Human Heredity* (Princeton: Princeton University Press, 2018). Already in 1899, Magnus Hirschfeld had developed a Psychobiological Questionnaire, which was revised several times and completed by thousands of patients; see Herrn, *Der Liebe und dem Leid*, 80–81, 210–11.

69. Wålinder, *Transsexualism*, V.

70. Wålinder, *Transsexualism*, 24. See Harry Benjamin, "Clinical Aspects of

Transsexualism in the Male and Female," *American Journal of Psychotherapy* 18 (1964), https://doi.org/10.1176/appi.psychotherapy.1964.18.3.458.

71. Wålinder, *Transsexualism*, 85.

72. Wålinder, *Transsexualism*, 34.

73. Wålinder, *Transsexualism*, 76–77.

74. Wålinder, *Transsexualism*, 88.

75. Wålinder, *Transsexualism*, 87.

76. Binder, "Geschlechtsumwandlung"; Max Marcuse, "Ein Fall von Geschlechtsumwandlungstrieb," *Zeitschrift für Psychotherapie und medizinische Psychologie* 6 (1914). For a discussion of the diagnostic entities, see Bättig, "Beitrag zur Frage des Transvestitismus," 27.

Chapter Four

1. Lecture by Trygve Braatøy in the Norwegian Medical Society, March 4, 1942, PA-0855 Braatøy, Trygve, H, Box 6, Folder 2, NAN.

2. Lecture by Trygve Braatøy in the Norwegian Medical Society, March 4, 1942.

3. Carl R. Moore and Dorothy Price, "Gonad Hormone Functions, and the Reciprocal Influence Between Gonads and Hypophysis with Its Bearing on the Problem of Sex Hormone Antagonism," *American Journal of Anatomy* 50, no. 1 (1932), https://doi.org/10.1002/aja.1000500103. See also Sengoopta, *Secret Quintessence*, 117–51. For the challenging of the hormonal sex specificity, see Stoff, *Ewige Jugend*, 483.

4. Nelly Oudshoorn, *Beyond the Natural Body: An Archeology of Sex Hormones* (London: Routledge, 1994), 28.

5. Fredrik Grøn, *Det norske medicinske selskab, 1833–1933* (Oslo: Steenske Boktrykkeri Johannes Bjørnstad, 1933), 173.

6. Sengoopta, *Secret Quintessence*, 153–204; Stoff, *Ewige Jugend*, 469–502.

7. Ovaropan, RA/S-4944/D/Da/L0086, Spesialitetskontrollen, Da, box 86 Ovarium–Ovoculin, NAN.

8. Oudshoorn, *Beyond the Natural Body*, 109. See also M. Holm and Morten Hillgaard Bülow, *Det stof, mænd er gjort af—Konstruktionen af maskulinitetsbegreber i forskningsprojekter om testosteron i Danmark fra 1910'erne til 1980'erne*, vol. 10 (Copenhagen: Varia, 2013), 10, 126.

9. Annemarie Mol, Ingunn Moser, and Jeannette Pols, eds., *Care in Practice: On Tinkering in Clinics, Homes and Farms* (Bielefeld: transcript Verlag, 2010).

10. Oudshoorn, *Beyond the Natural Body*, 41–63.

11. Karl Fredrik Støa, "Om hormon og hormongransking," *Syn og segn* 66 (1960).

12. Hormonlaboratoriet, Aker Hospital, the annual report of 1965, Aker årsmeldinger, OCA.

13. Egil Haug, *Hormonlaboratoriet—50 år i vekst og utvikling* (Oslo: Oslo universitetssykehus HF, Aker 2009), 7.

14. Hormonlaboratoriet, Aker Hospital, the annual report of 1965, Aker årsmeldinger, OCA.

15. Psychiatrist to Christian Hamburger, 1954, Psychiatric Department for Men, Ullevål Hospital, case 1013, OUHA.

16. Endocrine Society, "First International Congress of Endocrinology," *Journal of Clinical Endocrinology & Metabolism* 20, no. 2 (1960).

17. Christian Hamburger, "Endocrine Treatment of Male and Female Transsexualism," in *Transsexualism and Sex Reassignment*, ed. Richard Green and John Money (Baltimore: Johns Hopkins University Press, 1969).

18. Harry Benjamin to Christian Hamburger, February 19, 1969, Series IV B, Box 24, Folder 2, HBC.

19. Bättig, "Beitrag zur Frage des Transvestitismus," 39–40.

20. Oudshoorn, *Beyond the Natural Body*, 55–58.

21. Sengoopta, *Secret Quintessence*, 186–92.

22. See Holm, "Fleshing Out the Self," 313.

23. Karl Evang to NN, November 20, 1963, Evang seksualkorrespondanse, NAN.

24. Hans H. Bassøe to Karl Evang, November 18, 1966, Evang seksualkorrespondanse, NAN.

25. Hans H. Bassøe to Karl Evang, April 26, 1967, Evang seksualkorrespondanse, NAN.

26. Karl Evang to Hans H. Bassøe, May 3, 1967, Evang seksualkorrespondanse, NAN.

27. Hans H. Bassøe to Karl Evang, August 23, 1968, Evang seksualkorrespondanse, NAN.

28. Hans H. Bassøe, July 15, 1968, Evang seksualkorrespondanse, NAN.

29. For a Danish case, see, for example, Ostenfeld, "Paradoks kønsindstilling—Genuin transvestisme."

30. Anchersen, "Problems of Transvestism," 254.

31. Villars Lunn to the Ministry of Justice, August 3, 1955, Retslægerådet, Kasager (1929–1968), Box L66, Folder 798, DNA.

32. Jørgen H. Vogt to the Director General of Health, undated, received July 28, 1959, S-4736, Folder Kjønnsskifte, NAN.

33. Jørgen H. Vogt to the Director General of Health, undated 1959, S-4736, Folder Kjønnsskifte, NAN.

34. Jørgen H. Vogt to the Director General of Health, undated, received July 28, 1959, S-4736, Folder Kjønnsskifte, NAN.

35. Villars Lund, discussion paper for the Medico-Legal Council, May 31, 1954, Retslægerådet, Ka-sager (1929–1968), Box L64, Folder 774, DNA.

36. Bättig, "Beitrag zur Frage des Transvestitismus," 44.

37. Georg K. Stürup to Johan Bremer, April 6, 1956, quoted in Minority report, September 3, 1956, S-4736, Folder Kjønnsskifte, NAN, pp. 18–20.

38. Christine Jorgensen to Gen and Dr. Joe, October 30, 1950, Christine Jorgensens efterladte papirer, Box 28, DRL.

39. Christian Hamburger and Mogens Sprechler, "The Influence of Steroid Hormones on the Hormonal Activity of the Adenohypophysis in Man," *Acta Endocrinologica* 7, 1–4 (1951), https://doi.org/10.1530/acta.0.0070167.

40. Hertoft and Ritzau, *Paradiset*, 96–97.

41. Christine Jorgensen to Gen and Dr. Joe, October 30, 1950, Christine Jorgensens efterladte papirer, Box 28, DRL.

42. Christian Hamburger to Chris Jorgensen, November 9, 1955, Christine Jorgensens efterladte papirer, DRL.

43. Jørgen H. Vogt, case 1014, medical record, 1962, Aker Hospital, OUHA.

44. Sanna Wallin, *När jag letar efter Max* (Stockholm: Normal förlag, 2007). I was made aware of this source in a master's thesis; see Devin Baaring, "Könsgränsen. Om samproduktionen av könstillhörighetslagen (1972:119)" (Master's thesis, Lund University, 2019).

45. Wallin, *När jag letar efter Max*, 65.

46. Gill-Peterson, *Transgender Child*, 97.

47. Oslo Health Council, Transseksualitet, December 5, 1979, S-1286, Folder Transseksualitet, NAN.

48. Jørgen H. Vogt to N. V. Organon, November 9, 1956, case 1012, OUHA.

49. N. V. Organon to Jørgen H. Vogt, November 27, 1956, case 1012, OUHA.

50. Bård Hobæk and Anne Kveim Lie, "Less Is More: Norwegian Drug Regulation, Antibiotic Policy, and the 'Need Clause,'" *Milbank Quarterly* 97, no. 3 (2019), https://doi.org/10.1111/1468-0009.12405.

51. Jakob Bjerg Larsen et al., "Dynamics of Pharmacy Regulation in Denmark, 1932–1994: A Study of Profession-State Relations," *Pharmacy in History* 46, no. 2 (2004).

52. See Olav Hamran, *Riktig medisin? En historie om apotekvesenet* (Oslo: Pax forlag, 2010); Larsen et al., "Dynamics of Pharmacy Regulation in Denmark."

53. Christian Hamburger, January 7, 1955, Retslægerådet, Ka-sager (1929–1968), Box L65, Folder 789, DNA.

54. Etifollin, Spesialitetskontrollens råd, November 3, 1952, Preparatarkivet, NMAA.

55. Harry Benjamin, "For the Practicing Physician: Suggestions and Guidelines for the Management of Transsexuals," in *Transsexualism and Sex Reassignment*, ed. Richard Green and John Money (Baltimore: Johns Hopkins University Press, 1969), 305.

56. Christian Hamburger, "Endocrine Treatment of Male and Female Transsexualism," 300–2.

57. Depo-testosterone, April 20, 1953, S-4944 Spesialitetskontrollen, D, Box 38 Dakinol–Depotestosterone, NAN.

58. Hans G. Dedichen, Depo Testosterone Cyclopentylpropionate, April 20, 1953, S-4944 Spesialitetskontrollen, Da, box 38, Dakinol—Depotestosterone.

59. Norges Apotekerforening, *Resepthåndboken* (Oslo: A. W. Brøggers Boktrykkeri A/S, 1959).

60. CIBA, "Triolandren Spritzampullen 250 mg," February 24, 1965, Lab.nr. 12887, Preparatarkivet, Norwegian Medicines Agency Archive (hereafter cited as NMAA), Oslo.

61. Christian Hamburger and K. Pedersen-Bjergaard, "Om testosteron-suppositorier med særligt henblik på suppositorie-massen," *Archiv for pharmaci og chemi* 68 (1961).

62. Asbjørn Aakvaag and Jørgen H. Vogt, "Plasma Testosterone Values in Different Forms of Testosterone Treatment," *Acta Endocrinologica* 60, no. 3 (1969), https://doi.org/10.1530/acta.0.0600537.

63. Jørgen H. Vogt, medical note, 1964, case 1012, OUHA.

64. Aakvaag and Vogt, "Plasma Testosterone Values."

65. Claude J. Migeon, Marco A. Rivarola, and Maguelone G. Forest, "Studies of Androgens in Male Transsexual Subjects: Effects of Estrogen Therapy," in *Transsexualism and Sex Reassignment*, ed. Richard Green and John Money (Baltimore: Johns Hopkins University Press, 1969).

66. See Peter Keating and Alberto Cambrosio, *Biomedical Platforms: Realigning the Normal and the Pathological in Late-Twentieth-Century Medicine* (Cambridge: MIT Press, 2003).

67. J. R. Latham, "(Re)making Sex: A Praxiography of the Gender Clinic," *Feminist Theory* 18, no. 2 (2017), https://doi.org/10.1177/1464700117700051; Mak, *Doubting Sex*; Mol, *Body Multiple*; Annemarie Mol and John Law, "Embodied Action, Enacted Bodies: The Example of Hypoglycaemia," *Body & Society* 10, no. 2–3 (2004), https://doi.org/10.1177/1357034X04042932; Jörg Niewöhner et al., "Phenomenography: Relational Investigations into Modes of Being-in-the-World," *Cyprus Review* 28, no. 1 (2016); Stoff, "Identität und Differenz."

68. For the argument that care is more than "tender love," see Annemarie Mol, *The Logic of Care: Health and the Problem of Patient Choice* (London: Routledge, 2008), 5.

69. Minority report, September 3, 1956, pp. 6–7, S-4736, Folder Kjønnsskifte, NAN.

70. Hamburger, Stürup, and Dahl-Iversen, "Transvestism," 395.

71. Jørgen H. Vogt to Tollef Bredal A/S, February 13, 1962, case 1014, OUHA.

72. Jørgen H. Vogt to the chair of Domsarkivet, May 27, 1968, case 1014, OUHA.

73. Jørgen H. Vogt to Den Norske Creditbank, January 30, 1967, case 1014, OUHA.

74. For how medical care creates *collectives*, see Mol, Moser, and Pols, *Care*.

75. Myriam Winance, "Care and Disability: Practices of Experimenting, Tinkering with, and Arranging People and Technical Aids," in *Care in Practice: On*

Tinkering in Clinics, Homes and Farms, ed. Annemarie Mol, Ingunn Moser, and Jeannette Pols (Bielefeld: transcript Verlag, 2010), 95.

Chapter Five

1. Sophia interview, September 17, 2020.

2. See Brian Conroy, "The History of Facial Prostheses," in *Clinics in Plastic Surgery: An International Quarterly*, ed. Sharon Romm (Philadelphia: W. B. Saunders, 1983); Thomas Schlich, *The Origins of Organ Transplantation: Surgery and Laboratory Science*, 1880–1930 (Rochester: University of Rochester Press, 2010), 14–17; Antony F. Wallace, "The Influence of the Battle of Jutland on Plastic Surgery," in *Clinics in Plastic Surgery: An International Quarterly*, ed. Sharon Romm (Philadelphia: W. B. Saunders, 1983); Judith B. Zacher, "Plastic Surgery in the Late 1920s: Three Points of View," in *Clinics in Plastic Surgery: An International Quarterly*, ed. Sharon Romm (Philadelphia: W. B. Saunders, 1983). However, many of the procedures taken over by plastic surgeons in the late twentieth century have a much longer history. Plastic surgeons have often emphasized the long traditions of plastic surgery when writing the history of their profession: see Frederick Strange Kolle, *Plastic and Cosmetic Surgery* (New York: D. Appleton, 1911), 1–8; Maxwell Maltz, *Evolution of Plastic Surgery* (New York: Froben Press, 1946).

3. Maltz, *Evolution of Plastic Surgery*, 268.

4. Elizabeth Haiken, *Venus Envy: A History of Cosmetic Surgery* (Baltimore: Johns Hopkins University Press, 1997), 17–29.

5. Sander L. Gilman, *Creating Beauty to Cure the Soul: Race and Psychology in the Shaping of Aesthetic Surgery* (Durham: Duke University Press, 1998), 19.

6. Gunnar Eskeland, "Plastikkirurgi," in *Det norske medicinske selskab 150 år* (Oslo: Det norske medicinske selskab, 1983); Leiv M. Hove, *Fra håndkirurgiens historie* (Bergen: Molvik Grafisk, 2004).

7. Allan Ragnell, "The Development of Plastic Surgery in Sweden," *British Journal of Plastic Surgery* 8 (1955), https://doi.org/10.1016/S0007-1226(55)80022-7.

8. The specialty did not have specific educational requirements or qualifications until 1965: see Per Haave, *I medisinens sentrum—Den norske legeforening og spesialistregimet gjennom hundre år* (Oslo: Unipub, 2011), 41.

9. Halfdan Schjelderup, "The Gillies Memorial Lecture 1975," *British Journal of Plastic Surgery* 30 (1977).

10. Henrik Borchgrevink, "Centralized Surgical Treatment of Transsexuals in Norway," 1988, IHPA; patients were treated by different doctors in various hospitals, however, which means that the total number of operations is higher.

11. For the specter of litigation shaping doctors' decision in the United States, see Velocci, "Standards of Care."

12. Statement on the Establishment of a Clinic for Transsexuals at the Johns Hopkins Medical Institutions, The Johns Hopkins University and The Johns

Hopkins Hospital, box 503.602, Gender Identity Clinic Part II Archives 1966–1994, JMC.

13. Jacob B. Natvig and S. Normann, Plastisk kirurgisk avdeling ved Rikshospitalet 25 år, October 11, 1978, 311, Plastisk kirurgisk avdeling, 1977–1981, Rikshospitalet, OUHA.

14. Gunnar Eskeland to the Directorate of Health, June 24, 1986, Rikshospitalet, 300 Plastisk kirurgisk avdeling, Wergelandsveien 27, 1982–1986, OUHA.

15. Hanne Gamnes, "Prehistoriske forhold på Rikshospitalet," *Dagbladet*, May 9, 1980, p. 8.

16. John Law, "Care and Killing: Tensions in Veterinary Practice," in *Care in Practice: On Tinkering in Clinics, Homes and Farms*, ed. Annemarie Mol, Ingunn Moser, and Jeannette Pols (Bielefeld: transcript Verlag, 2010), 67.

17. Isak interview, November 14, 2019.

18. Astrid interview, October 1, 2019.

19. Astrid interview.

20. Medical record, 1981, case 1007, OUHA.

21. Isak interview.

22. Isak interview.

23. Isak interview.

24. Henrik Borchgrevink, "Transsexualisme som landsfunksjon," August 10, 1990, no. 1991/1410, The Norwegian Board of Health Supervision Archive, Oslo (hereafter cited as NBHSA).

25. Sophia interview.

26. Sophia interview.

27. Eric Plemons and Chris Straayer, "Introduction: Reframing the Surgical," *TSQ: Transgender Studies Quarterly* 5, no. 2 (2018), https://doi.org/10.1215/23289252-4348605.

28. Milton T. Edgerton, "The Surgical Treatment of Male Transsexuals," *Clinics in Plastic Surgery* 1, no. 2 (1974).

29. Eric Plemons, *The Look of a Woman: Facial Feminization Surgery and the Aims of Trans-Medicine* (Durham: Duke University Press, 2017).

30. Benjamin, "Transvestism and Transsexualism"; Benjamin, "Transsexualism and Transvestism as Psycho-Somatic and Somato-Psychic Syndromes."

31. Oslo Health Council, Transseksualitet, December 5, 1979, Oslo helseråd, S-1286, Folder Transseksualitet, NAN.

32. Wålinder, *Transsexualism*, 87.

33. Sophia interview.

34. Sara interview, December 12, 2019.

35. Sandra Mesics, "When Building a Better Vulva, Timing Is Everything: A Personal Experience with the Evolution of MTF Genital Surgery," *TSQ: Transgender Studies Quarterly* 5, no. 2 (2018), https://doi.org/10.1215/23289252-4348672.

36. Mesics, "When Building a Better Vulva, Timing Is Everything."

37. Sophia interview.

38. Henrik Borchgrevink, "Centralized Surgical Treatment of Transsexuals,"

abstract for Trans-Actions of the IX International Congress of Plastic and Reconstructive Surgery March 1987, 1986, IHPA.

39. Henrik Borchgrevink to the Ministry of Justice, October 7, 1970, S-4250 Justisdepartementet, Den administrative avdeling, A, D, Box 2455, NAN (hereafter cited as "Box sivilstandsregister").

40. Henrik Borchgrevink, Centralized Surgical Treatment of Transsexuals, abstract for Trans-Actions of the IX International Congress of Plastic and Reconstructive Surgery March 1987, 1986, IHPA.

41. Medical record, 1982, case 1007, OUHA.

42. Medical record, 1982, case 1007, OUHA.

43. Howard W. Jones, "Operative Treatment of the Male Transsexual," in *Transsexualism and Sex Reassignment*, ed. Richard Green and John Money (Baltimore: Johns Hopkins University Press, 1969), 316–17.

44. Herrn, *Der Liebe und dem Leid*, 423–27. There are several earlier documented cases of castration performed on trans women.

45. Felix Abraham, "Genitalumwandlung an zwei männlichen Transvestiten," *Zeitschrift für Sexualwissenschaft und Sexualpolitik* 18 (1931).

46. Archibald H. McIndoe and J. Bright Banister, "An Operation for the Cure of Congenital Absence of the Vagina," *Journal of Obstetrics and Gynaecology of the British Empire* 45, no. 3 (1938), https://doi.org/10.1111/j.1471-0528.1938.tb11141.x.

47. Archibald McIndoe, "The Treatment of Congenital Absence and Obliterative Conditions of the Vagina," *British Journal of Plastic Surgery* 2, no. 4 (1950).

48. Surgical description by Erling Dahl-Iversen, November 20, 1952, Retslægerådet, Ka-sager (1929–1968), Box L55, Folder 641 "Chris," DNA.

49. Alex Bakker, "In the Shadows of Society: Trans People in the Netherlands in the 1950s," in *Others of My Kind: Transatlantic Transgender Histories*, ed. Alex Bakker et al. (Calgary, AB: University of Calgary Press, 2020), 152.

50. Holm, "Fleshing Out the Self," 368.

51. Milton T. Edgerton, Norman J. Knorr, and James R. Callison, "The Surgical Treatment of Transsexual Patients: Limitations and Indications," *Plastic and Reconstructive Surgery* 45, no. 1 (1970).

52. The files have been lost and it is unclear if these included the construction of a neo-vagina; see H. Henrik Chr. Borchgrevink, "Kirurgisk konstruksjon av vagina," *Tidsskrift for Den Norske Lægeforening* 93 (1973).

53. Medical record, 1981, case 1007, OUHA.

54. Law, "Care and Killing," 62.

55. Borchgrevink, "Kirurgisk konstruksjon av vagina."

56. Sophia interview.

57. Patricia Hadley Thompson, "Apparatus to Maintain Adequate Vaginal Size Postoperatively in the Male Transsexual Patient," in *Transsexualism and Sex Reassignment*, ed. Richard Green and John Money (Baltimore: Johns Hopkins University Press, 1969), 323–24.

58. Anne Kveim Lie, "Blodprøver og piller er også omsorg," *Tidsskrift for Den norske legeforening* 133 (2013), https://doi.org/10.4045/tidsskr.12.1389; Mol, Moser, and Pols, *Care*.

59. Law, "Care and Killing," 67.

60. Patient to Jørgen H. Vogt, case 1014, OUHA.

61. Patient to Jørgen H. Vogt, case 1014, OUHA.

62. John E. Hoopes, "Operative Treatment of the Female Transsexual," in *Transsexualism and Sex Reassignment*, ed. Richard Green and John Money (Baltimore: Johns Hopkins University Press, 1969), 340.

63. Henrik Borchgrevink, surgical note, amputatio ammae bilat. Transplantation areolae, 1968, case 1019, OUHA. The 50 øre coin had a diameter of 22 mm.

64. J. Hertz, K. G. Tillinger, and A. Westman, "Transvestitism. Report on Five Hormonally and Surgically Treated Cases," *Acta Psychiatr Scand* 37, no. 4 (1961), https://doi.org/10.1111/j.1600-0447.1961.tb07363.x.

65. Thorkil Sørensen, "A Follow-Up Study of Operated Transsexual Females," *Acta Psychiatrica Scandinavica* 64 (1981).

66. Henrik Borchgrevink to Dr NN, 1968, case 1019, OUHA.

67. Henrik Borchgrevink, surgical note, construction pediculi cutis abdominis, 1969, case 1019, OUHA.

68. Nikolaj A. Bogoras, "Über die volle plastische Wiederherstellung eines zum Koitus fähigen Penis (Peniplastica totalis)," *Zentralblatt für Chirurgie* 63, no. 22 (1936).

69. Maltz, *Evolution of Plastic Surgery*, 269–77.

70. T. Roderick Hester, H. Louis Hill, and M. J. Jurkiewicz, "One-Stage Reconstruction of the Penis," *British Journal of Plastic Surgery* 31, no. 4 (1978), https://doi.org/10.1016/S0007-1226(78)90110-8; Miguel Orticochea, "A New Method of Total Reconstruction of the Penis," *British Journal of Plastic Surgery* 25, no. 4 (October 1972), https://doi.org/10.1016/s0007-1226(72)80077-8. See also, Bakker, *The Dutch Approach*, 62–63.

71. Eskeland, "Plastikkirurgi," 236.

72. Henrik Borchgrevink to Alv A. Dahl, description of surgical techniques conducted at Rikshospitalet, 23 September 1988, IHPA.

73. Rikshospitalet, medical record, 1982, case 1005, OUHA.

74. Rikshospitalet, medical record, 1982, case 1005, OUHA.

75. Medical record, 1982, case 1005, OUHA.

76. Jonas interview, November 14, 2019.

77. Isak interview.

78. John E. Hoopes, "Surgical Construction of the Male External Genitalia," *Clinics in Plastic Surgery* 1, no. 2 (1974).

79. Historians and plastic surgeons have made this argument before. See Gilman, *Creating Beauty to Cure the Soul*, 3–10; Mario González-Ulloa, ed., *The Creation of Aesthetic Plastic Surgery* (New York: Springer-Verlag, 1976). Elizabeth Haiken,

on the other hand, argued that cosmetic surgery lies "at the nexus of medicine and consumer culture" enabled by "medical knowledge, leisure, and money": see Haiken, *Venus Envy*, 12.

80. Karin Haugen, "Forandrer livet for mange," *VG*, 1974, 16–17.

81. Sophia interview.

82. Sophia interview.

83. Protocol for Alv Dahl's research project, 1988, IHPA.

Chapter Six

1. Anne Kveim Lie and Per Haave, "Social Medicine in Social Democracy," in *Medicine on a Larger Scale: Global Histories of Social Medicine*, ed. Anne Kveim Lie, Jeremy Greene, and Warwick Anderson (Cambridge: Cambridge University Press, 2025).

2. Slagstad, *Nasjonale strateger*, 387.

3. Knud Sand, November 24, 1954, Retslægerådet, Ka-sager (1929–1968), Box L64, Folder 774, DNA.

4. Letter to Knud Sand, November 8, 1954, Retslægerådet, Ka-Kv-sager (1935–1956), 1–5, Box 85, DNA.

5. Ebbe Brandstrup to Knud Sand, January 7, 1955, in ibid.

6. B. Vestergaard to Knud Sand, 12 December 1958, Retslægerådet, Ka-Kv-sager (1935–1956), 1–5, Box 85, DNA.

7. Jan Wålinder and Inga Thuwe, "A Law Concerning Sex Reassignment of Transsexuals in Sweden," *Archives of Sexual Behavior* 5, no. 3 (1976); see also Erika Alm, "'Ett emballage för inälvor och emotioner': föreställningar om kroppen i statliga utredningar från 1960- & 1970-talen" (PhD thesis, University of Gothenburg, 2006); Daniela Alaattinoğlu, *Grievance Formation, Rights and Remedies: Involuntary Sterilisation and Castration in the Nordics, 1930s–2020s* (Cambridge: Cambridge University Press, 2023).

8. Alm, "Emballage"; Julian Honkasalo, "In the Shadow of Eugenics: Transgender Sterilisation Legislation and the Struggle for Self-Determination," in *The Emergence of Trans: Culture, Politics and Everyday Lives*, ed. Ruth Pearce et al. (Oxon: Routledge, 2020); Rolf Luft et al., "Transsexualism," *Läkartidningen* 74, no. 44 (1977); Amy Rappole, "Trans People and Legal Recognition: What the U.S. Federal Government Can Learn from Foreign Nations," *Maryland Journal of International Law* 30, no. 1 (2015).

9. Erika Alm, "What Constitutes an In/Significant Organ? The Vicissitudes of Juridical and Medical Decision-Making Regarding Genital Surgery for Intersex and Trans People in Sweden," in *Body, Migration, Re/Constructive Surgeries*, ed. Gabriele Griffin and Malin Jordal (London: Routledge, 2018), 227.

10. The Ministry of Justice to the Directorate of Health, December 7, 1953, S-3212 Justisdepartementet, Lovavdelingen, Ba, Box 56, NAN.

11. Per Anchersen to the Ministry of Social Affairs, July 31, 1974, S-1286, Folder Transseksualitet, NAN.

12. Per Anchersen to the Ministry of Justice, October 12, 1970, Box sivilstandsregister, NAN.

13. Henrik Borchgrevink to the Ministry of Justice, October 7, 1970, Box sivilstandsregister, NAN.

14. See Holm, "Fleshing Out the Self," 372–76.

15. Jørgen H. Vogt to Oslo Folkeregister, December 8, 1961, case 1014, OUHA.

16. Jørgen H. Vogt to the 3. Civil Office in the Ministry of Justice, June 20, 1963, case 1014, OUHA.

17. Jørgen H. Vogt to Chris Dalen, case 1014, 1967, OUHA.

18. Jørgen Herman Vogt, "Five Cases of Transsexualism in Females," *Acta Psychiatrica Scandinavica* 44, no. 1 (1968): 72, https://doi.org/10.1111/j.1600-0447.1968.tb07636.x.

19. Vogt, "Five Cases of Transsexualism in Females," 72.

20. Medical record, 1982, case 1005, the Radium Hospital, OUHA.

21. Bauer, "Epidemiologist's Dream."

22. From January 1, 1983, the National Population Register handled requests for change of personal identification number and first name; see Sentralkontoret for folkeregistering to Henrik Borchgrevink, 28 June 1983, IHPA. Concomitantly, the register took over processing name change applications; see Jan Furseth and Olav Ljones, "50-årsjubilant med behov for oppgradering," *Samfunnsspeilet* 1 (2015), https://www.ssb.no/befolkning/artikler-og-publikasjoner/_attachment/224332?_ts=14cb7c21460.

23. Statistisk sentralbyrå, Sivilstandsendring for transseksuelle, 2 July 1969, Box sivilstandsregister, NAN.

24. Conrad Clemetsen, 3. sivilkontor/Administrativ avdeling to the Legislation Department in the Ministry of Justice, August 26, 1977, Box sivilstandsregister, NAN; Chief of Bureau to NN, May 3, 1972, Den administrative avdeling, Justisdepartementet, 311.10 Personrett alm, j.nr. 2011/93 E, MJPS.

25. Sigrid Sandal, "Kirurgi og byråkrati," in *Frihet, likhet og mangfold*, ed. Anne Hellum and Anniken Sørlie (Oslo: Gyldendal, 2021), 56.

26. Chief county psychiatrist to the Ministry of Justice, March 26, 1969, S-1286, Folder Transseksualitet, NAN.

27. Jørgen H. Vogt, medical statement, 1972, S-1286, Folder Transseksualitet, NAN.

28. The Ministry of Justice to Jørgen H. Vogt, July 5, 1973, S-1286, Folder Transseksualitet, NAN.

29. Undecipherable signature, handwritten note by the Ministry of Justice, July 13, 1973, Box sivilstandsregister, NAN.

30. Ole Herman Fisknes by the Ministry of Justice to Jørgen H. Vogt, December 10, 1973, Box sivilstandsregister, NAN.

31. Letter to the Ministry of Justice, July 16, 1973, S-1286, Folder Transseksualitet, NAN.

32. Jørgen H. Vogt to the Ministry of Justice, January 2, 1974, Box sivilstandsregister, NAN. I have not been able to find out if the marriages were dissolved.

33. The Ministry of Justice to Jørgen H. Vogt, July 5, 1973, S-1286, Folder Transseksualitet, NAN.

34. Jørgen H. Vogt to the Ministry of Justice, July 26, 1973, S-1286, Folder Transseksualitet, NAN.

35. Christofer Lohne Knudsen, handwritten note on memo, February 16, 1974, S-1286, Folder Transseksualitet, NAN.

36. Johan Bremer to the Office of Psychiatry, July 22, 1974, S-1286, Folder Transseksualitet, NAN.

37. Psychiatrist to the Ministry of Justice, application for change of name, 1977, S-1286, Folder Transseksualitet, NAN.

38. The Ministry of Justice to NN, October 3, 1977, S-1286, Folder Transseksualitet, NAN.'

39. Psychiatrist to the Ministry of Justice, February 23, 1979, S-1286, Folder Transseksualitet, NAN.

40. Principal Officer of the Ministry of Church and Educational Affairs Ole Herman Fisknes to Per Anchersen, July 25, 1979, S-1286, Folder Transseksualitet, NAN.

41. Per Anchersen to the Ministry of Church and Educational Affairs, August 7, 1979, S-1286, Folder Transseksualitet, NAN.

42. The Directorate of Health by Torbjørn Mork to the Ministry of Justice by Leif Thomas Eldring, November 9, 1979, S-1286, Folder Transseksualitet, NAN.

43. The Office of Psychiatry in the Directorate at Health to the 3rd Civil Office in the Ministry of Justice, memo, February 20, 1980, S-1286, Folder Transseksualitet, NAN.

44. Application for the change of name, undated, unaddressed, S-1286, Folder Transseksualitet, NAN.

45. KBH, memo, April 29, 1972, Justisdepartementet, Lovavdelingen, box 311.10, The Archive of the Ministry of Justice and Public Security, Oslo.

46. S.S., handwritten note, June 7, 1963, 3. sivilkontor, the Ministry of Justice, Box Sivilstandsregister, NAN.

47. Unsigned handwritten note, July 20, 1979, S-4711, Kirke-, utdannings- og forskningsdepartementet, Kirkeavdelingen, Da, Box 330, NAN.

48. Memo by Elsbeth Bergsland in the Legislation Department to the Department of Administration, February 26, 1980, Justisdepartementet, Lovavdelingen, box 311.10, the Archives of the Ministry of Justice and Public Security.

49. Ibid., memo signed I.S.

50. Sandal, "Kirurgi og byråkrati," 56–58.

51. Holm, "Fleshing Out the Self," 373–74. Similarly, Julian Honkasalo argued that the Swedish law of 1972 must be seen in light of longer eugenic traditions in Scandinavia: see Honkasalo, "Shadow of Eugenics."

52. Alm, "What Constitutes an In/Significant Organ?" 231.

Chapter Seven

1. In 1977, the Norwegian Psychiatric Association recommended that its members avoid using this diagnosis. In 1981, it was removed from the public list of diagnoses; see Torbjørn Mork to Det Norske Forbundet av 1948, Diagnoselistenes punkt 302, August 17, 1981. S-1286, Sosialdepartementet, Helsedirektoratet, Kontoret for psykiatri, H4, Dc box 611, Homofili, NAN.

2. Lie and Haave, "Social Medicine in Social Democracy."

3. Fredrik Mellbye, "Embetet som stadsfysikus i Oslo," in *Samfunnsmedisin i praksis. Oslo Helseråd i 80-årene*, ed. Harald Siem, Kåre Berg, and Berthold Grünfeld (Oslo: Universitetsforlaget, 1987), 26.

4. Inge A. Dahl and Erling Viksjø, "Oslo Helseråd," *Byggekunst* 51, no. 6 (1969).

5. See, for example, Pentecost et al., "Revitalising Global Social Medicine"; Porter and Porter, "Social Medicine"; Porter, *Social Medicine and Medical Sociology in the Twentieth Century*.

6. Ohlsson Al Fakir, "Socialt medborgarskap," 95–99.

7. Sunniva Engh, "The Complexities of Postcolonial International Health: Karl Evang in India 1953," *Medical History* 67, no. 1 (2023), https://doi.org/10.1017/mdh.2023.12.

8. Berg, *Helsebegrepet*, 15.

9. Lie and Haave, "Social Medicine in Social Democracy."

10. Axel Strøm, *Lærebok i sosialmedisin* (Oslo: Liv og helses forlag, 1956), 12.

11. Ohlsson Al Fakir, "Socialt medborgarskap," 96–97.

12. Axel Strøm, Per Sundby, and Yngvar Løchen, *Lærebok i sosialmedisin*, 4th ed. (Oslo: Fabritius Forlag, 1973), 12.

13. See, for example, Donna J. Drucker, *The Classification of Sex: Alfred Kinsey and the Organization of Knowledge* (Pittsburgh: University of Pittsburgh Press, 2014). For West Germany, see Moritz Liebeknecht, *Wissen über Sex. Die Deutsche Gesellschaft für Sexualforschung im Spannungsfeld westdeutscher Wandlungsprozesse* (Göttingen: Wallstein Verlag, 2020).

14. Preben Hertoft, *Undren og befrielse. Erindringer* (Copenhagen: Hans Reitzels Forlag, 2001), 271–77.

15. Steven Epstein, *The Quest for Sexual Health: How an Elusive Ideal Has Transformed Science, Politics, and Everyday Life* (Chicago: University of Chicago Press, 2022).

16. WHO, *Education and Treatment in Human Sexuality: The Training of Health Professionals* (Geneva: WHO, 1975), 6.

17. Lars Ullerstam, *De seksuelle minoriteter* (Oslo: Pax Forlag, 1964).

18. Bodil Solberg interview, Oslo, January 20, 2020.

19. Hans Døvik, Rådgivningstjenesten for homofile—egen seksjon for medisinsk sexologi, July 3, 1979, Oslo helseråd, Box 122, Folder homofile-transseksualitet, OCA.

20. Preben Hertoft, *Undersøgelser over unge mænds seksuelle adfærd, viden og holdning. Bind 1–3* (Copenhagen: Akademisk forlag, 1968).

21. Christine Jorgensen to Iriving Lazar, undated, probably 1984, Christine Jorgensens efterladte papirer, Box 28, DRL.

22. Teit Ritzau to Christine Jorgensen, June 9, 1983, Christine Jorgensens efterladte papirer, Box 9, DRL.

23. Christine Jorgensen to Christian Hamburger, September 14, 1983,

24. Preben Hertoft, "Tilblivelsen af Sexologisk Klinik. De første år," in *25 år med Sexologisk Klinik*, ed. Ellids Kristensen and Annamaria Giraldi (Copenhagen: Psykiatrisk Center København, 2011).

25. Preben Hertoft, *Klinisk sexologi. En introduktion* (Copenhagen: Munksgaard, 1976); Ellids Kristensen, "Sexologisk Klinik 1989–2011," in *25 år med Sexologisk Klinik*, ed. Ellids Kristensen and Annamaria Giraldi (Copenhagen: Psykiatrisk Center København, 2011); Rikke Kildevæld Simonsen and Ellids Kristensen, "Transseksuelle i Sexologisk Klinik," in *25 år med Sexologisk Klinik*, ed. Ellids Kristensen and Annamaria Giraldi (Copenhagen: Psykiatrisk Center København, 2011), 69.

26. Preben Hertoft, "Lægen må overvinde egne fordomme og tabuer for at løse patienters sexuelle problemer," *Nordisk Medicin* 94, no. 4 (1979).

27. Preben Hertoft and Thorkil Sørensen, "Transsexuality: Some Remarks Based on Clinical Experience," *Ciba Foundation Symposium*, no. 62 (1978): 170, https://doi.org/10.1002/9780470720448.ch9.

28. 81 MfF and 29 FtM operations. See Hertoft and Sørensen, "Transsexuality: Some Remarks Based on Clinical Experience," 170.

29. Hertoft, *Klinisk sexologi. En introduktion*, 215–16.

30. Hertoft, *Undersøgelser over unge mænds seksuelle adfærd, viden og holdning. Bind 1–3* (Copenhagen: Akademisk forlag, 1968), 369.

31. Berthold Grünfeld, *Vårt seksuelle liv* (Oslo: Gyldendal, 1979), 168.

32. Grünfeld, *Vårt seksuelle liv*, 114.

33. Lisa Downing, Iain Morland, and Nikki Sullivan, *Fuckology: Critical Essays on John Money's Diagnostic Concepts* (Chicago: University of Chicago Press, 2014); Sandra Eder, *How the Clinic Made Gender: The Medical History of a Transformative Idea* (Chicago: University of Chicago Press, 2022), 110–40; Gill-Peterson, *Transgender Child*, 97–127.

34. Eder, *How the Clinic Made Gender*, 132–35.

35. Eder, *How the Clinic Made Gender*, 211–12.

36. For the reception and integration of the Baltimore model in East and West Germany, see Klöppel, *XXoXY ungelöst*, 531–35.

37. Gunhild Grünfeld interview, November 16, 2020.

38. Berthold Grünfeld, "Homofili og transseksualitet—psykiatriske og sosialmedisinske betraktninger," in *NOU 1979: 46 Strafferettslig vern for homofile* (Oslo: Justis og politidepartemenet, 1979); see John Money and Anke A. Ehrhardt, *Man & Woman, Boy & Girl* (Baltimore: Johns Hopkins University Press, 1972), 3.

39. Eder, *How the Clinic Made Gender*, 169–86.

40. Grünfeld, *Seksuelle liv*, 26. In Norwegian, there is only one term for sex/gender, *kjønn*.

41. Hertoft and Ritzau, *Paradiset*, 130.

42. Money and Ehrhardt, *Man & Woman*, 1.

43. Money and Ehrhardt, *Man & Woman*, 14.

44. Jennifer Germon, *Gender: A Genealogy of an Idea* (New York: Palgrave Macmillan, 2009), 55.

45. Grünfeld, *Seksuelle liv*, 21.

46. Robert J. Stoller, "A Contribution to the Study of Gender Identity," *International Journal of Psychoanalysis* 45, nos. 2–3 (1964). As Ulrikke Klöppel has pointed out, Robert Stoller, Harold Garfinkel, and Alexander C. Rosen coined the term "core sexual identity" in an article published in 1962 about the psychiatric management of children with intersex conditions; see Robert J. Stoller, Harold Garfinkel, and Alexander C. Rosen, "Psychiatric Management of Intersexed Patients," *California Medicine* 96, no. 1 (1962). For a discussion of Stoller's concept of "core gender identity," see Klöppel, *XXoXY ungelöst*, 499–503.

47. For a discussion of Stoller's concept, see Eder, *How the Clinic Made Gender*, 196–201. For Stoller's theories and the practices at the UCLA clinic, see Gill-Peterson, *Transgender Child*, 143–50.

48. Robert J. Stoller, *Sex and Gender: The Development of Masculinity and Femininity* (London: Karnac, 1968), 263–64.

49. Hertoft and Ritzau, *Paradiset*, 132.

50. Sorensen and Hertoft, "Sex Modifying Operations on Transsexuals in Denmark in the Period 1950–1977," 64.

51. Jemima Repo, *The Biopolitics of Gender* (New York: Oxford University Press, 2016), 24–48.

52. Stoller, *Sex and Gender*, xi; see also Germon, *Gender*, 65–66.

53. Grünfeld, *Seksuelle liv*, 24.

54. Grünfeld, *Seksuelle liv*, 25.

55. John Forrester, *Thinking in Cases* (Cambridge: Polity Press, 2017), 138–39.

56. Fredrik Mellbye to Torbjørn Mork, February 5, 1979, Oslo helseråd, Box 122, Folder homofile-transseksualitet, OCA; Kirsti Malterud, Georg Petersen and Berthold Grünfeld to the Office of Psychiatry, the Directorate of Health, February 15, 1980, Oslo helseråd, Box 122, Folder homofile-transseksualitet, OCA.

57. Torbjørn Mork to Fredrik Mellbye, February 16, 1979, Oslo helseråd, Box 122, Folder homofile-transseksualitet, OCA.

58. Oslo Health Council, December 1979, Utredning om transseksualitet [Report on transsexuality], 2, Oslo helseråd, Box 122, Folder homofile-transseksualitet, OCA.

59. Oslo Health Council, December 1979, Utredning om transseksualitet [Report on transsexuality], 6.

60. Medical record, Oslo Health Council, case 0006, Bodil Solberg private

archive (hereafter cited as BSA). The patients were examined by various psychologists in the Psychiatric Department of the Oslo Health Council.

61. Harry Benjamin, "Transvestism and Transsexualism in the Male and Female," *Journal of Sex Research* 3, no. 2 (1967).

62. Benjamin, "Transvestism and Transsexualism in the Male and Female," 5.

63. Benjamin, "Transvestism and Transsexualism in the Male and Female," 5.

64. Benjamin, "Transvestism and Transsexualism in the Male and Female," 7.

65. Psychologist, medical record, case 0007, BSA.

66. Kirsti Malterud interview, October 24, 2019.

67. Oslo Health Council, December 1979, Utredning om transseksualitet, 2, Oslo helseråd, Box 122, Folder homofile-transseksualitet, OCA.

68. Jan-Henrik Pederstad interview, Oslo, June 17, 2019.

69. Wintherbauer and Stokholm, *Janni*, 74–75.

70. Kirsti Malterud and Georg Petersen, "Blant sexologer i Mexico," *Fritt Fram* (August 1980), 22–23; "Skal snakke kjønnsskifte," *VG* (December 8, 1979), 5.

71. Ketil Slagstad, "How the Idea of Social Contagion Shaped Trans Medicine," *New England Journal of Medicine* 391, no. 16, https://doi.org/10.1056/NEJMms2407430.

72. Kirsti Malterud and Georg Petersen, Rapport fra den 4. verdenskongress i sexologi, Mexico City 16.–21. December 1979, Oslo helseråd, box 122, homofile—transseksualitet, OCA.

73. Jesse R. Pitts, "Social Control: The Concept," in *International Encyclopaedia of the Social Sciences*, ed. David L. Sills (New York: Macmillan, 1968); Irving Kenneth Zola, "Medicine as an Institution of Social Control," *Sociological Review* 20, no. 4 (1972), https://doi.org/10.1111/j.1467-954X.1972.tb00220.x; see also Peter Conrad, "The Discovery of Hyperkinesis: Notes on the Medicalization of Deviant Behavior," *Social Problems* 23, no. 1 (1975). Medicalization was originally suggested as a critique of psychiatry without referring specifically to the term "medicalization"; see Michel Foucault, *Folie et déraison: Histoire de la folie à l'âge classique* (Paris: Plon, 1961); Erving Goffman, *Asylums: Essays on the Social Situation of Mental Patients and Other Inmates* (Garden City, New York: Anchor Books, 1961); Ronald David Laing, *The Divided Self* (New York: Pantheon Books, 1960); Thomas S. Szasz, *The Myth of Mental Illness: Foundations of a Theory of Personal Conduct* (New York: Dell, 1961).

74. Janice G. Raymond, *The Transsexual Empire: The Making of the She-Male* (New York: Teachers College Press, 1994 [1979]), 119.

75. Kirsti Malterud and Bodil Solberg, "Kliniske erfaringer med kjønnsskifteklienter," *Tidsskrift for Den norske lægeforening* 103, no. 1 (1983): 6; see Thomas S. Szasz, *Sex by Prescription* (New York: Anchor Press/Doubleday, 1980).

76. Astrid Fiskvik, "Å være kvinne er ingen sykdom," *Nationen*, February 27, 1982, p. 10.

77. Malterud interview.

78. Kirsti Malterud, "Sexologi er en politisk vitenskap," *Kontrast*, no. 5–6 (1980).

79. Cf. Hobæk and Lie, "Less Is More."

80. Malterud and Solberg, "Kjønnsskifteklienter," 6.

81. Malterud interview.

82. Malterud interview.

83. Malterud interview.

84. Letter from Isak to Oslo Health Council, 1981, case 1005, OUHA.

85. See, for example, the medical statement, 1980, S-4711 Kirke-, utdannings- og forskningsdepartementet, Kirkeavdelingen, Da, Box L0330, Folder 453.2, NAN.

Chapter Eight

1. Per Anthi and Erik Stänicke, "Psykoanalysens inntog i norsk psykologi og psykiatri. En historisk beretning," *Tidsskrift for Norsk psykologforening* 56, no. 6 (2018); Gunn Johansson, ed., *Historien om svensk psykologisk forskning. Utvecklingen från perception och psykofysik* (Stockholm: Kungl. vitterhets historie och antikvitets akademien, 2020); Andrew Scull, "Contending Professions: Sciences of the Brain and Mind in the United States, 1850–2013," *Science in Context* 28, no. 1 (2015), https://doi.org/10.1017/S0269889714000350. Traditionally, psychiatrists were critical of the entry of psychologists and psychoanalysts into the Norwegian health system, but a new generation of psychiatrists were more open to psychoanalytic thinking. In 1969, psychotherapeutic training became a mandatory part of psychiatric training; see Per Haave, "Striden om nevrosene: Psykiatriens møte med psykoanalysen," in *Den mangfoldige velferden*, ed. Edgeir Benum et al. (Oslo: Gyldendal, 2003), 209–10.

2. For a history of psychoanalysis in Norway, see Randolf Alnæs, "Psychoanalysis in Norway: History, Training, Treatment, Research," *Nordic Journal of Psychiatry* 48, 32 (1994); Haave, "Striden om nevrosene: Psykiatriens møte med psykoanalysen"; Håvard Friis Nilsen, "Da psykoanalysen ble en profesjon," *Arr*, no. 2–3 (2010); Håvard Nilsen, *Du må ikke sove. Wilhelm Reich og psykoanalysen i Norge* (Oslo: Aschehoug, 2022).

3. Håvard Friis Nilsen, "Psykologene," in *Profesjonshistorier*, ed. Rune Slagstad and Jan Messel (Oslo: Pax forlag, 2014). In 1948, the University of Oslo introduced a dedicated cand. psychol. exam; see also Haave, "Striden om nevrosene: Psykiatriens møte med psykoanalysen," 205.

4. Nilsen, "Psykologene." See also, Karl Halvor Teigen, *En psykologihistorie*, 2nd ed. (Fagbokforlaget, 2015), 406–7.

5. Solberg interview.

6. Solberg interview.

7. The Draw-a-Man test is sometimes referred to as the Draw-a-Person test.

8. Nomran D. Sundberg, "The Practice of Psychological Testing in Clinical Services in the United States," *American Psychologist* 16, no. 2 (1961), https://doi.org/10.1037/h0040647.

9. Ruth Rae Doorbar, "Psychological Testing of Male Transsexuals: A Brief Report of Results from the Wechsler Adult Intelligence Scale, the Thematic Apperception Test, and the House-Tree-Person Test," in *Transsexualism and Sex Reassignment*, ed. Richard Green and John Money (Baltimore: Johns Hopkins University Press, 1969).

10. Hertz, Tillinger, and Westman, "Transvestitism: Report on Five Hormonally and Surgically Treated Cases."

11. Wålinder, *Transsexualism*, 51–57; see also minutes from the Oslo Health Council, December 5, 1979, 1, Oslo helseråd, Box 122, Folder homofile-transseksualitet, OCA.

12. Oslo Health Council, Transseksualitet, December 5, 1979, Oslo helseråd, Box 122, Folder homofile-transseksualitet, OCA.

13. Thematic Apperception Test, 1954, case 1013, Ullevål Hospital Department XIII, OUHA.

14. Thematic Apperception Test, 1954, case 1013, Ullevål Hospital Department XIII, OUHA.

15. Thematic Apperception Test, 1954, case 1013, Ullevål Hospital Department XIII, OUHA. Underlining retained.

16. Rorschach test, 1954, case 1013, Psychiatric Department for Men, Ullevål Hospital, OUHA.

17. Rorschach test, 1954, case 1013, Psychiatric Department for Men, Ullevål Hospital, OUHA.

18. Admission note, 1954, case 1013, Psychiatric Department for Men, Ullevål Hospital, OUHA.

19. Malterud and Solberg, "Kjønnsskifteklienter."

20. Solberg interview.

21. Malterud and Solberg, "Kjønnsskifteklienter."

22. Rebecca Lemov, "X-rays of Inner Worlds: The Mid-Twentieth-Century American Projective Test Movement," *Journal of the History of the Behavioral Sciences* 47, no. 3 (2011), https://doi.org/10.1002/jhbs.20510.

23. Hermann Rorschach, *Psychodiagnostik. Methodik und Ergebnisse eines Wahrnehmungsdiagnostischen Experiments* (Bern: Verlag Hans Huber, 1937 [1921]), 16–17.

24. On the emergence of "trained judgment" as a new scientific virtue in the early twentieth century, see Lorraine Daston and Peter Galison, *Objectivity* (New York: Zone Books, 2010), 309–61.

25. Rebecca Lemov, *Database of Dreams: The Lost Quest to Catalog Humanity* (New Haven, CT: Yale University Press, 2015), 34–35.

26. Peter Galison, "Image of Self," in *Things That Talk: Object Lessons from Art and Science*, ed. Lorraine Daston (New York: Zone Books, 2004), 291.

27. Ellen Hartmann and Geir Nielsen, *Kompendium 1 i Rorschach-metodikk* (Bergen: Universitetet i Bergen, 1976), 2.

28. Solberg interview.

29. Kirsti Malterud interview, October 24, 2019.

30. Psychologist note, case 0003, BSA.

31. Bjørn Killingmo, *Rorschachmetode og psykoterapi. En egopsykologisk studie* (Oslo: Universitetsforlaget, 1980), 18.

32. Killingmo, *Rorschachmetode og psykoterapi*, 25.

33. Killingmo, *Rorschachmetode og psykoterapi*, 26.

34. Florence L. Goodenough, *Measurement of Intelligence by Drawings* (Yonkers-on-Hudson: World Book, 1926).

35. Karen Machover, *Personality Projection in the Drawing of the Human Figure* (Springfield: Charles C. Thomas, 1949), 4.

36. Machover, *Personality Projection*, 35.

37. Machover, *Personality Projection*, 5.

38. Machover, *Personality Projection*, 101.

39. Daniel G. Brown and Alexander Tolor, "Human Figure Drawings as Indicators of Sexual Identification and Inversion," *Perceptual and Motor Skills* 7 (1957).

40. Michael Fleming, Gerald Koocher, and Judith Nathans, "Draw-A-Person Test: Implications for Gender Identification," *Archives of Sexual Behavior* 8, no. 1 (1979).

41. Machover, *Personality Projection*, 35–36.

42. Wålinder, *Transsexualism*, 56.

43. Psychologist note, case 0004, BSA.

44. Psychologist, medical record, case 0004, BSA.

45. Psychologist, medical record, case 0001, BSA.

46. Solberg interview.

47. Doorbar, "Psychological Testing of Male Transsexuals," 192.

48. Isak interview.

49. Isak interview.

50. Isak interview.

51. Isak to the team at Oslo Health Council, case 1005, OUHA.

52. Psychologist note, case 0001, BSA.

53. Malterud and Solberg, "Kjønnsskifteklienter."

54. Psychologist note, case unnumbered, BSA.

55. Solberg interview.

56. Psychologist, medical record, case 0002, BSA.

57. Astri Eidsbø Lindholm second interview, February 8, 2019.

58. Lindholm second interview.

59. Sonja was a traditional Norwegian name, very popular after Sonja Haraldsen married Crown Prince Harald of Norway in 1968. Silvia was a less common name, but its popularity also surged after 1976, when Silvia Renate Sommerlath married Swedish Crown Prince Carl Gustaf; see "Historisk utvikling av jentenavn," https://www.ssb.no/a/navn/historisk-utvikling-av-jentenavn/, accessed June 23, 2022.

60. Lindholm second interview.

61. Letter from patient, case 1005, OUHA.

62. Psychologist, medical record, case 0002, BSA.

63. For *inscription devices*, see Cornelius Borck and Armin Schäfer, eds., *Das psychiatrische Aufschreibesystem* (Paderborn: Wilhelm Fink, 2015).

64. Cornelia Vismann, *Akten: Medientechnik und Recht* (Frankfurt am Main: Fischer Taschenbuch Verlag, 2000), 47.

65. See Knut Ebeling and Stephan Günzel, eds., *Archivologie: Theorien des Archivs in Phiolosophie, Medien und Künsten* (Berlin: Kulturverlag Kadmos Berlin, 2009); Arlette Farge, *The Allure of the Archives*, trans. Thomas Scott-Railton (New Haven: Yale University Press, 2013 [1989]); Anja Horstmann and Vanina Kopp, eds., *Archiv—Macht—Wissen. Organisation und Konstruktion von Wissen und Wirklichkeit in Archiven* (Frankfurt a.M.: Campus-Verlag, 2010); Vismann, *Akten*.

66. Heike Bauer, *The Hirschfeld Archives: Violence, Death, and Modern Queer Culture* (Philadelphia: Temple University Press, 2017), 9.

67. Solberg interview.

68. Malterud interview.

69. For a critique of traditional pathologizing psychoanalytic approaches to transgender life, especially the concept of "core gender identity" and a new approach to thinking about "development dynamically, without falling into developmentalism," i.e., as gender without a final destination, but as a "matter of self-theorizing," see Avgi Saketopoulou and Ann Pellegrini, *Gender Without Identity* (New York: Unconscious in Translation, 2023); quotes from pp. 29–30.

70. Carlo Ginzburg, "Signes, traces, pistes," *Le Débat* 6, no. 6 (1980): 12.

71. Paul Wenzel Geissler and Guillaume Lachenal, "Brief Instructions for Archaeologists of African Futures," in *Traces of the Future: An Archaeology of Medical Science in Twenty-First-Century Africa*, ed. Paul Wenzel Geissler et al. (Bristol: Intellect, 2016), 16.

72. Wålinder, *Transsexualism*, 57.

73. This information is based on interviews; see also Malterud and Solberg, "Kjønnsskifteklienter." Another document states that, among nine patients, two started hormone treatment: see Georg Petersen, Kirsti Malterud, and Berthold Grünfeld to the Directorate of Health, Behandling av transseksuelle i Norge, February 15, 1980, Oslo helseråd, Box 122, Folder homofile-transseksualitet, OCA.

74. Allan Young, *The Harmony of Illusions: Inventing Post-Traumatic Stress Disorder* (Princeton: Princeton University Press, 1996); see, in particular, chapters 4 and 5.

75. Lemov, *Database*, 34.

Chapter Nine

1. In the following, I reproduce the actors' category of "transvestites," since this was the self-description in use at that time. Even if some people today find the term

pejorative, I seek to give voice to a community that has often been overlooked in trans historiography.

2. Anette Hall, "Phi Pi Epsilon—Northern Europe," undated, 1966, *Feminform*, SA.

3. Esben Benestad, Berthold Grünfeld, and Bernt Krøvel, "Heterofil transvestisme," *Tidsskrift for Den norske lægeforening* 106, no. 25 (1986). According to another article from 1988, it had 450 members in the Nordic region and 70 in Norway: see Turid Eikvam and Trine Brekke, "Transvestittisme—en livsberikelse . . . !?," *Løvetann*, 1988.

4. Susan Stryker, "Foreword," *International Journal of Transgenderism* 8, no. 4 (2005), https://doi.org/10.1300/J485v08n04_a.

5. Richard Ekins and Dave King, "Virginia Prince: Transgender Pioneer," *International Journal of Transgenderism* 8, no. 4 (2005), https://doi.org/10.1300/J485v08n04_02; C. V. Prince, "Homosexuality, Transvestism and Transsexuality: Reflections on Their Etiology and Differentiation," *International Journal of Transgenderism* 8, no. 4 (2005), https://doi.org/10.1300/J485v08n04_03.

6. Second International Symposium on Gender Identity, September 12–14, 1971, folder, Brev- og materialesamling, Christine Jorgensen 1926–1989 fotograf, kps. 3 Materialer, RDL.

7. Gill-Peterson, *Transgender Child*, 19.

8. Hans W. Kristiansen, *Masker og motstand: diskré homoliv i Norge 1920–1970* (Oslo: Unipub, 2008), 203–10.

9. Aase Schibsted Knudsen interview with Skeivt arkiv, February 18, 2019, SA.

10. *Feminform* had published a total of 163 issues by the time of its last issue in 2001.

11. Bjørn Ellen Brudal interview with Bjørn André Widvey, February 19, 2019, Skeivt arkiv.

12. This chapter draws on a close reading of the comprehensive collection of Scandinavian transvestite magazines at Skeivt arkiv at the University of Bergen. This includes the magazines *Feminform* (1967–2001, 163 issues), *Intermezzo* (1989–2002, 44 issues), and *Attitude* (2002–2004, 9 issues).

13. Thorkil Sørensen, "Tilblivelsen af Sexologisk Klinik—Samlivsgruppen," in *25 år med Sexologisk Klinik*, ed. Ellids Kristensen and Annamaria Giraldi (Copenhagen: Psykiatrisk Center København, 2011), 16; see also Hertoft and Ritzau, *Paradiset*, 21–24.

14. Hertoft and Ritzau, *Paradiset*, 21–22.

15. Hertoft, *Klinisk sexologi. En introduktion*, 210–14.

16. Erna, *Feminform*, no. 58 (1976), 9.

17. Erna, *Feminform*, no. 58 (1976), 10.

18. Åse Eriksen, *Feminform*, no. 59 (1977), 16.

19. Erna, *Feminform*, no. 58 (1976), 12.

20. Britt-Eva, "Med dig ikke ut av FPE!" *Feminform*, no. 85 (1981), 10.

21. "Transvestisme—kvindebevægelse—uni-sex," *Feminform*, no. 67 (1978), 10.

22. Preben Hertoft, *Feminform*, no. 60 (1977), 6–7; Hertoft and Ritzau, *Paradiset*, 7, 15.

23. Stryker, *Transgender History*, 147–53.

24. Erika Alm, "A State Affair? Notions of the State in Discourses on Trans Rights in Sweden," in *Pluralistic Struggles in Gender, Sexuality and Coloniality: Challenging Swedish Exceptionalism*, ed. Erika Alm et al. (Cham: Palgrave Macmillan, 2021).

25. Timm, "Trans Circles of Knowledge."

26. Queers mot kapitalism, "Transvestia söker medlemmar. Intervju med Eva-Lisa Begntsson," *Queers mot kapitalism* (December 10, 2010), https://queersmotkapitalism.wordpress.com/2010/12/10/transvestia-soker-medlemmar-intervju-med-eva-lisa-bengtsson/; "Eva-Lisa's Monument: Interview with Sam Hultin," C-print (August 19, 2019), https://www.c-print.se/post/eva-lisa-s-monument.

27. Sam Hultin, "Resistance and care," *Kunstkritikk: Nordic Art Review* (May 26, 2021), https://kunstkritikk.com/resistance-and-care/.

28. Aleksa Lundberg, "Eva-Lisa Bengtson—Sveriges första transaktivist" (Sweden: Sveriges radio, 2020), https://sverigesradio.se/avsnitt/1482205.

29. By the mid-1980s, only one member was assigned female at birth. Eikvam and Brekke, "Transvestittisme."

30. For an explanation of the acronym "FFs," see note 33 below.

31. Anette Hall, "F.P.E. och sex-frågan," *Feminform*, no. 2 (1967), 5.

32. Eva-Karin Rydberg, "Hur manga har det så pas bra??," *Feminform*, no. 64 (1978), 4.

33. Åse Eriksen, *Feminform*, no. 65 (1978), 5. "FP" was a person who expressed their personality freely; "FF" (*FlickFlicka/födt flicka*) was the abbreviation for "girl-girl" or "born girl." The American organization used "GG" for Genuine Girl and Genetic Girl.

34. *Feminform*, no. 66 (1978), 1.

35. *Feminform*, no. 66 (1978), 1.

36. Virginia Prince, "Transvestisme og sex," translated into Danish, *Feminform*, no. 66 (1978), 4.

37. Åse Eriksen, "Skal FPE's love revideres? Skal medlemskredsen udvides?" *Feminform*, no. 66 (1978), 6.

38. Prince, "Transvestisme og sex."

39. Virginia Prince, "Transvestisme og sex," intercept in Danish, *Feminform*, no. 66 (1978), 5.

40. Erna, "Nogle betragtninger over TV-isme og dette fænomens relation til sexualitet," *Feminform*, no. 66 (1978), 13.

41. Genesis 2:23.

42. Sigmund Raanes, "En Transvestitt Står Fram," *Fritt Fram*, nos. 2–3 (1979), 14.

43. Hertoft and Ritzau, *Paradiset*, 49.

44. Rikke Bjurstrøm, "—Kona syr kjolene mine," *VG* (July 30, 1987), 4.

45. Grünfeld, *Seksuelle liv*, 87–88.

46. Åse Eriksen, *Feminform*, no. 66 (1978), 6.

47. Jan Elisabeth Lindvik interview, January 15, 2021. As leader of Foreningen for Transpersoner, Norge Lindvik later helped change the membership rules so that all trans people were welcome.

48. Esben Esther Pirelli Benestad, November 2, 2020.

49. Ragnhild interview, October 23, 2019.

50. Ragnhild interview.

51. Jeanette Solstad interview, January 13, 2021.

52. Åse Eriksen, *Feminform*, no. 66 (1978), 6.

53. Gunhild Grünfeld interview.

54. Benestad interview.

55. Benestad interview.

56. Benestad, Grünfeld, and Krøvel, "Heterofil transvestisme," 2071.

57. Benestad, Grünfeld, and Krøvel, "Heterofil transvestisme," 2070.

58. Berthold Grünfeld, Esben Benestad, and Bernt Krøvel, "Transvestisme og transvestitter," *Nordisk Sexologi* 4 (1986).

59. Esben Benestad, "Heterofil transvestisme," *Feminform*, no. 106 (1987), 32–38.

60. Kadji Amin, "Temporality," *TSQ: Transgender Studies Quarterly* 1, nos. 1–2 (2014), https://doi.org/10.1215/23289252-2400073.

Chapter Ten

1. Aina Schiøtz, *Doktoren. Distriktslegenes historie 1900–1984* (Oslo: Pax Forlag, 2003), 351–60; Aina Schiøtz, *Folkets helse—landets styrke, 1850–2003, Det offentlige helsevesen i Norge 1603–2003* (Oslo: Universitetsforlaget, 2003), 371–74; Per Haave, "Velferdskommunen i støpeskjeen," in *Den nye velferdsstatens historie. Ekspansjon og omdanning etter 1966*, ed. Anne Lise Ellingsæter et al. (Oslo: Gyldendal akademisk, 2020).

2. Halvard Vike, Anette Fagertun, and Heidi Haukelien, "Introduction: Welfare State Capitalism, Universalism, and Social Reproduction in Scandinavia," in *The Political Economy of Care: Welfare State Capitalism, Universalism, and Social Reproduction*, ed. Halvard Vike, Anette Fagertun, and Heidi Haukelien (Oslo: Scandinavian University Press, 2024).

3. Kirsti Malterud, Bodil Solberg, and Kåre Duckert to Stadsfysikus, June 3, 1981, homophile—transseksualitet folder, OCA.

4. Berthold Grünfeld to Stadsfysikus Mellbye, June 25, 1981, homofile—transseksualitet folder, OCA.

5. Berthold Grünfeld, "Seksualitet som helseproblem," in *Samfunnsmedisin i*

praksis. Oslo Helseråd i 80-årene, ed. Harald Siem, Kåre Berg, and Berthold Grünfeld (Oslo: Universitetsforlaget, 1987), 203.

6. Velocci, "Standards of Care."

7. Knut Chr. Skolleborg, "Transsexualisme," annual report 1996, Plastisk kirurgisk avdeling, Rikshospitalet, no. 1997/4045, NBHSA.

8. Henrik Borchgrevink, "Transsexualisme som landsfunksjon," August 10, 1990, Rikshospitalet, no. 1991/1410, NBHSA.

9. Knut Skolleborg interview, January 14, 2021.

10. Borchgrevink, "Transsexualisme som landsfunksjon."

11. Grünfeld, case 1005, medical record, Oslo University Hospital Archive.

12. Borchgrevink, "Transsexualisme som landsfunksjon," NBHSA.

13. Isak interview.

14. Bengt Lundström and Jan Wålinder, "Evaluation of Candidates for Sex Reassignment," *Nordisk Psykiatrisk Tidsskrift* 39, no. 3 (1985), https://doi.org/10.3109/08039488509106160.

15. Hertoft and Sørensen, "Transsexuality: Some Remarks Based on Clinical Experience," 70, 166.

16. Adeen-Wintherbauer interview.

17. "St. meld. nr. 9 (1974–75). Sykehusutbygging m.v. i et regionalisert helsevesen" (Oslo: Sosialdepartementet), 41–42, 53.

18. Jonas interview.

19. Alv A. Dahl interview, September 25, 2019.

20. Eviatar Zerubavel, "Timetables and Scheduling: On the Social Organization of Time," *Sociological Inquiry* 46, no. 2 (1976): 89–90, https://doi.org/10.1111/j.1475-682X.1976.tb00753.x.

21. Todd Sekuler, "Un/Certain Care: From a Diagnostic to a Somatechnic Regime of Care for Medical Transition in Public Hospitals in France" (PhD thesis, Humboldt University of Berlin, 2018), 139.

22. Hanna interview, November 13, 2019.

23. Hanna interview.

24. Hanna interview.

25. Rådgivningstjenesten for homofile, 1977, case 1007, medical records, OUHA.

26. See Eira Bjørvik, "Conceiving Infertility: Infertility Treatment and Assisted Reproductive Technologies in 20th Century Norway" (PhD thesis, University of Oslo, 2018), 80–83.

27. Bjørvik, "Conceiving Infertility," 212.

28. Knut Bjøro to Berthold Grünfeld, 1981, case 1005, OUHA.

29. Berthold Grünfeld to Fridtjof Jervel, 1981, case 1005, OUHA.

30. Money and Ehrhardt, *Man & Woman*, 14.

31. Hanna interview.

32. Harry Benjamin, *The Transsexual Phenomenon* (New York: Julian Press, 1966), 137.

33. Robert J. Stoller, "Male Transsexualism: Uneasiness," *American Journal of Psychiatry* 130, no. 5 (1973): 251.

34. Astri Eidsbø Lindholm interview, October 31, 2019.

35. Daisy Hafstad interview with André Widvey, Skeivt arkiv, April 19, 2018.

36. Sandal, "En særlig trang," 93, 98, 108.

37. Benjamin, "Transvestism and Transsexualism in the Male and Female."

38. J. Hoenig, J. Kenna, and Ann Youd, "A Follow-Up Study of Transsexualists: Social and Economic Aspects," *Psychopathology* 3, no. 2 (1970), https://doi.org/10.1159/000278595; J. Hoenig, J. C. Kenna, and A. Youd, "Surgical Treatment for Transsexualism," *Acta Psychiatrica Scandinavica* 47, no. 1 (1971); B. D. Hore, F. V. Nicolle, and J. S. Calnan, "Male Transsexualism in England: Sixteen Cases with Surgical Intervention," *Archives of Sexual Behavior* 4, no. 1 (1975), https://doi.org/10.1007/bf01541889; Jon K. Meyer and Donna J. Reter, "Sex Reassignment: Follow-Up," *Archives of General Psychiatry* 36, no. 9 (1979), https://doi.org/10.1001/archpsyc.1979.01780090096010; John Money and Anke A. Ehrhardt, "Transsexuelle nach Geschlechtswechsel. Erfahrungen und Befunde am Johns Hopkins Hospital," *Beitrage zur Sexualforschung* 49 (1970); Georg K. Stürup, "Male Transsexuals: A Long-term Follow-Up After Sex Reassignment Operations," *Acta Psychiatrica Scandinavica* 53, no. 1 (1976), https://doi.org/10.1111/j.1600-0447.1976.tb00058.x.

39. Sekuler, "Un/Certain Care," 102.

40. Bauer, "Epidemiologist's Dream."

41. Jan Wålinder and Inga Thuwe, *A Social-Psychiatric Follow-Up Study of 24 Sex-Reassigned Transsexuals* (Gothenborg: Akademiförlaget, 1975), 10.

42. Wålinder and Thuwe, *Social-Psychiatric Follow-Up Study*, 21–22.

43. Wålinder and Thuwe, *Social-Psychiatric Follow-Up Study*, 27–29.

44. Jan Wålinder, Bengt Lundström, and Inga Thuwe, "Prognostic Factors in the Assessment of Male Transsexuals for Sex Reassignment," *British Journal of Psychiatry* 132 (1978).

45. Harry Benjamin to Bengt Lundström, April 13, 1979, Box 6, Series II C, HBC.

46. Harry Benjamin to Bengt Lundström, May 29, 1982, Box 6, Series II C, HBC.

47. Bengt Lundström, Ira Pauly, and Jan Wålinder, "Outcome of Sex Reassignment Surgery," *Acta Psychiatrica Scandinavica* 70 (1984); Lundström and Wålinder, "Evaluation of Candidates for Sex Reassignment."

48. See, for instance, Oslo Health Council, Utredning om transseksualitet, December 1979, OCA.

49. Gunnar Lindemalm, Dag Körlin, and Nils Uddenberg, "Long-Term Follow-Up of 'Sex Change' in 13 Male-to-Female Transsexuals," *Archives of Sexual Behavior* 15, no. 3 (1986): 188.

50. Luft et al., "Transsexualism."

51. Astrid interview.

52. D. Daniel Hunt and John L. Hampson, "Transsexualism: A Standardized Psychosocial Rating Format for the Evaluation of Results of Sex Reassignment Surgery," *Arch Sex Behav* 9, no. 3 (June 1980), https://doi.org/10.1007/bf01542251.

53. Alv A. Dahl, Cato A. Guldberg, Helge Hansen, Ellen Kjelsberg, Henrik Borchgrevink, and Berthold Grünfeld, *The Norwegian Outcome Study of Patients with Transsexualism Treated with Sex Reassignment Surgery,* draft of research manuscript for publication, undated, p. 5, IHPA.

54. Del av Alv Dahl's forskningsopplegg 1988, IHPA.

55. Alv A. Dahl to the working group on transsexuality, Vedr. etterundersøkelsen, 21 May 1993, IHPA.

56. Dahl et al., *The Norwegian Outcome Study*, IHPA.

Chapter Eleven

1. Bjørn interview, March 9, 2020.

2. Anne Lise Ellingsæter et al., *Den nye velferdsstatens historie. Ekspansjon og omdanning etter 1966* (Oslo: Gyldendal Akademisk, 2020); Maren Skaset, "Reformtid og markedsgløtt: det offentlige helsevesen etter 1985," in *Folkets helse—landets styrke, 1850–2003*, ed. Aina Schiøtz (Oslo: Universitetsforlaget, 2003). For a comparison of this period in Norway and Sweden, see Sejersted, *Social Democracy*, 388–430.

3. Schiøtz, *Folkets helse*, 374–79.

4. St. meld. nr. 9 (1974–75).

5. St. meld. nr. 9 (1974–75), 95. On the institutionalization and centralization of infertility treatment, see Bjørvik, "Conceiving Infertility."

6. "St. meld. nr. 41 (1987–88). Helsepolitikken mot år 2000, Nasjonal Helseplan" (Oslo: Sosialdepartementet).

7. "St. meld. nr. 41 (1987–88). Helsepolitikken mot år 2000, Nasjonal Helseplan," 192. A report on the centralization of highly specialized somatic health care was published in 1992. Helsedirektoratet, *Styring av høyspesialiserte somatiske funksjoner i helsetjenesten* (Oslo: Helsedirektoratet, 1992).

8. At the time, Rikshospitalet was still a state-owned hospital, while other hospitals were under the authority of the counties. In 2001, the Storting decided that, beginning in 2002, the ownership of specialized health care services, including the hospitals, would be transferred to the state. In addition, five regional health authorities were established under state ownership.

9. Schiøtz, *Folkets helse*, 376–86.

10. Statens helsetilsyn, *Styring av høyspesialiserte funksjoner innen psykisk helsevern for voksne* (Oslo, 1995).

11. Statens helsetilsyn, *Styring av høyspesialiserte funksjoner innen psykisk helsevern for voksne*, 32.

12. Director Odd Arild Haugen and Medical Director Dagfinn Abrechtsen, Rikshospitalet, to Directorate of Health, July 15, 1992, no 1991/1410, NBHSA.

13. Borchgrevink, "Transsexualisme som landsfunksjon."

14. Rapportskjema for lands- og flerregionale helsetjenester, July 15, 1992, no. 1991/1410, NBHSA.

15. Stadsfysikus in Oslo Fredrik Mellbye to Director General of Health, February 5, 1979, Oslo helseråd, Box 122, Homofile-transseksualitet folder, OCA.

16. Harald Frey, Aker Hospital, to Otto W. Steenfeldt-Foss, Directorate of Health, November 23, 1979, S-1286, Folder Transseksualitet, NAN.

17. Torbjørn Mork to Fredrik Mellbye, February 16, 1979, Homofile-transseksualitet folder, OCA.

18. Alv A. Dahl and Berthold Grünfeld to the Norwegian Board of Health Supervision, December 30, 1994, no. 1995/94, NBHSA.

19. Berthold Grünfeld to Chief County Medical Officer in Oslo, "Innhenting av uttalelse vedr. klage på avslag for kjønnsskifteoperasjon ved Rikshospitalet for NN," no. 1995/6160, NBHSA.

20. See Steven Epstein, *Impure Science: AIDS, Activism, and the Politics of Knowledge* (Berkeley: University of California Press, 1996); Wendy Kline, *Bodies of Knowledge: Sexuality, Reproduction, and Women's Health in the Second Wave* (Chicago: University of Chicago Press, 2010); Alex Mold, *Making the Patient-Consumer: Patient Organisations and Health Consumerism in Britain* (Manchester: Manchester University Press, 2015); Nancy Tomes, *Remaking the American Patient: How Madison Avenue and Modern Medicine Turned Patients into Consumers* (Chapel Hill: University of North Carolina Press, 2015), esp. chapter 8.

21. Asbjørn Kjønstad, *Kompendium i helserett* (Oslo: Gyldendal, 2003), 53; see also David J. Rothman, *Strangers at the Bedside: A History of How Law and Bioethics Transformed Medical Decision Making* (New York: Basic Books, 1991).

22. Seip, *Veiene til velferdsstaten*, 381.

23. "NOU 1992:8. Lov om pasientrettigheter" (Oslo: Sosialdepartementet), 86.

24. Alv A. Dahl interview, Oslo, September 25, 2019.

25. See Council of Europe, "Recommendation 1117 (1989)—Condition of Transsexuals"; and European Court of Human Rights, "Case of B. v. France, Application no. 13343/87" (1992).

26. Dahl and Grünfeld to Norwegian Board of Health Supervision, December 30, 1994, NBHSA. This case history is based on numerous letters exchanged between the working group, the patient's lawyers, and the Norwegian Board of Health Supervision. To avoid compromising the person's anonymity, the archival material is not described in any further detail. All materials originate from no. 1995/6160, NBHSA.

27. Working Group to Chief County Medical Officer in Oslo, 1995/6160, NBHSA.

28. Director Åge Danielsen and Medical Director Arnt Jackobsen, Rikshospitalet, to Ministry of Health and Social Affairs, Vedlegg 1, June 7, 2001, p. 3, no. 1999/2202, NBHSA.

29. Helsedirektoratet, *Kjønnsinkongruens—Nasjonal faglig retningslinje, Helsedirektoratet* (Oslo, 2020), 27.

30. Berthold Grünfeld to Chief County Medical Officer in Oslo, 1995/6160, NBHSA.

31. Health Ombudsman in Hordaland Grethe Brundtland to Ministry of Health and Social Affairs, July 20, 2000, no. 1999/2202, NBHSA. For the patient letters, see no. 1999/2202—pasientbrev, NBHSA.

32. Danielsen and Jackobsen to Ministry of Health and Social Affairs, May 11, 2000, no. 1999/2202, NBHSA.

33. Skolleborg interview.

34. Petter Frode Amland et al., "Plastikkirurger og yrkesetikk," *Tidsskrift for Den norske lægeforening* 123, no. 21 (2003); Knut Skolleborg, "Plastikkirurger og etikk," *Tidsskrift for Den norske lægeforening* 123, no. 21 (2003).

35. Danielsen and Jackobsen to Ministry of Health and Social Affairs, August 16, 2000, no. 1999/2202, NBHSA.

36. For the history of DRGs, see Rick Mayes, "The Origins, Development, and Passage of Medicare's Revolutionary Prospective Payment System," *Journal of the History of Medicine and Allied Sciences* 62, no. 1 (2007), https://doi.org/10.1093/jhmas/jrj038.

37. Danielsen and Jackobsen to the Ministry for Social Affairs and Health, May 11, 2000, no. 2000/00824, NBHSA.

38. Danielsen and Jackobsen to the Ministry of Health and Social Affairs, "Vedlegg 1," NBHSA.

39. Danielsen and Jackobsen to the Ministry of Health and Social Affairs, May 11, 2000, no. 2000/00824, NBHSA.

40. Stryker, *Transgender History*, 212–74; Meyerowitz, *How Sex Changed*.

41. Christina Støp, "En anelse rødt," *TV2* (May 19, 1994); Aslaug Tangvald Pedersen, "—Føler meg som mann og kvinne," *VG* (November 13, 1994), 22–23.

42. Esben Esther Pirelli Benestad and Elsa Almås, *Kjønn i bevegelse* (Oslo: Universitetsforlaget, 2001). For the media discourse, see, for instance, Tone Foss Aspevoll, "43 millioner kjønn—og enda noen til," *Klassekampen* (June 21, 2001), 14–15; Annicken Vargel, "Alias pappa," *Dagbladet Magasinet* (July 7, 2001), 40–43.

43. Roderick A. Ferguson, *One-Dimensional Queer* (Cambridge: Polity Press, 2019); Victor Silverman and Susan Stryker, *Screaming Queens: The Riot at Compton's Cafeteria* (USA: Kanopy Streaming, 2005); Stryker, *Transgender History*, 119–65.

44. "Mann ble kvinne," *VG* (August 29, 1974), 20–21.

45. "'Jon' blir Linda," *Dagbladet* (July 29, 1983), 1; "Eva (23) var mann," *Dagbladet* (July 30, 1983).

46. "Norsk kvinne ble mann og er lykkelig gift," *Dagbladet* (October 20, 1979), 1; "Eva (23) var mann."

47. "Nytt kjønn like ensom," *VG* (February 17, 1988), 13; "–Ikke skift kjønn!" *VG* (November 22, 1975), 1.

48. Thomas Ergo, "Snart kvinne," *Dagbladet Magasinet* (May 26, 2001), 16–22.

49. The figure on patient suicides is from a statement by Ira Haraldsen quoted in Ergo, "Snart kvinne." Haraldsen has confirmed the figure in private correspondence. Unfortunately, it has not been possible to confirm the exact number of people who took their own lives, as the hospital has not kept a formal register and because I was denied access to such recent medical records by the research ethics committees. For media reports on the deadlock at the Rikshospitalet, see, e.g., "Tar livet sitt i operasjonskø," *Bergens Tidene*, December 24, 2000, p. 6; "Transseksuelle nektes behandling," *Klassekampen*, November 15, 2000, p. 7; and the documentary *Fakta på lørdag: Manndomsprøven*, NRK, May 6, 2000.

50. Arbeidsgruppen for transseksuelle, memo, January 28, 1999, Ira Haraldsen private archive (hereafter cited as IHA).

51. Charles L. Briggs and Daniel C. Hallin, *Making Health Public: How News Coverage Is Remaking Media, Medicine, and Contemporary Life* (Oxon: Routledge, 2016), 47–49. Highlighting the role of media in health communication, not merely as mirrors of science and medicine, but as coproducers of knowledge, is a central component of biomediatization theory.

52. Berthold Grünfeld and Ira Haraldsen to Olav Gunnar Ballo and Annelise Høegh, April 12, 2000, no. 1999/2202, NBHSA.

53. Minister of Health Tore Tønne to Annelise Høegh and Olav Gunnar Ballo, May 29, 2000, no. 1999/2202, NBHSA.

54. The Ministry of Health and Social Affairs to Rikshospitalet, November 3, 2000, no. 1999/2202, NBHSA.

55. Harald Stabell to the Ministry of Health and Social Affairs, June 11, 2001, no. 1999/2202, NBHSA.

56. The Ministry of Health and Social Affairs to Rikshospitalet, May 28, 2001, no. 1999/2202, NBHSA.

57. Dagfinn Høybråten, *Drivkraft* (Oslo: Cappelen Damm, 2012), 53.

58. Gyri Aure, "Transer feiret Tønne," *VG*, March 19, 2001, p. 25. Jens Stoltenberg, from the Social Democratic Party, became prime minister on March 17, 2000. Tore Tønne was appointed minister of health, replacing Dagfinn Høybråten from the Christian Democratic Party.

59. Landsforeningen for lesbiske og homofile foreslår å endre navn til Skeive folk, *Transskript* no. 1 (2004), 10–11, SKA/AKT-0063/Fa/LFTS 1, Stein Wolff Frydenlund, SA.

60. "LFTS vil ut av homoplan," *Transskript* no. 4 (2007), 7 and "En oppklaring fra sentralstyret i LLH og landsstyret i LFTS," *Transskript* no. 4 (2007), 10, both in SKA/AKT-0063/Fa/LFTS 1, Stein Wolff Frydenlund, Skeivt arkiv.

61. Stein Wolff Frydenlund interview with Bjørn André Widvey, January 17, 2018, SA.

62. On matters of public concern, see Bruno Latour, "From Realpolitik to Dingpolitik: How to Make Things Public. An Introduction," in *Making Things Public: Atmospheres of Democracy*, ed. Bruno Latour and Peter Weibel (Cambridge, MA: MIT Press, 2005).

63. Jasanoff, *Designs on Nature: Science and Democracy in Europe and the United States*.

64. Alm, "Emballage"; Holm, "Fleshing Out the Self"; Luft et al., "Transsexualism"; Katariina Parhi, "Boyish Mannerisms and Womanly Coquetry: Patients with the Diagnosis of Transvestitismus in the Helsinki Psychiatric Clinic in Finland, 1954–68," *Medical History* 62, no. 1 (2018), https://doi.org/10.1017/mdh.2017.73.

65. Borchgrevink, "Transsexualisme som landsfunksjon," NBHSA.

66. Øyvind Borch Bugge and Frank Åbyholm, "Transseksualisme, Årsrapport 2000, Plastisk kirurgisk avdeling, Rikshospitalet," no. 1997/4054, NBHSA.

67. Danielsen and Jackobsen to the Ministry of Health and Social Affairs, "Vedlegg 1," NBHSA.

68. Bugge and Åbyholm, "Transseksualisme, Årsrapport 2000," no. 1997/4054, NBHSA.

69. Danielsen and Jackobsen to the Ministry of Health and Social Affairs, "Vedlegg 1," NBHSA.

70. Harry Benjamin International Gender Dysphoria Association, *The Standards of Care for Gender Identity Disorders*, 5th ed. (Düsseldorf: Symposion, 1998), The World Professional Association of Transgender Health Archive.

71. Danielsen and Jackobsen to the Ministry of Health and Social Affairs, "Vedlegg 1," NBHSA.

72. See Haave, *Medisinens sentrum*, 333–60.

73. The Rikshospitalet had one position for a psychiatrist, Spesiallegen i psykiatri, established in 1958. The psychiatrist was supposed to work within psychosomatics and deliver liaison services for the somatic departments; see Instruks for spesiallegen i psykiatri, Sosialdepartementet December 27, 1957, RA/S-4134/F/Fd/L0360, Rikshospitalet, Administrasjonen, 1922–1971, Psykiatrisk avdeling 1951–1971, NAN.

74. Ira Haraldsen interview, Oslo, October 9, 2019.

75. LFTS info folder, "Hva er transseksualisme?" undated, probably early 2000s, SKA/AKT-0063/Fa/LFTS 2, Stein Wolff Frydenlund, Skeivt arkiv.

76. The LFTS to the Minister of Health and Social Affairs, undated, archived October 31, 2000, no. 1999/2202, NBHSA.

77. LFTS to the Minister of Health and Social Affairs, undated, archived October 31, 2000, no. 1999/2202.

78. Hansen interview.

79. Haraldsen interview.

80. ICD-9 is the *International Classification of Diseases*, 9th ed., published by the World Health Organization. For the history of the classification of these diagnoses, see Jack Drescher, Peggy Cohen-Kettenis, and Sam Winter, "Minding the Body: Situating Gender Identity Diagnoses in the ICD-11," *International Review of Psychiatry* 24, no. 6 (2012), https://doi.org/10.3109/09540261.2012.741575.

81. Dahl interview.

82. Fritz Handerer et al., "How Did Mental Health Become So Biomedical? The Progressive Erosion of Social Determinants in Historical Psychiatric Admission Registers," *History of Psychiatry* 32, no. 1 (2020), https://doi.org/10.1177/0957154X20968522; Stefan Timmermans and Marc Berg, *The Gold Standard: The Challenge of Evidence-Based Medicine and Standardization in Health Care* (Philadelphia: Temple University Press, 2003).

83. Stuart A. Kirk and Herb Kutchins, *The Selling of DSM: The Rhetoric of Science in Psychiatry* (New York: Aldine de Gruyter, 1992); Sing Lee, "Diagnosis Postponed: Shenjing Shuairuo and the Transformation of Psychiatry in Post-Mao China," *Culture, Medicine and Psychiatry* 23 (1999), https://doi.org/10.1023/a:1005586301895; Jackie Orr, *Panic Diaries: A Genealogy of Panic Disorder* (Durham: Duke University Press, 2006); Jackie Orr, "Biopsychiatry and the Informatics of Diagnosis: Governing Mentalities," in *Biomedicalization: Technoscience, Health, and the Illness in the U.S.*, ed. Adele E. Clarke et al. (Durham: Duke University Press, 2010); Martyn Pickersgill, "Standardising Antisocial Personality Disorder: The Social Shaping of a Psychiatric Technology," *Sociology of Health & Illness* 34, no. 4 (2012), https://doi.org/10.1111/j.1467-9566.2011.01404.x; Josef Parnas and Pierre Bovet, "Psychiatry Made Easy: Operation(al)ism and Some of Its Consequences," in *Philosophical Issues in Psychiatry III: The Nature and Sources of Historical Change*, ed. Kenneth S. Kendler and Josef Parnas (Oxford: Oxford University Press, 2015). The processes of standardization were not uncontested and should not be seen as a singular revolutionary process: see Nicolas Henckes, "Magic Bullet in the Head? Psychiatric Revolutions and Their Aftermath," in *Therapeutic Revolutions: Pharmaceuticals and Social Change in the Twentieth Century*, ed. Jeremy A. Greene, Flurin Condrau, and Elizabeth Siegel Watkins (Chicago: University of Chicago Press, 2016).

84. Yaron Ezrahi, *The Descent of Icarus: Science and the Transformation of Contemporary Democracy* (Cambridge, MA: Harvard University Press, 1990), esp. 29–30.

85. Porter, *Trust in Numbers*, 229.

86. Sheila Jasanoff, "The Idiom of Co-Production," in *States of Knowledge: The Co-Production of Science and Social Order*, ed. Sheila Jasanoff (London: Routledge, 2004), 2.

87. Georges Canguilhem, *On the Normal and the Pathological*, trans. Carolyn R. Fawcett, ed. Robert S. Cohen (Dordrecht: D. Reidel, 1978), 147.

Conclusion

1. Psychiatrist to Maria, August 10, 2016, a complaint about the rejection of appointment, private possession. Maria is a pseudonym. To avoid compromising the anonymity of patients, the case stories are based on several similar stories from the same period.

2. Geoffrey M. Reed et al., "Disorders Related to Sexuality and Gender Identity in the ICD-11: Revising the ICD-10 Classification Based on Current Scientific Evidence, Best Clinical Practices, and Human Rights Considerations," *World Psychiatry* 15, no. 3 (2016), https://doi.org/10.1002/wps.20354; Steven Epstein, "Cultivated Co-Production: Sexual Health, Human Rights, and the Revision of the ICD," *Social Studies of Science* 51, no. 5 (2021), https://doi.org/10.1177/03063127211014283.

3. Lovisenberg DPS to Felix, September 2021, private possession.

4. "St. meld. nr. 9 (1974–75)," 41–42, 53.

5. The NBTS to Felix's primary care physician, September 2021, private possession.

6. Socialstyrelsen, "Beslut om nationell högspecialiserad vård—viss vård vi könsdysfori" (December 1, 2020), https://www.socialstyrelsen.se/globalassets/sharepoint-dokument/dokument-webb/ovrigt/nationell-hogspecialiserad-vard-konsdysfori-beslut.pdf.

7. Geoffrey C. Bowker and Susan Leigh Star, *Sorting Things Out: Classification and Its Consequences* (Cambridge, MA: MIT Press, 1999).

8. Annemarie Jutel, "Sociology of Diagnosis: A Preliminary Review," *Sociology of Health & Illness* 32, no. 2 (2009), https://doi.org/10.1111/j.1467-9566.2008.01152.x; Anne Kveim Lie and Jeremy A. Greene, "From Ariadne's Thread to the Labyrinth Itself—Nosology and the Infrastructure of Modern Medicine," *New England Journal of Medicine* 382, no. 13 (2020), https://doi.org/10.1056/NEJMms1913140.

9. Charles E. Rosenberg, "The Tyranny of Diagnosis: Specific Entities and Individual Experience," *Milbank Quarterly* 80, no. 2 (2002), https://doi.org/10.1111/1468-0009.t01-1-00003.

10. Annemarie Mol and Marc Berg, "Differences in Medicine: An Introduction," in *Differences in Medicine: Unraveling Practices, Techniques, and Bodies*, ed. Marc Berg and Annemarie Mol (Durham: Duke University Press, 1998), 3.

11. Stefan Timmermans and Marc Berg, "Standardization in Action: Achieving Local Universality Through Medical Protocols," *Social Studies of Science* 27, no. 2 (1997), https://doi.org/10.1177/030631297027002003.

12. Gill-Peterson, *Transgender Child*, 129–43; Jacob Moses, "Medical Regret Without Remorse: A Moral History of Harm, Responsibility, and Emotion in American Surgery since 1945" (PhD thesis, Harvard University, 2020); Stryker, *Transgender History*, 168–70; Velocci, "Standards of Care." For the "legalization" of medicine more generally, see Carol Heimer, Juleigh Petty, and Rebecca Culyba, "Risk and Rules: The 'Legalization' of Medicine," in *Organizational Encounters with Risk*, ed. Bridget Hutter and Michael Power (Cambridge: Cambridge University Press, 2005).

13. Clarke et al., *Biomedicalization*.

14. Meyerowitz, *How Sex Changed*; Afsaneh Najmabadi, *Professing Selves:*

Transsexuality and Same-Sex Desire in Contemporary Iran (Durham, NC: Duke University Press, 2014).

15. Paisley Currah, *Sex Is as Sex Does: Governing Transgender Identity* (New York: NYU Press, 2022).

16. Florence Ashley, "Thinking an Ethics of Gender Exploration: Against Delaying Transition for Transgender and Gender Creative Youth," *Clinical Child Psychology and Psychiatry* 24, no. 2 (2019), https://doi.org/10.1177/1359104519836462; Florence Ashley, "Gatekeeping Hormone Replacement Therapy for Transgender Patients Is Dehumanising," *Journal of Medical Ethics* 45, no. 7 (2019), https://doi.org/10.1136/medethics-2018-105293; Kinnon R. MacKinnon et al., "Preventing Transition 'Regret': An Institutional Ethnography of Gender-Affirming Medical Care Assessment Practices in Canada," *Social Science & Medicine* 291 (2021); stef m. shuster, *Trans Medicine: The Emergence and Practice of Treating Gender* (New York: New York University Press, 2021); Sandy Stone, "The Empire Strikes Back: A Posttransexual Manifesto," in *Body Guards: The Cultural Politics of Gender Ambiguity*, ed. Julia Epstein and Kristina Straub (New York: Routledge, 1991).

17. Andrea Long Chu, "My New Vagina Won't Make Me Happy," *New York Times*, November 24, 2018.

18. Andreas Long Chu, "On Liking Women," *n+1*, no. 30 (2018).

19. Hertoft and Ritzau, *Paradiset*, 132.

20. Mol, *Logic of Care*, xi.

21. See also Mol, Moser, and Pols, *Care*; Jeannette Pols, "Enforcing Patient Rights or Improving Care? The Interference of Two Modes of Doing Good in Mental Health Care," *Sociology of Health & Illness* 25, no. 4 (2003), https://doi.org/10.1111/1467-9566.00349.

22. Cooter, "Inside the Whale." See also Nancy D. Campbell and Laura Stark, "Making Up 'Vulnerable' People: Human Subjects and the Subjective Experience of Medical Experiment," *Social History of Medicine* 28, no. 4 (2015), https://doi.org/10.1093/shm/hkv031.

23. François Delaporte, *Figures of Medicine: Blood, Face Transplants, Parasites*, ed. Stefanos Geroulanos and Todd Meyers, trans. Nils F. Schott (New York: Fordham University Press, 2013), 65.

24. See Kadji Amin, "Trans Negative Affect," in *The Routledge Companion to Gender and Affect*, ed. Todd Reeser (New York: Routledge University Press, 2022).

25. Hanna interview.

26. Sophia interview.

27. Avgi Saketopoulou, "Thinking Psychoanalytically, Thinking Better: Reflections on Transgender," *International Journal of Psychoanalysis* 101, no. 5 (2020), https://doi.org/10.1080/00207578.2020.1810884.

28. Georges Canguilhem, "Introduction: Thought and the Living," in *Knowledge of Life*, ed. Paola Marrati and Todd Meyers, trans. Stefanos Geroulanos and Daniela Ginsburg (New York: Fordham University Press, 2008), xx.

Interviews

By the author

Alv A. Dahl, September 25, 2019
Astri Lindholm, February 8 and October 31, 2019
Astrid, October 1, 2019
Bjørn, March 9, 2020
Bodil Solberg, January 20, 2020
Esben Esther Pirelli Benestad, November 2, 2020
Gunhild Grünfeld, November 16, 2020
Hanna, November 13, 2019
Ira Haraldsen, October 9, 2019
Isak, November 14, 2019
Jan Elisabeth Lindvik, January 15, 2021
Jan-Henrik Pederstad, June 17, 2019
Janni Christin Adeen-Wintherbauer, December 13, 2019
Jeanette Solstad, January 13, 2021
Jonas, November 14, 2019
Kirsti Malterud, October 24, 2019
Knut Skolleborg, January 14, 2021
Ragnhild, October 23, 2020
Sara, December 12, 2019
Sophia, September 17, 2020
Thomas Schreiner, November 12, 2019
Thore Langfeldt, January 29, 2020
Tone Maria Hansen, October 1, 2019

By Skeivt arkiv, Bergen

Aase Schibsted Knudsen, February 18, 2019
Bjørn Ellen Brudal, February 19, 2019
Daisy Hafstad, April 19, 2018
Janni Christin Adeen-Wintherbauer, June 3, 2016
Jeanette Solstad Remø, November 23, 2018
Per Halvor Kaja Mostad, October 18, 2017
Sara Claes Schmidt, February 21, 2019
Stein Wolff Frydenlund, January 17, 2018

Bibliography

Aakvaag, Asbjørn, and Jørgen H. Vogt. "Plasma Testosterone Values in Different Forms of Testosterone Treatment." *Acta Endocrinologica* 60, no. 3 (1969): 537–42. https://doi.org/10.1530/acta.0.0600537.

Aas, Ingeborg. *Hvordan kan samfundet beskytte sig mot åndssvake og sedelighetsforbrytere.* Oslo: Olaf Norlis forlag, 1932.

Abraham, Felix. "Genitalumwandlung an zwei männlichen Transvestiten." *Zeitschrift für Sexualwissenschaft und Sexualpolitik* 18 (1931): 223–26.

Ackerknecht, Erwin H. "A Plea for a 'Behaviorist' Approach in Writing the History of Medicine." *Journal of the History of Medicine and Allied Sciences* XXII, no. 3 (1967): 211–14. https://doi.org/10.1093/jhmas/XXII.3.211.

Alaattinoğlu, Daniela. *Grievance Formation, Rights and Remedies: Involuntary Sterilisation and Castration in the Nordics, 1930s–2020s.* Cambridge: Cambridge University Press, 2023.

Alm, Erika. "A State Affair? Notions of the State in Discourses on Trans Rights in Sweden." In *Pluralistic Struggles in Gender, Sexuality and Coloniality: Challenging Swedish Exceptionalism*, edited by Erika Alm, Linda Berg, Mikela Lundahl Hero, Anna Johansson, Pia Laskar, Lena Martinsson, Diana Mulinari, and Cathrin Wasshede. Cham: Palgrave Macmillan, 2021, pp. 209–37.

Alm, Erika. "'Ett emballage för inälvor och emotioner': föreställningar om kroppen i statliga utredningar från 1960- & 1970-talen." PhD thesis, University of Gothenburg, 2006.

Alm, Erika. "What Constitutes an In/Significant Organ? The Vicissitudes of Juridical and Medical Decision-Making Regarding Genital Surgery for Intersex and Trans People in Sweden." In *Body, Migration, Re/Constructive Surgeries*, edited by Gabriele Griffin and Malin Jordal. London: Routledge, 2018, pp. 225–40.

Alnæs, Randolf. "Psychoanalysis in Norway: History, Training, Treatment, Research." *Nordic Journal of Psychiatry* 48, 32 (1994): 1–103.

Amin, Kadji. "Temporality." *TSQ: Transgender Studies Quarterly* 1, nos. 1–2 (2014): 219–22. https://doi.org/10.1215/23289252-2400073.

Amin, Kadji. "Trans Negative Affect." In *The Routledge Companion to Gender and Affect*, edited by Todd Reeser. New York: Routledge, 2022, pp. 33–42.

Amland, Petter Frode, Kjell Andenæs, Marius Barstad, Gorm Bretteville, Einar Gjessing, Helge Einar Roald, Frode Samdal, Knut Skolleborg, and Kjell Aass. "Plastikkirurger og yrkesetikk." *Tidsskrift for Den norske lægeforening* 123, no. 21 (2003): 3086–87.

Anchersen, Per. "Problems of Transvestism." *Acta Psychiatrica et Neurologica Scandinavica* 106 (1957): 249–56.

Anchersen, Per. "Sammenfatning og utsyn." In *Nervøse lidelser og sinnets helse*, edited by Per Anchersen and Leo Eitinger. Oslo: H. Aschehough, 1955, pp. 225–33.

Andenæs, Johs. *Alminnelig strafferett*. Oslo: Akademisk forlag, 1956.

Andenæs, Johs. *Strafferettens alminnelige del*. Oslo: Universitetets studentkontor, 1952.

Andersson, Jenny. "Mellan tillväxt och trygghet. Idéer om produktiv socialpolitik i socialdemokratisk socialpolitisk ideologi under efterkrigstiden." PhD thesis, Uppsala University, 2003.

Anthi, Per, and Erik Stänicke. "Psykoanalysens inntog i norsk psykologi og psykiatri. En historisk beretning." *Tidsskrift for Norsk psykologforening* 56, no. 6 (2018): 446–57.

Apotekerforening, Norges. *Resepthåndboken*. Oslo: A. W. Brøggers Boktrykkeri A/S, 1959.

Asdal, Kristin, and Tone Druglitrø. "Modifying the Biopolitical Collective: The Law as a Moral Technology." Chap. 4 in *Humans, Animals and Biopolitics: The More-Than-Human Condition*, edited by Kristin Asdal, Tone Druglitrø, and Steve Hinchliffe. London: Routledge, 2017, pp. 66–84.

Asdal, Kristin, and Christoph Gradmann. "Introduction: Science, Technology, Medicine—and the State: The Science-State Nexus in Scandinavia, 1850–1980." *Science in Context* 27, no. 2 (2014): 177–86. https://doi.org/10.1017/S0269889714000039.

Asdal, Kristin, and Bård Hobæk. "The Modified Issue: Turning Around Parliaments, Politics as Usual and How to Extend Issue-Politics with a Little Help from Max Weber." *Social Studies of Science* 50, no. 2 (2020): 252–70. https://doi.org/10.1177/0306312720902847.

Ashley, Florence. "Gatekeeping Hormone Replacement Therapy for Transgender Patients Is Dehumanising." *Journal of Medical Ethics* 45, no. 7 (2019): 480–82. https://doi.org/10.1136/medethics-2018-105293.

Ashley, Florence. "Thinking an Ethics of Gender Exploration: Against Delaying Transition for Transgender and Gender Creative Youth." *Clinical Child Psychology and Psychiatry* 24, no. 2 (2019): 223–36. https://doi.org/10.1177/1359104519836462.

Aubert, Vilhelm. *The Hidden Society*. Totowa, NJ: Bedminster Press, 1965.

Baaring, Devin. "Könsgränsen. Om samproduktionen av könstillhörighetslagen (1972:119)." Master's thesis, Lund University, 2019.

Bakker, Alex. *The Dutch Approach: Fifty Years of Transgender Health Care at the CVU Amsterdam Gender Clinic*. Amsterdam: Boom, 2021.

Bakker, Alex. "In the Shadows of Society: Trans People in the Netherlands in the 1950s." In *Others of My Kind: Transatlantic Transgender Histories*, edited by Alex Bakker, Rainer Herrn, Michael Thomas Taylor, and Annette F. Timm. Calgary, AB: University of Calgary Press, 2020, pp. 133–75.

Bakker, Alex, Rainer Herrn, Michael Thomas Taylor, and Annette F. Timm, eds. *Others of My Kind: Transatlantic Transgender Histories*. Calgary, AB: University of Calgary Press, 2020.

Bashford, Alison, and Philippa Levin, eds. *The Oxford Handbook of the History of Eugenics*. New York: Oxford University Press, 2010.

Bättig, Fritz. "Beitrag zur Frage des Transvestitismus." Dissertation, Buchdrückerei Fluntern, 1952.

Bauer, Heike. *The Hirschfeld Archives: Violence, Death, and Modern Queer Culture*. Philadelphia: Temple University Press, 2017.

Bauer, Susanne. "From Administrative Infrastructure to Biomedical Resource: Danish Population Registries, the 'Scandinavian Laboratory,' and the 'Epidemiologist's Dream.'" *Science in Context* 27, no. 2 (2014): 187–213. https://doi.org/10.1017/S0269889714000040.

Bendixsen, Synnøve, Mary Bente Bringslid, and Halvard Vike, eds. *Egalitarianism in Scandinavia: Historical and Contemporary Perspectives*. Cham: Palgrave Macmillan, 2018.

Benestad, Esben Esther Pirelli, and Elsa Almås. *Kjønn i bevegelse*. Oslo: Universitetsforlaget, 2001.

Benestad, Esben, Berthold Grünfeld, and Bernt Krøvel. "Heterofil transvestisme." *Tidsskrift for Den norske lægeforening* 106, no. 25 (1986): 2069–71.

Benjamin, Harry. "Clinical Aspects of Transsexualism in the Male and Female." *American Journal of Psychotherapy* 18 (1964): 458–69. https://doi.org/10.1176/appi.psychotherapy.1964.18.3.458.

Benjamin, Harry. "For the Practicing Physician: Suggestions and Guidelines for the Management of Transsexuals." In *Transsexualism and Sex Reassignment*, edited by Richard Green and John Money. Baltimore: Johns Hopkins University Press, 1969, pp. 305–7.

Benjamin, Harry. "Transsexualism and Transvestism as Psycho-Somatic and Somato-Psychic Syndromes." *American Journal of Psychotherapy* 8, no. 2 (1954): 219–30. https://doi.org/10.1176/appi.psychotherapy.1954.8.2.219.

Benjamin, Harry. *The Transsexual Phenomenon*. New York: Julian Press, 1966.

Benjamin, Harry. "Transvestism and Transsexualism." *International Journal of Sexology* VII, no. 1 (1953): 12–14.

Benjamin, Harry. "Transvestism and Transsexualism in the Male and Female." *Journal of Sex Research* 3, no. 2 (1967): 107–27.

Berg, Annika. *De samhällsbesvärliga. Förhandlingar om psykopati och kverulans i 1930- och 1940-talens Sverige*. Gothenburg: Makadam förlag, 2018.

Berg, Marc, and Annemarie Mol, eds. *Differences in Medicine: Unraveling Practices, Techniques, and Bodies*. Durham, NC: Duke University Press, 1998.

Berg, Ole. *Spesialisering og profesjonalisering. En beretning om den sivile norske helseforvaltnings utvikling fra 1809 til 2009. Del 1: 1809–1983—Den gamle helseforvaltning*. Oslo: Statens helsetilsyn, 2009.

Berg, Siv Frøydis. *Den unge Karl Evang og utvidelsen av helsebegrepet*. Oslo: Solum Forlag, 2002.

Binder, Hans. "Das Verlangen nach Geschlechtsumwandlung." *Zeitschrift für die gesamte Neurologie und Psychiatrie* 110, no. 4/6 (1933): 84.

Bjørvik, Eira. "Conceiving Infertility: Infertility Treatment and Assisted Reproductive Technologies in 20th Century Norway." PhD thesis, University of Oslo, 2018.

Bogoras, Nikolaj A. "Über die volle plastische Wiederherstellung eines zum Koitus fähigen Penis (Peniplastica totalis)." *Zentralblatt für Chirurgie* 63, no. 22 (1936): 1271–76.

Borchgrevink, H. Henrik Chr. "Kirurgisk konstruksjon av vagina." *Tidsskrift for Den Norske Lægeforening* 93 (1973): 1042–43.

Borck, Cornelius, and Armin Schäfer, eds. *Das psychiatrische Aufschreibesystem*. Paderborn: Wilhelm Fink, 2015.

Bowker, Geoffrey C., and Susan Leigh Star. *Sorting Things Out: Classification and Its Consequences*. Cambridge, MA: MIT Press, 1999.

Brändli, Sibylle, Barbara Lüthi, and Gregor Spuhler, eds. *Zum Fall machen, zum Fall werden: Wissensproduktion und Patientenerfahrung in Medizin und Psychiatrie des 19. und 20. Jahrhunderts*. Frankfurt am Main: Campus Verlag, 2009.

Bremer, Johan. *Asexualization: A Follow-Up Study of 244 Cases*. Oslo: Oslo University Press, 1958.

Bremer, Johan. "Mutilerende behandling av transseksualisme?" *Tidsskrift for Den Norske Lægeforening* 68, no. 13–14 (1961): 921–23.

Bremer, Signe. "'Jag söker gemenskap': Svenska sextidningar som transmöjliggörande rum under 1960-talet." *Lamda Nordica* 28, no. 1 (2023): 48–75. https://doi.org/10.34041/ln.v28.867.

Briggs, Charles L., and Daniel C. Hallin. *Making Health Public: How News Coverage Is Remaking Media, Medicine, and Contemporary Life*. Oxon: Routledge, 2016.

Broberg, Gunnar, and Mattias Tydén. "Eugenics in Sweden: Efficient Care." In *Eugenics and the Welfare State: Sterilization Policy in Denmark, Sweden, Norway, and Finland*, edited by Gunnar Broberg and Nils Roll-Hansen. East Lansing: Michigan State University Press, 2005, pp. 77–149.

Brown, Daniel G., and Alexander Tolor. "Human Figure Drawings as Indicators of Sexual Identification and Inversion." *Perceptual and Motor Skills* 7 (1957): 199–211.

Bürger-Prinz, Hans, Heinrich Albrecht, and Hans Giese. "Zur Phänomenologie des Transvestitismus bei Männern." *Beiträge zur Sexualforschung* 3 (1953): 1–43.

C-print. "Eva-Lisa's Monument: Interview with Sam Hultin." C-print, August 19, 2019. https://www.c-print.se/post/eva-lisa-s-monument.

Campbell, Nancy D., and Laura Stark. "Making Up 'Vulnerable' People: Human Subjects and the Subjective Experience of Medical Experiment." *Social History of Medicine* 28, no. 4 (2015): 825–48. https://doi.org/10.1093/shm/hkv031.

Canguilhem, Georges. "Introduction: Thought and the Living." Translated by Stefanos Geroulanos and Daniela Ginsburg. In *Knowledge of Life*, edited by Paola Marrati and Todd Meyers. New York: Fordham University Press, 2008, pp. xvii–xx.

Canguilhem, Georges. *On the Normal and the Pathological.* Translated by Carolyn R. Fawcett. Edited by Robert S. Cohen. Dordrecht: D. Reidel, 1978.

Christiansen, Niels Finn, and Pirjo Markkola. "Introduction." In *The Nordic Model of Welfare—a Historical Reappraisal*, edited by Niels Finn Christiansen, Klaus Petersen, Nils Edling, and Per Haave. Copenhagen: Museum Tusculanum Press, 2006, pp. 9–29.

Chu, Andrea Long. "My New Vagina Won't Make Me Happy." *New York Times*, November 24, 2018.

Chu, Andrea Long. "On Liking Women." *n+1*, no. 30 (2018).

Clarke, Adele E., Laura Mamo, Jennifer Ruth Fosket, Jennifer R. Fishman, and Janet K. Shim, eds. *Biomedicalization: Technoscience, Health, and Illness in the US*. Durham, NC: Duke University Press, 2010.

Conrad, Peter. "The Discovery of Hyperkinesis: Notes on the Medicalization of Deviant Behavior." *Social Problems* 23, no. 1 (1975): 12–21.

Conroy, Brian. "The History of Facial Prostheses." In *Clinics in Plastic Surgery: An International Quarterly*, edited by Sharon Romm. Philadelphia: W. B. Saunders, 1983, pp. 689–707.

Cooter, Roger. "Inside the Whale: Bioethics in History and Discourse." *Social History of Medicine* 23, no. 3 (2010): 662–72. https://doi.org/10.1093/shm/hkq058.

Currah, Paisley. *Sex Is as Sex Does: Governing Transgender Identity*. New York: NYU Press, 2022.

Dahl, Inge A., and Erling Viksjø. "Oslo Helseråd." *Byggekunst* 51, no. 6 (1969): 232–33.

Das, Veena, and Deborah Poole. "State and Its Margins: Comparative Ethnographies." In *Anthropology in the Margins of the State*, edited by Veena Das and Deborah Poole. New Delhi: Oxford University Press, 2004, pp. 3–33.

Daston, Lorraine, and Peter Galison. *Objectivity*. New York: Zone Books, 2010.

Daston, Lorraine, and Katharine Park. "The Hermaphrodite and the Orders of Nature." *GLQ* 1 (1995): 419–38.

Delaporte, François. *Figures of Medicine: Blood, Face Transplants, Parasites*. Translated by Nils F. Schott. Edited by Stefanos Geroulanos and Todd Meyers. New York: Fordham University Press, 2013.

DeVun, Leah. *The Shape of Sex: Nonbinary Gender from Genesis to the Renaissance*. Columbia University Press, 2021. https://doi.org/10.7312/devu19550.

Doorbar, Ruth Rae. "Psychological Testing of Male Transsexuals: A Brief Report of Results from the Wechsler Adult Intelligence Scale, the Thematic Apperception Test, and the House-Tree-Person Test." In *Transsexualism and Sex Reassignment*, edited by Richard Green and John Money. Baltimore: Johns Hopkins University Press, 1969, pp. 189–201.

Downing, Lisa, Iain Morland, and Nikki Sullivan. *Fuckology: Critical Essays on John Money's Diagnostic Concepts*. Chicago: University of Chicago Press, 2014.

Drager, Emmett H., and Lucas Platero. "At the Margins of Time and Place: Transsexuals and the Transvestites in Trans Studies." *TSQ: Transgender Studies Quarterly* 8, no. 4 (2021). https://doi.org/10.1215/23289252-9311018.

Dreger, Alice Domurat. *Hermaphrodites and the Medical Invention of Sex*. Cambridge, MA: Harvard University Press, 1998.

Drescher, Jack, Peggy Cohen-Kettenis, and Sam Winter. "Minding the Body: Situating Gender Identity Diagnoses in the ICD-11." *International Review of Psychiatry* 24, no. 6 (2012): 568–77. https://doi.org/10.3109/09540261.2012.741575.

Drucker, Donna J. *The Classification of Sex: Alfred Kinsey and the Organization of Knowledge*. Pittsburgh: University of Pittsburgh Press, 2014.

Dukor, Benno. "Probleme um den Transvestitismus." *Schweizerische Medizinische Wochenschrift* 81, no. 22 (1951): 516–19.

Ebeling, Knut, and Stephan Günzel, eds. *Archivologie: Theorien des Archivs in Philosophie, Medien und Künsten*. Berlin: Kulturverlag Kadmos Berlin, 2009.

Eder, Sandra. *How the Clinic Made Gender: The Medical History of a Transformative Idea*. Chicago: University of Chicago Press, 2022.

Edgerton, Milton T. "The Surgical Treatment of Male Transsexuals." *Clinics in Plastic Surgery* 1, no. 2 (1974): 285–323.

Edgerton, Milton T., Norman J. Knorr, and James R. Callison. "The Surgical Treatment of Transsexual Patients: Limitations and Indications." *Plastic and Reconstructive Surgery* 45, no. 1 (1970): 38–50.

Eggen, Bernt. *Bastard! En sannferdig roman om kjønnsskifte i Norge*. Oslo: Aschehoug, 1979.

Eikvam, Turid, and Trine Brekke. "Transvestittisme—en livsberikelse . . . !?" *Løvetann* (1988), 22–26.

Ekins, Richard, and Dave King. "Virginia Prince: Transgender Pioneer." *International Journal of Transgenderism* 8, no. 4 (2005): 5–15. https://doi.org/10.1300/J485v08n04_02.

Elbe, Lili. *Fra Mand til Kvinde: Lili Elbes Bekendelser*, a compilation of autobiographic material by Niels Hoyer. Copenhagen: Hage & Clausens Forlag, 1931.

Elbe, Lili. *Man into Woman: The First Sex Change. A Portrait of Lili Elbe*. Edited by Niels Hoyer. London: Blue Boat Books, 2004 [1933].

Ellingsæter, Anne Lise, Aksel Hatland, Per Haave, and Steinar Stjernø. *Den nye velferdsstatens historie. Ekspansjon og omdanning etter 1966*. Oslo: Gyldendal Akademisk, 2020.

Ellis, Havelock. "Eonism and Other Supplementary Studies." In *Studies in the Psychology of Sex*. Philadelphia: F. A. Davis, 1928.

Ellis, Havelock. "Sexo-ästhetische Inversion." *Zeitschrift für Psychotherapie und medizinische Psychologie* V (1914): 134–62.

Endocrine Society. "First International Congress of Endocrinology." *Journal of Clinical Endocrinology & Metabolism* 20, no. 2 (1960): 334–35.

Engh, Sunniva. "The Complexities of Postcolonial International Health: Karl Evang in India 1953." *Medical History* 67, no. 1 (2023): 23–41. https://doi.org/10.1017/mdh.2023.12.

Epstein, Julia. "Historiography, Diagnosis, and Poetics." *Literature and Medicine* 11, no. 1 (1992): 23–44.

Epstein, Steven. "Cultivated Co-Production: Sexual Health, Human Rights, and the Revision of the ICD." *Social Studies of Science* 51, no. 5 (2021): 657–82. https://doi.org/10.1177/03063127211014283.

Epstein, Steven. *Impure Science: AIDS, Activism, and the Politics of Knowledge*. Berkeley: University of California Press, 1996.

Epstein, Steven. *The Quest for Sexual Health: How an Elusive Ideal Has Transformed Science, Politics, and Everyday Life*. Chicago: University of Chicago Press, 2022.

Eskeland, Gunnar. "Plastikkirurgi." In *Det norske medicinske selskab 150 år*. Oslo: Det norske medicinske selskab, 1983, pp. 232–37.

Etzemüller, Thomas. "Rationalizing the Individual—Engineering Society: The Case of Sweden." In *Engineering Society: The Role of the Human and Social Sciences in Modern Societies, 1880–1980*, edited by Kerstin Brückweh, Dirk Schumann, Richard F. Wetzell, and Benjamin Ziemann. Houndmills: Palgrave Macmillan, 2012, pp. 97–118.

Ezrahi, Yaron. *The Descent of Icarus: Science and the Transformation of Contemporary Democracy*. Cambridge, MA: Harvard University Press, 1990.

Farge, Arlette. *The Allure of the Archives*. Translated by Thomas Scott-Railton. New Haven: Yale University Press, 2013 [1989].

Fassin, Didier. "Introduction: Governing Precarity." Translated by Patrick Brown and Didier Fassin. In *At the Heart of the State: The Moral World of Institutions*, edited by Didier Fassin, Yasmine Bouagga, Isabelle Coutant, Jean-Sébastien Eideliman, Fabrice Fernandez, Nicolas Fischer, Carolina Kobelinsky, et al. London: Pluto Press, 2015, pp. 1–14.

Ferguson, Roderick A. *One-Dimensional Queer*. Cambridge: Polity Press, 2019.

Fleck, Ludwik. *Entstehung und Entwicklung einer wissenschaftlichen Tatsache*. Frankfurt am Main: Suhrkamp, 1980 [1935].

Fleming, Michael, Gerald Koocher, and Judith Nathans. "Draw-A-Person Test: Implications for Gender Identification." *Archives of Sexual Behavior* 8, no. 1 (1979): 55–61.

Forrester, John. "If *p*, Then What? Thinking in Cases." *History of the Human Sciences* 9, no. 3 (1996): 1–25. https://doi.org/10.1177/095269519600900301.

Forrester, John. *Thinking in Cases*. Cambridge: Polity Press, 2017.

Foucault, Michel. *Folie et déraison. Histoire de la folie à l'âge classique*. Paris: Plon, 1961.

Furseth, Jan, and Olav Ljones. "50-årsjubilant med behov for oppgradering." *Samfunnsspeilet* 1 (2015): 23–28. https://www.ssb.no/befolkning/artikler-og-publikasjoner/_attachment/224332?_ts=14cb7c21460.

Galison, Peter. "Image of Self." In *Things That Talk: Object Lessons from Art and Science*, edited by Lorraine Daston. New York: Zone Books, 2004, pp. 257–94.

Geissler, Paul Wenzel, and Guillaume Lachenal. "Brief Instructions for Archaeologists of African Futures." In *Traces of the Future: An Archaeology of Medical Science in Twenty-First-Century Africa*, edited by Paul Wenzel Geissler, Guillaume Lachenal, John Manton, and Noémi Tousignant. Bristol: Intellect, 2016, pp. 14–30.

Geisthövel, Alexa, and Volker Hess. "Handelndes Wissen: Die Praxis des Gutachtens." In *Medizinisches Gutachten: Geschichte einer neuzeitlichen Praxis*, edited by Alexa Geisthövel and Volker Hess. Göttingen: Wallstein Verlag, 2017, pp. 9–39.

Germon, Jennifer. *Gender: A Genealogy of an Idea*. New York: Palgrave Macmillan, 2009.

Gill-Peterson, J. *Histories of the Transgender Child*. Minneapolis: University of Minnesota Press, 2018.

Gill-Peterson, Jules. *A Short History of Trans Misogyny*. London: Verso Books, 2024.

Gilman, Sander L. *Creating Beauty to Cure the Soul: Race and Psychology in the Shaping of Aesthetic Surgery*. Durham: Duke University Press, 1998.

Ginzburg, Carlo. "Signes, traces, pistes." *Le Débat* 6, no. 6 (1980): 3–44.

Goffman, Erving. *Asylums: Essays on the Social Situation of Mental Patients and Other Inmates*. Garden City, NY: Anchor Books, 1961.

Golan, Tal. *Laws of Men and Laws of Nature*. Cambridge, MA: Harvard University Press, 2004.

González-Ulloa, Mario, ed. *The Creation of Aesthetic Plastic Surgery*. New York: Springer-Verlag, 1976.

Goodenough, Florence L. *Measurement of Intelligence by Drawings*. Yonkers-on-Hudson: World Book, 1926.

Graugaard, Christian. "Professor Sands høns—om sexualbiologi i mellemkrigstidens Danmark." PhD thesis, University of Copenhagen, 1997.

Greene, Jeremy A. "Knowledge in Medias Res: Toward a Media History of Science, Medicine, and Technology." *History and Theory* 59, no. 4 (2020): 48–66. https://doi.org/10.1111/hith.12181.

Greene, Jeremy A., Flurin Condrau, and Elizabeth Siegel Watkins. "Medicine Made Modern by Medicines." In *Therapeutic Revolutions: Pharmaceuticals and Social Change in the Twentieth Century*, edited by Jeremy A. Greene, Flurin

Condrau, and Elizabeth Siegel Watkins. Chicago: University of Chicago Press, 2016, pp. 1–17.

Grøn, Fredrik. *Det norske medicinske selskab, 1833–1933*. Oslo: Steenske Boktrykkeri Johannes Bjørnstad, 1933.

Grünfeld, Berthold. "Homofili og transseksualitet—psykiatriske og sosialmedisinske betraktninger." In *NOU 1979: 46 Strafferettslig vern for homofile*. Oslo: Justis og politidepartemenet, 1979, pp. 68–79.

Grünfeld, Berthold. "Seksualitet som helseproblem." In *Samfunnsmedisin i praksis. Oslo Helseråd i 80-årene*, edited by Harald Siem, Kårc Berg, and Berthold Grünfeld. Oslo: Universitetsforlaget, 1987, pp. 200–206.

Grünfeld, Berthold. *Vårt seksuelle liv*. Oslo: Gyldendal, 1979.

Grünfeld, Berthold, Esben Benestad, and Bernt Krøvel. "Transvestisme og transvestitter." *Nordisk Sexologi* 4 (1986): 195–207.

Gullestad, Marianne. *Det norske sett med nye øyne. Kristisk analyse av norsk innvandringsdebatt*. Oslo: Universitetsforlaget, 2002.

Gullestad, Marianne. "The Scandinavian Version of Egalitarian Individualism." *Ethnologia Scandinavica: A Journal for Nordic Ethnology* 21 (1991): 3–17.

Gustainis, Emily R. Novak, and Phoebe Evans Letocha. "The Practice of Privacy." In *Innovation, Collaboration, and Models: Proceedings of the CLIR Cataloging Hidden Special Collections and Archives*, edited by Cheryl Oestreicher. Washington: Council on Library and Information Resources, 2015.

Gutheil, Emil. "Analysis of a Case of Transvestism." In *Sexual Aberrations: The Phenomena of Fetishism in Relations to Sex*, edited by Wilhelm Stekel. New York: Horace Liveright, 1930, pp. 281–318.

Haave, Per. *I medisinens sentrum—Den norske legeforening og spesialistregimet gjennom hundre år*. Oslo: Unipub, 2011.

Haave, Per. *Sterilisering av tatere 1934–1977: en historisk undersøkelse av lov og praksis*. Oslo: Norges forskningsråd, 2000.

Haave, Per. "Striden om nevrosene: Psykiatriens møte med psykoanalysen." In *Den mangfoldige velferden*, edited by Edgeir Benum, Per Haave, Hilde Ibsen, Aina Schiøtz, and Ellen Schrumpf. Oslo: Gyldendal, 2003, pp. 193–212.

Haave, Per. "Velferdskommunen i støpeskjeen." Chap. 16 in *Den nye velferdsstatens historie. Ekspansjon og omdanning etter 1966*, edited by Anne Lise Ellingsæter, Aksel Hatland, Per Haave, and Steinar Stjernø. Oslo: Gyldendal akademisk, 2020.

Hacking, Ian. *Historical Ontology*. Cambridge, MA: Harvard University Press, 2002.

Hacking, Ian. "Kinds of People: Moving Targets: British Academy Lecture." *Proceedings of the British Academy* 151 (2007): 285–318. https://doi.org/10.5871/bacad/9780197264249.003.0010.

Hacking, Ian. "The Looping Effect of Human Kinds." In *Causal Cognition: A Multi-Disciplinary Debate*, edited by D. Sperber, D. Premack and A. J. Premack. New York: Oxford University Press, 1995, pp. 351–83.

Haiken, Elizabeth. *Venus Envy: A History of Cosmetic Surgery*. Baltimore: Johns Hopkins University Press, 1997.

Hamburger, Christian. "The Desire for Change of Sex as Shown by Personal Letters from 465 Men and Women." *Acta Endocrinologica* 14, no. 4 (1953): 361–75. https://doi.org/10.1530/acta.0.0140361.

Hamburger, Christian. "Endocrine Treatment of Male and Female Transsexualism." In *Transsexualism and Sex Reassignment*, edited by Richard Green and John Money. Baltimore: Johns Hopkins University Press, 1969, pp. 291–307.

Hamburger, Christian. "Intersexualität." In *Mensch, Geschlecht, Gesellschaft: Das Geschlechtsleben unserer Zeit gemeinverständlich dargestellt*, edited by Hans Giese and A. Willy. Paris: Guillaume Aldor; 1954, pp. 816–25.

Hamburger, Christian, and K. Pedersen-Bjergaard. "Om testosteron-suppositorier med særligt henblik på suppositorie-massen." *Archiv for pharmaci og chemi* 68 (1961): 146–51.

Hamburger, Christian, and Mogens Sprechler. "The Influence of Steroid Hormones on the Hormonal Activity of the Adenohypophysis in Man." *Acta Endocrinologica* 7, 1–4 (1951): 167–95. https://doi.org/10.1530/acta.0.0070167.

Hamburger, Christian, Georg K. Stürup, and E. Dahl-Iversen. "Transvestisme. Hormonal, psykiatrisk og kirurgisk behandling af et tilfælde." *Nordisk Medicin* 12, no. VI (1953): 844–48.

Hamburger, Christian, Georg K. Stürup, and Erling Dahl-Iversen. "Transvestism: Hormonal, Psychiatric, and Surgical Treatment." *Journal of the American Medical Association* 152, no. 5 (1953): 391–96. https://doi.org/10.1001/jama.1953.0369 0050015006.

Hamran, Olav. *Riktig medisin? En historie om apotekvesenet*. Oslo: Pax forlag, 2010.

Handerer, Fritz, Peter Kinderman, Carsten Timmermann, and Sara J. Tai. "How Did Mental Health Become So Biomedical? The Progressive Erosion of Social Determinants in Historical Psychiatric Admission Registers." *History of Psychiatry* 32, no. 1 (2020): 37–51. https://doi.org/10.1177/0957154X20968522.

Hartmann, Ellen, and Geir Nielsen. *Kompendium 1 i Rorschach-metodikk*. Bergen: Universitetet i Bergen, 1976.

Haug, Egil. *Hormonlaboratoriet—50 år i vekst og utvikling*. Oslo: Oslo universitetssykehus HF, Aker, 2009.

Haugen, Karin. "Forandrer livet for mange." *VG* (1974), 16–17.

Hausman, Bernice L. *Changing Sex: Transsexualism, Technology, and the Idea of Gender*. Durham, NC: Duke University Press, 1995.

Heaney, Emma. *The New Woman: Literary Modernism, Queer Theory, and the Trans Feminine Allegory*. Evanston: Northwestern University Press, 2017.

Heimer, Carol, Juleigh Petty, and Rebecca Culyba. "Risk and Rules: The 'Legalization' of Medicine." In *Organizational Encounters with Risk*, edited by Bridget Hutter and Michael Power. Cambridge: Cambridge University Press, 2005, pp. 92–131.

Helsedirektoratet. *Kjønnsinkongruens—Nasjonal faglig retningslinje*. Oslo: Helsedirektoratet, 2020.

Helsedirektoratet. *Styring av høyspesialiserte somatiske funksjoner i helsetjenesten*. Oslo: Helsedirektoratet, 1992.

Henckes, Nicolas. "Magic Bullet in the Head? Psychiatric Revolutions and Their Aftermath." In *Therapeutic Revolutions: Pharmaceuticals and Social Change in the Twentieth Century*, edited by Jeremy A. Greene, Flurin Condrau, and Elizabeth Siegel Watkins. Chicago: University of Chicago Press, 2016, pp. 65–96.

Hernes, Helga Maria. *Staten—kvinner ingen adgang?* Oslo: Universitetsforlaget, 1982.

Hernes, Helga Maria. *Welfare State and Woman Power: Essays in State Feminism*. Oslo: Norwegian University Press, 1987.

Herrn, Rainer. *Der Liebe und dem Leid. Das Institut für Sexualwissenschaft 1919–1933*. Berlin: Suhrkamp, 2022.

Herrn, Rainer. "Die falsche Hofdame vor Gericht: Transvestitismus in Psychiatrie und Sexualwissenschaftoder die Regulierung der öffentlichen Kleiderordnung." *Medizinhistorisches Journal* 49, no. 3 (2014): 199–236.

Herrn, Rainer. "Geschlecht als Option: Selbstversuche und medizinische Experimente zur Geschlechtsumwandlung im frühen 20. Jahrhundert." In *Sexualität als Experiment: Identität, Lust und Reproduktion zwischen Science und Fiction*, edited by Nicolas Pethes and Silke Schicktanz. Frankfurt: Campus Verlag, 2008, pp. 45–70.

Herrn, Rainer. *Schnittmuster des Geschlechts: Transvestismus und Transsexualität in der frühen Sexualwissenschaft*. Gießen: Psychosozial-Verlag, 2005.

Hertoft, Preben. *Klinisk sexologi. En introduktion*. Copenhagen: Munksgaard, 1976.

Hertoft, Preben. "Lægen må overvinde egne fordomme og tabuer for at løse patienters sexuelle problemer." *Nordisk Medicin* 94, no. 4 (1979): 100–104.

Hertoft, Preben. "Tilblivelsen af Sexologisk Klinik. De første år." In *25 år med Sexologisk Klinik*, edited by Ellids Kristensen and Annamaria Giraldi. Copenhagen: Psykiatrisk Center København, 2011, pp. 18–23.

Hertoft, Preben. *Undersøgelser over unge mænds seksuelle adfærd, viden og holdning. Bind 1-3*. Copenhagen: Akademisk forlag, 1968.

Hertoft, Preben. *Undren og befrielse. Erindringer*. Copenhagen: Hans Reitzels Forlag, 2001.

Hertoft, Preben, and Teit Ritzau. *Paradiset er ikke til salg. Trangen til at være begge køn*. Viborg: Lindhardt og Ringhof, 1984.

Hertoft, Preben, and Thorkil Sørensen. "Transsexuality: Some Remarks Based on Clinical Experience." *Ciba Foundation Symposium*, no. 62 (1978): 165–81. https://doi.org/10.1002/9780470720448.ch9.

Hertz, J., K. G. Tillinger, and A. Westman. "Transvestitism. Report on Five Hormonally and Surgically Treated Cases." *Acta Psychiatr Scand* 37, no. 4 (1961): 283–94. https://doi.org/10.1111/j.1600-0447.1961.tb07363.x.

Hess, Volker. "Formalisierte Beobachtung. Die Genese der modernen Krankenakte am Beispiel der Berliner und Pariser Medizin (1725–1830)." *Medizinhistorisches Journal* 45, no. 3/4 (2010): 293–340.

Hess, Volker. "Observatio und Casus. Status und Funktion der medizinischen Fallgeschichte." In *Fall, Fallgeschichte, Fallstudie: Theorie und Geschichte einer Wissensform*, edited by Susanne Düwell and Nicolas Pethes, 34–59. Frankfurt am Main: Campus Verlag, 2014.

Hess, Volker. "A Paper Machine of Clinical Research in the Early Twentieth Century." *Isis* 109, no. 3 (2018): 473–93. https://doi.org/10.1086/699619.

Hess, Volker, and J. Andrew Mendelsohn. "Case and Series: Medical Knowledge and Paper Technology, 1600–1900." *History of Science* 48, no. 3–4 (2010): 287–314. https://doi.org/10.1177/007327531004800302.

Hess, Volker, and J. Andrew Mendelsohn. "Fallgeschichte, Historia, Klassifikation." *N.T.M* 21 (2013): 61–92. https://doi.org/10.1007/s00048-013-0086-0.

Hester, T. Roderick, H. Louis Hill, and M. J. Jurkiewicz. "One-Stage Reconstruction of the Penis." *British Journal of Plastic Surgery* 31, no. 4 (1978): 279–85. https://doi.org/10.1016/S0007-1226(78)90110-8.

Hilson, Mary. *The Nordic Model: Scandinavia Since 1945*. London: Reaktion Books, 2008.

Hirdman, Yvonne. *Att lägga livet til rätta. Studier i svensk folkhemspolitik*. Stockholm: Carlssons, 2018 [1989].

Hirschauer, Stefan. *Die soziale Konstruktion der Transsexualität: Über die Medizin und den Geschlechtswechsel*. Frankfurt: Suhrkamp, 1993.

Hirschfeld, Magnus. "Der Transvestitismus." In *Sexualpathologie*, 139–78. Bonn: A. Marcus & E. Webers Verlag, 1918.

Hirschfeld, Magnus. "Die intersexuelle Konstitution." In *Jahrbuch für sexuelle Zwischenstufen*, 3–27. Stuttgart: Julius Püttmann, 1923.

Hirschfeld, Magnus. *Die Transvestiten: Eine Untersuchung über den erotischen Verkleidungstrieb*. Berlin: Alfred Pulvermacher, 1910.

Hirschfeld, Magnus. *Was soll das Volk vom dritten Geschlecht wissen! Eine Aufklärungsschrift*. Leipzig: Verlag von Max Spohr, 1901.

Hobæk, Bård, and Anne Kveim Lie. "Less Is More: Norwegian Drug Regulation, Antibiotic Policy, and the 'Need Clause.'" *Milbank Quarterly* 97, no. 3 (2019): 762–95. https://doi.org/10.1111/1468-0009.12405.

Hoenig, J., J. Kenna, and Ann Youd. "A Follow-Up Study of Transsexualists: Social and Economic Aspects." *Psychopathology* 3, no. 2 (1970): 85–100. https://doi.org/10.1159/000278595.

Hoenig, J., J. C. Kenna, and A. Youd. "Surgical Treatment for Transsexualism." *Acta Psychiatrica Scandinavica* 47, no. 1 (1971): 106–33.

Holm, M., and Morten Hillgaard Bülow. *Det stof, mænd er gjort af—Konstruktionen af maskulinitetsbegreber i forskningsprojekter om testosteron i Danmark fra 1910'erne til 1980'erne*. Vol. 10. Copenhagen: Varia, 2013. https://koensforskning.ku.dk/nyeudgivelser/varia/Det_stof_m_nd_er_gjort_af_-_Varia_2013.pdf.

Holm, Marie-Louise (now Sølve M. Holm). "Fleshing Out the Self: Reimagining Intersexed and Trans Embodied Lives Through (Auto)Biographical Accounts of the Past." PhD thesis, Linköping University, 2017.

Honkasalo, Julian. "In the Shadow of Eugenics: Transgender Sterilisation Legislation and the Struggle for Self-Determination." In *The Emergence of Trans: Culture, Politics and Everyday Lives*, edited by Ruth Pearce, Igi Moon, Kat Gupta, and Deborah Lynn Steinberg. Oxon: Routledge, 2020, pp. 17–33.

Honkasalo, Julian. "Transfeminine Letter Clubs, Community Care and the Radical Politics of the Erotic." *European Journal of Women's Studies* 30, no. 2 (2023): 274–89. https://doi.org/10.1177/13505068231164215.

Hoopes, John E. "Operative Treatment of the Female Transsexual." In *Transsexualism and Sex Reassignment*, edited by Richard Green and John Money. Baltimore: Johns Hopkins University Press, 1969, pp. 335–52.

Hoopes, John E. "Surgical Construction of the Male External Genitalia." *Clinics in Plastic Surgery* 1, no. 2 (1974): 325–33.

Hore, B. D., F. V. Nicolle, and J. S. Calnan. "Male Transsexualism in England: Sixteen Cases with Surgical Intervention." *Archives of Sexual Behavior* 4, no. 1 (1975): 81–88. https://doi.org/10.1007/bf01541889.

Horstmann, Anja, and Vanina Kopp, eds. *Archiv—Macht—Wissen. Organisation und Konstruktion von Wissen und Wirklichkeit in Archiven.* Frankfurt a.M.: Campus-Verlag, 2010.

Hove, Leiv M. *Fra håndkirurgiens historie*. Bergen: Molvik Grafisk, 2004.

Høybråten, Dagfinn. *Drivkraft*. Oslo: Cappelen Damm, 2012.

Hultin, Sam. "Resistance and Care." *Kunstkritikk: Nordic Art Review* (May 26, 2021). https://kunstkritikk.com/resistance-and-care/.

Hunt, D. Daniel, and John L. Hampson. "Transsexualism: A Standardized Psychosocial Rating Format for the Evaluation of Results of Sex Reassignment Surgery." *Arch Sex Behav* 9, no. 3 (June 1980): 255–63. https://doi.org/10.1007/bf01542251.

Jasanoff, Sheila. *Designs on Nature: Science and Democracy in Europe and the United States*. Princeton, NJ: Princeton University Press, 2005.

Jasanoff, Sheila. "The Idiom of Co-Production." In *States of Knowledge: The Co-Production of Science and Social Order*, edited by Sheila Jasanoff. London: Routledge, 2004, pp. 1–12.

Jasanoff, Sheila. "Science, Common Sense & Judicial Power in U.S. Courts." *Daedalus* 147, no. 4 (2018): 15–27. https://doi.org/10.1162/daed_a_00517.

Johansson, Gunn, ed. *Historien om svensk psykologisk forskning. Utvecklingen från perception och psykofysik*. Stockholm: Kungl. vitterhets historie och antikvitets akademien, 2020.

Jones, Howard W. "Operative Treatment of the Male Transsexual." In *Transsexualism and Sex Reassignment*, edited by Richard Green and John Money. Baltimore: Johns Hopkins University Press, 1969, pp. 313–17.

Jordåen, Runar. "Inversjon og perversjon: Homoseksualitet i norsk psykiatri og psykologi frå slutten av 1800-talet til 1960." PhD thesis, University of Bergen, 2010.

Jutel, Annemarie. "Sociology of Diagnosis: A Preliminary Review." *Sociology of Health & Illness* 32, no. 2 (2009): 278–99. https://doi.org/10.1111/j.1467-9566.2008.01152.x.

Keating, Peter, and Alberto Cambrosio. *Biomedical Platforms: Realigning the Normal and the Pathological in Late-Twentieth-Century Medicine*. Cambridge, MA: MIT Press, 2003.

Killingmo, Bjørn. *Rorschachmetode og psykoterapi. En egopsykologisk studie*. Oslo: Universitetsforlaget, 1980.

Kirk, Stuart A., and Herb Kutchins. *The Selling of DSM: The Rhetoric of Science in Psychiatry*. New York: Aldine de Gruyter, 1992.

Kjønstad, Asbjørn. *Kompendium i helserett*. Oslo: Gyldendal, 2003.

Kline, Wendy. *Bodies of Knowledge: Sexuality, Reproduction, and Women's Health in the Second Wave*. Chicago: University of Chicago Press, 2010.

Klöppel, Ulrike. *XXoXY ungelöst: Hermaphroditismus, Sex und Gender in der deutschen Medizin. Eine historische Studie zu Intersexualität*. Bielefeld: transcript Verlag, 2010.

Koch, Lene. "Eugenic Sterilisation in Scandinavia." *European Legacy* 11, no. 3 (2006): 299–309. https://doi.org/10.1080/10848770600668340.

Koch, Lene. *Racehygiejne i Danmark 1920–56*. Copenhagen: Informations Forlag, 2010.

Kolle, Frederick Strange. *Plastic and Cosmetic Surgery*. New York: D. Appleton, 1911.

Krafft-Ebing, Richard von. *Psychopathia Sexualis. Eine klinisch-forensiche Studie*. Stuttgart: Verlag von Ferdinand Enke, 1886.

Kristensen, Ellids. "Sexologisk Klinik 1989–2011." In *25 år med Sexologisk Klinik*, edited by Ellids Kristensen and Annamaria Giraldi. Copenhagen: Psykiatrisk Center København, 2011, pp. 24–37.

Kristiansen, Hans W. *Masker og motstand: diskré homoliv i Norge 1920–1970*. Oslo: Unipub, 2008.

Kvaale, Reidun. *Kvinner i norsk presse gjennom 150 år*. Oslo: Gyldendal, 1986.

Laing, Ronald David. *The Divided Self*. New York: Pantheon Books, 1960.

Laqueur, Thomas. *Making Sex: Body and Gender from the Greeks to Freud*. Cambridge: MA: Harvard University Press, 1990.

Larsen, Jakob Bjerg, Jeanine K. Mount, Poul R. Kruse, and Karsten Vrangbäk. "Dynamics of Pharmacy Regulation in Denmark, 1932–1994: A Study of Profession-State Relations." *Pharmacy in History* 46, no. 2 (2004): 43–61.

Latham, J. R. "(Re)making Sex: A Praxiography of the Gender Clinic." *Feminist Theory* 18, no. 2 (2017): 177–204. https://doi.org/10.1177/1464700117700051.

Latour, Bruno. "From Realpolitik to Dingpolitik: How to Make Things Public. An Introduction." In *Making Things Public: Atmospheres of Democracy*, edited by Bruno Latour and Peter Weibel. Cambridge, MA: MIT Press, 2005, pp. 14–41.

Latour, Bruno. *We Have Never Been Modern*. Cambridge, MA: Harvard University Press, 1993.

Latour, Bruno, and Steve Woolgar. *Laboratory Life: The Construction of Scientific Facts*. Princeton, NJ: Princeton University Press, 1986.

Law, John. "Care and Killing: Tensions in Veterinary Practice." In *Care in Practice: On Tinkering in Clinics, Homes and Farms*, edited by Annemarie Mol, Ingunn Moser, and Jeannette Pols. Bielefeld: transcript Verlag, 2010, pp. 57–71.

Lawrence, Susan C. "Access Anxiety: HIPAA and Historical Research." *Journal of the History of Medicine and Allied Sciences* 62, no. 4 (2007): 422–60.

Le Maire, Louis. "Danish Experiences Regarding the Castration of Sexual Offenders." *Journal of Criminal Law, Criminology, and Police Science* 47, no. 3 (1956): 294–310. https://doi.org/10.2307/1140320.

Lee, Sing. "Diagnosis Postponed: Shenjing Shuairuo and the Transformation of Psychiatry in Post-Mao China." *Culture, Medicine and Psychiatry* 23 (1999): 349–80. https://doi.org/10.1023/a:1005586301895.

Lemov, Rebecca. *Database of Dreams: The Lost Quest to Catalog Humanity*. New Haven, CT: Yale University Press, 2015.

Lemov, Rebecca. "X-rays of Inner Worlds: The Mid-Twentieth-Century American Projective Test Movement." *Journal of the History of the Behavioral Sciences* 47, no. 3 (2011): 251–78. https://doi.org/10.1002/jhbs.20510.

Lie, Anne Kveim. "Blodprøver og piller er også omsorg." *Tidsskrift for Den norske legeforening* 133 (2013): 988–89. https://doi.org/10.4045/tidsskr.12.1389.

Lie, Anne Kveim. "Producing Standards, Producing the Nordic Region: Antibiotic Susceptibility Testing, from 1950–1970." *Science in Context* 27, no. 2 (2014): 215–48. https://doi.org/10.1017/S0269889714000052.

Lie, Anne Kveim, and Jeremy A. Greene. "From Ariadne's Thread to the Labyrinth Itself—Nosology and the Infrastructure of Modern Medicine." *New England Journal of Medicine* 382, no. 13 (2020): 1273–77. https://doi.org/10.1056/NEJMms1913140.

Lie, Anne Kveim, and Per Haave. "Social Medicine in Social Democracy." In *Medicine on a Larger Scale: Global Histories of Social Medicine*, edited by Anne Kveim Lie, Jeremy Greene, and Warwick Anderson. Cambridge: Cambridge University Press, 2025.

Liebeknecht, Moritz. *Wissen über Sex. Die Deutsche Gesellschaft für Sexualforschung im Spannungsfeld westdeutscher Wandlungsprozesse*. Göttingen: Wallstein Verlag, 2020.

Lindemalm, Gunnar, Dag Körlin, and Nils Uddenberg. "Long-Term Follow-Up of 'Sex Change' in 13 Male-to-Female Transsexuals." *Archives of Sexual Behavior* 15, no. 3 (1986): 187–210.

Löwy, Ilana. *Between Bench and Bedside: Science, Healing and Interleukine-2 in a Cancer Ward*. Cambridge, MA: Harvard University Press, 1996.

Luft, Rolf, Jan Wålinder, Bengt Hult, Dan Sundberg, and Gunnar Fahlberg. "Transsexualism." *Läkartidningen* 74, no. 44 (1977): 3857–60.

Lundberg, Aleksa. "Eva-Lisa Bengtson—Sveriges första transaktivist." Sweden: Sveriges radio, 2020. https://sverigesradio.se/avsnitt/1482205.

Lundström, Bengt, Ira Pauly, and Jan Wålinder. "Outcome of Sex Reassignment Surgery." *Acta Psychiatrica Scandinavica* 70 (1984): 289–94.

Lundström, Bengt, and Jan Wålinder. "Evaluation of Candidates for Sex Reassignment." *Nordisk Psykiatrisk Tidsskrift* 39, no. 3 (1985): 225–28. https://doi.org/10.3109/08039488509106160.

Machover, Karen. *Personality Projection in the Drawing of the Human Figure*. Springfield: Charles C. Thomas, 1949.

MacKinnon, Kinnon R., Florence Ashley, Hannah Kia, June S. H. Lam, Yonah Krakowsky, and Lori E. Ross. "Preventing Transition 'Regret': An Institutional Ethnography of Gender-Affirming Medical Care Assessment Practices in Canada." *Social Science & Medicine* 291 (2021): art. 114477.

Mak, Geertje. *Doubting Sex: Inscriptions, Bodies and Selves in Nineteenth-Century Hermaphrodite Case Histories*. Manchester: Manchester University Press, 2012.

Malterud, Kirsti. "Sexologi er en politisk vitenskap." *Kontrast*, no. 5–6 (1980): 26–31.

Malterud, Kirsti, and Bodil Solberg. "Kliniske erfaringer med kjønnsskifteklienter." *Tidsskrift for Den norske lægeforening* 103, no. 1 (1983): 5–6.

Maltz, Maxwell. *Evolution of Plastic Surgery*. New York: Froben Press, 1946.

Marcuse, Max. "Ein Fall von Geschlechtsumwandlungstrieb." *Zeitschrift für Psychotherapie und medizinische Psychologie* 6 (1914): 176–92.

Marhoefer, Laurie. *Racism and the Making of Gay Rights: A Sexologist, His Student, and the Empire of Queer Love*. Toronto: University of Toronto Press, 2022.

Mayes, Rick. "The Origins, Development, and Passage of Medicare's Revolutionary Prospective Payment System." *Journal of the History of Medicine and Allied Sciences* 62, no. 1 (2007): 21–55. https://doi.org/10.1093/jhmas/jrj038.

McIndoe, Archibald. "The Treatment of Congenital Absence and Obliterative Conditions of the Vagina." *British Journal of Plastic Surgery* 2, no. 4 (1950): 254–67.

McIndoe, Archibald H., and J. Bright Banister. "An Operation for the Cure of Congenital Absence of the Vagina." *Journal of Obstetrics and Gynaecology of the British Empire* 45, no. 3 (1938): 490–94. https://doi.org/10.1111/j.1471-0528.1938.tb11141.x.

McKay, Richard Andrew. *Patient Zero and the Making of the AIDS Epidemic*. Chicago: University of Chicago Press, 2017.

Melby, Kari. "Husmortid. 1900–1950." In *Med kjønnsperspektiv på norsk historie: fra viktingtid til 2000-årsskiftet*, edited by Ida Blom and Søvi Sogner. Oslo: Cappelen akademisk forlag, 2005, pp. 255–331.

Mellbye, Fredrik. "Embetet som stadsfysikus i Oslo." In *Samfunnsmedisin i praksis. Oslo Helseråd i 80-årene*, edited by Harald Siem, Kåre Berg, and Berthold Grünfeld. Oslo: Universitetsforlaget, 1987, pp. 22–27.

Mendelsohn, J. Andrew. "Empiricism in the Library: Medicine's Case Histories." Chap. III in *Science in the Archives: Pasts, Presents, Futures*, edited by Lorraine Daston. Chicago: University of Chicago Press, 2017, pp. 85–109.

Mendelsohn, J. Andrew. "Public Practice: The European *Longue Durée* of Knowing for Health and Polity." In *Civic Medicine: Physician, Polity, and Pen in Early Modern Europe*, edited by John Andrew Mendelsohn, Annemarie Kinzelbach, and Ruth Schilling. London: Routledge, 2020, pp. 7–64.

Mendelsohn, J. Andrew, Annemarie Kinzelbach, and Ruth Schilling, eds. *Civic Medicine: Physician, Polity, and Pen in Early Modern Europe*. London: Routledge, 2020.

Mesics, Sandra. "When Building a Better Vulva, Timing Is Everything: A Personal Experience with the Evolution of MTF Genital Surgery." *TSQ: Transgender Studies Quarterly* 5, no. 2 (2018): 245–50. https://doi.org/10.1215/23289252-4348672.

Meyer, Jon K., and Donna J. Reter. "Sex Reassignment: Follow-Up." *Archives of General Psychiatry* 36, no. 9 (1979): 1010–15. https://doi.org/10.1001/archpsyc.1979.01780090096010.

Meyer, Sabine. *"Wie Lili zu einem richtingen Mädchen wurde." Lili Elbe: Zur Konstruktion von Geschlecht und Identität zwischen Medialisierung, Regulierung und Subjektivierung*. Bielefeld: Transcript Verlag, 2015.

Meyerowitz, Joanne. *How Sex Changed: A History of Transsexuality in the United States*. Cambridge, MA: Harvard University Press, 2002.

Meyerowitz, Joanne. "Sex Change and the Popular Press: Historical Notes on Transsexuality in the United States, 1930–1955." *GLQ: A Journal of Lesbian and Gay Studies* 4, no. 2 (1998): 159–87. https://doi.org/10.1215/10642684-4-2-159.

Migeon, Claude J., Marco A. Rivarola, and Maguelone G. Forest. "Studies of Androgens in Male Transsexual Subjects: Effects of Estrogen Therapy." In *Transsexualism and Sex Reassignment*, edited by Richard Green and John Money. Baltimore: Johns Hopkins University Press, 1969, pp. 203–11.

Mol, Annemarie. *The Body Multiple: Ontology in Medical Practice*. Durham: Duke University Press, 2002.

Mol, Annemarie. *The Logic of Care: Health and the Problem of Patient Choice*. London: Routledge, 2008.

Mol, Annemarie, and Marc Berg. "Differences in Medicine: An Introduction." In *Differences in Medicine: Unraveling Practices, Techniques, and Bodies*, edited by Marc Berg and Annemarie Mol. Durham: Duke University Press, 1998, pp. 1–12.

Mol, Annemarie, and John Law. "Embodied Action, Enacted Bodies: The Example of Hypoglycaemia." *Body & Society* 10, no. 2–3 (2004): 43–62. https://doi.org/10.1177/1357034X04042932.

Mol, Annemarie, Ingunn Moser, and Jeannette Pols, eds. *Care in Practice: On Tinkering in Clinics, Homes and Farms*. Bielefeld: transcript Verlag, 2010.

Mold, Alex. *Making the Patient-Consumer: Patient Organisations and Health Consumerism in Britain*. Manchester: Manchester University Press, 2015.

Money, John, and Anke A. Ehrhardt. *Man & Woman, Boy & Girl*. Baltimore: Johns Hopkins University Press, 1972.

Money, John, and Anke A. Ehrhardt. "Transsexuelle nach Geschlechtswechsel.

Erfahrungen und Befunde am Johns Hopkins Hospital." *Beitrage zur Sexualforschung* 49 (1970): 70–87.

Moore, Carl R., and Dorothy Price. "Gonad Hormone Functions, and the Reciprocal Influence Between Gonads and Hypophysis with Its Bearing on the Problem of Sex Hormone Antagonism." *American Journal of Anatomy* 50, no. 1 (1932): 13–71. https://doi.org/10.1002/aja.1000500103.

Moses, Jacob. "Medical Regret Without Remorse: A Moral History of Harm, Responsibility, and Emotion in American Surgery Since 1945." PhD thesis, Harvard University, 2020.

Mühsam, Richard. "Chirurgische Eingriffe bei Anomalien des Sexuallebens." In *Therapie der Gegenwart*. Berlin: Urban & Schwarzenberg, 1926, pp. 451–55.

Mulinari, Diana, Suvi Keskinen, Sari Irni, and Salla Tuori. "Introduction: Postcolonialism and the Nordic Models of Welfare and Gender." In *Complying with Colonialism: Gender, Race, and Ethnicity in the Nordic Region*, edited by Suvi Keskinen, Salla Tuori, Sari Irni, and Diana Mulinari. Farnham: Ashgate, 2009, pp. 1–16.

Myrdal, Alva, and Gunnar Myrdal. *Kris i befolkningsfrågan*. Stockholm: Albert Bonniers forlag, 1934.

Najmabadi, Afsaneh. *Professing Selves: Transsexuality and Same-Sex Desire in Contemporary Iran*. Durham, NC: Duke University Press, 2014.

Niewöhner, Jörg, Patrick Bieler, Maren Heibges, and Martina Klausner. "Phenomenography. Relational Investigations into Modes of Being-in-the-World." *Cyprus Review* 28, no. 1 (2016): 67–84.

Nilsen, Håvard. *Du må ikke sove. Wilhelm Reich og psykoanalysen i Norge*. Oslo: Aschehoug, 2022.

Nilsen, Håvard Friis. "Da psykoanalysen ble en profesjon." *Arr*, no. 2–3 (2010): 29–41.

Nilsen, Håvard Friis. "Psykologene." In *Profesjonshistorier*, edited by Rune Slagstad and Jan Messel. Oslo: Pax forlag, 2014, pp. 450–70.

Nordberg, Kari H. "Frigjøring gjennom vitenskap—Karl Evang som seksualopplyser." *Tidsskrift for Den Norske Legeforening* 141, no. 9 (2021). https://doi.org/10.4045/tidsskr.20.0558.

Nordby, Trond. *Karl Evang: en biografi*. Oslo: Aschehoug, 1989.

Nordli, Marianne. *Det sterke kjønn—min kamp for å bli kvinne*. Oslo: Juritzen forlag, 2017.

"NOU 1992:8. Lov om pasientrettigheter." Oslo: Sosialdepartementet.

Ohlsson Al Fakir, Ida. "Nya rum för socialt medborgarskap. Om vetenskap och politik i 'Zigenarundersökningen'—en socialmedicinsk studie av svenska romer 1962–1965." PhD thesis, Linnaeus University, 2015.

Olwig, Karen Fog, and Karsten Paerregaard, eds. *The Question of Integration: Immigration, Exclusion and the Danish Welfare State*. Newcastle: Cambridge Scholars Publishing, 2011.

Oosterhuis, Harry. *Stepchildren of Nature: Krafft-Ebing, Psychiatry, and the Making of Sexual Identity*. Chicago: University of Chicago Press, 2000.

Orr, Jackie. "Biopsychiatry and the Informatics of Diagnosis: Governing Mentalities." In *Biomedicalization: Technoscience, Health, and Illness in the U.S.*, edited by Adele E. Clarke, Laura Mamo, Jennifer Ruth Fosket, Jennifer R. Fishman, and Janet K. Shim. Durham: Duke University Press, 2010, pp. 353–79.

Orr, Jackie. *Panic Diaries: A Genealogy of Panic Disorder*. Durham: Duke University Press, 2006.

Orticochea, Miguel. "A New Method of Total Reconstruction of the Penis." *British Journal of Plastic Surgery* 25, no. 4 (October 1972): 347–66. https://doi.org/10.1016/s0007-1226(72)80077-8.

Ostenfeld, Ib. "Paradoks kønsindstilling—Genuin transvestisme." *Ugeskrift for læger* 121, no. 13 (1959): 488–96.

Oudshoorn, Nelly. *Beyond the Natural Body: An Archeology of Sex Hormones*. London: Routledge, 1994.

Parhi, Katariina. "Boyish Mannerisms and Womanly Coquetry: Patients with the Diagnosis of Transvestitismus in the Helsinki Psychiatric Clinic in Finland, 1954–68." *Medical History* 62, no. 1 (2018): 50–66. https://doi.org/10.1017/mdh.2017.73.

Parnas, Josef, and Pierre Bovet. "Psychiatry Made Easy: Operation(al)ism and Some of its Consequences." In *Philosophical Issues in Psychiatry III: The Nature and Sources of Historical Change*, edited by Kenneth S. Kendler and Josef Parnas. Oxford: Oxford University Press, 2015, pp. 190–212.

Pentecost, Michelle, Vincanne Adams, Rama Baru, Carlo Caduff, Jeremy A. Greene, Helena Hansen, David S. Jones, Junko Kitanaka, and Francisco Ortega. "Revitalising Global Social Medicine." *The Lancet* 398, no. 10300 (2021): 573–74. https://doi.org/10.1016/S0140-6736(21)01003-5.

Pickersgill, Martyn. "Standardising Antisocial Personality Disorder: The Social Shaping of a Psychiatric Technology." *Sociology of Health & Illness* 34, no. 4 (2012): 544–59. https://doi.org/10.1111/j.1467-9566.2011.01404.x.

Pitts, Jesse R. "Social Control: The Concept." In *International Encyclopaedia of the Social Sciences*, edited by David L. Sills. New York: Macmillan, 1968, pp. 381–96.

Plemons, Eric. *The Look of a Woman: Facial Feminization Surgery and the Aims of Trans-Medicine*. Durham: Duke University Press, 2017.

Plemons, Eric, and Chris Straayer. "Introduction: Reframing the Surgical." *TSQ: Transgender Studies Quarterly* 5, no. 2 (2018): 164–73. https://doi.org/10.1215/23289252-4348605.

Pols, Jeannette. "Enforcing Patient Rights or Improving Care? The Interference of Two Modes of Doing Good in Mental Health Care." *Sociology of Health & Illness* 25, no. 4 (2003): 320–47. https://doi.org/10.1111/1467-9566.00349.

Porter, Dorothy. "How Did Social Medicine Evolve, and Where Is It Heading?". *PLOS Medicine* 3, no. 10 (2006): e399. https://doi.org/10.1371/journal.pmed.0030399.

Porter, Dorothy, ed. *Social Medicine and Medical Sociology in the Twentieth Century*. Vol. 43, Clio Medica. Amsterdam: Rodopi, 1997.

Porter, Dorothy, and Roy Porter. "What Was Social Medicine? An Historiographical Essay." *Journal of Historical Sociology* 1, no. 1 (1988): 90–106.

Porter, Theodore M. *Genetics in the Madhouse: The Unknown History of Human Heredity*. Princeton: Princeton University Press, 2018.

Porter, Theodore M. *Trust in Numbers: The Pursuit of Objectivity in Science and Public Life*. Princeton, NJ: Princeton University Press, 1995.

Prince, C. V. "Homosexuality, Transvestism and Transsexuality: Reflections on Their Etiology and Differentiation." *International Journal of Transgenderism* 8, no. 4 (2005): 17–20. https://doi.org/10.1300/J485v08n04_03.

Prosser, Jay. *Second Skins: The Body Narrative of Transsexuality*. New York: Columbia University Press, 1998.

Queers mot kapitalism. "Transvestia söker medlemmar. Intervju med Eva-Lisa Begntsson." *Queers mot kapitalism* (2010). https://queersmotkapitalism.wordpress.com/2010/12/10/transvestia-soker-medlemmar-intervju-med-eva-lisa-bengtsson/.

Ragnell, Allan. "The Development of Plastic Surgery in Sweden." *British Journal of Plastic Surgery* 8 (1955): 118–35. https://doi.org/10.1016/S0007-1226(55)80022-7.

Rappole, Amy. "Trans People and Legal Recognition: What the U.S. Federal Government Can Learn From Foreign Nations." *Maryland Journal of International Law* 30, no. 1 (2015): 191–216.

Raymond, Janice G. *The Transsexual Empire: The Making of the She-Male*. New York: Teachers College Press, 1994 [1979].

Reed, Geoffrey M., Jack Drescher, Richard B. Krueger, Elham Atalla, Susan D. Cochran, Michael B. First, Peggy T. Cohen-Kettenis, et al. "Disorders Related to Sexuality and Gender Identity in the ICD-11: Revising the ICD-10 Classification Based on Current Scientific Evidence, Best Clinical Practices, and Human Rights Considerations." *World Psychiatry* 15, no. 3 (2016): 205–21. https://doi.org/10.1002/wps.20354.

Reis, Elizabeth. *Bodies in Doubt: An American History of Intersex*. Baltimore: Johns Hopkins University Press, 2009.

Repo, Jemima. *The Biopolitics of Gender*. New York: Oxford University Press, 2016.

Retslægerådet. *Retslægerådet 1909–2009*. Copenhagen: Retslægerådet, 2009.

Rheinberger, Hans-Jörg. *Toward a History of Epistemic Things: Synthetizing Proteins in the Test Tube*. Stanford, CA: Stanford University Press, 1997.

Risse, Guenter B., and John Harley Warner. "Reconstructing Clinical Activities: Patient Records in Medical History." *Social History of Medicine* 5, no. 2 (1992): 183–205. https://doi.org/10.1093/shm/5.2.183.

Roll-Hansen, Nils. "Norwegian Eugenics: Sterilization as Social Reform." In *Eugenics and the Welfare State: Sterilization Policy in Denmark, Sweden, Norway, and Finland*, edited by Gunnar Broberg and Nils Roll-Hansen. East Lansing: Michigan State University Press, 2005, pp. 151–94.

Roll-Hansen, Nils, and Gunnar Broberg (eds.). *Eugenics and the Welfare State: Norway, Sweden, Denmark, and Finland*. East Lansing: Michigan State University Press, 2005.

Rorschach, Hermann. *Psychodiagnostik. Methodik und Ergebnisse eines Wahrnehmungsdiagnostischen Experiments*. Bern: Verlag Hans Huber, 1937 [1921].

Rosenberg, Charles E. "Meanings, Policies, and Medicine: On the Bioethical Enterprise and History." *Daedalus* 128, no. 4 (1999): 27–46.

Rosenberg, Charles E. "The Tyranny of Diagnosis: Specific Entities and Individual Experience." *Milbank Quarterly* 80, no. 2 (2002): 237–60. https://doi.org/10.1111/1468-0009.t01-1-00003.

Rothman, David J. *Strangers at the Bedside: A History of How Law and Bioethics Transformed Medical Decision Making*. New York: Basic Books, 1991.

Rydström, Jens. "Into the Wild and Back Again: Pornographic Discourse and Sexual Liberation in Sweden, 1954–1986." In *Gender, Materiality, and Politics: Essays on the Making of Power*, edited by Anna Nilsson Hammar, Daniel Nyström, and Martin Almbjär. Lund: Nordic Academic Press, 2022, pp. 157–78.

Rydström, Jens, and David Tjeder. *Kvinnor, män och alla andra—En svensk genushistoria*. Lund: Studentlitteratur, 2009.

Ryymin, Teemu, and Kari Ludvigsen. "From Equality to Equivalence? Norwegian Health Policies Towards Immigrants and the Sámi, 1970–2009." *Nordic Journal of Migration Research* 3, no. 1 (2013): 10–18. https://doi.org/10.2478/v10202-012-0011-y.

Saketopoulou, Avgi. "Thinking Psychoanalytically, Thinking Better: Reflections on Transgender." *International Journal of Psychoanalysis* 101, no. 5 (2020): 1019–30. https://doi.org/10.1080/00207578.2020.1810884.

Saketopoulou, Avgi, and Ann Pellegrini. *Gender Without Identity*. New York: Unconscious in Translation, 2023.

Sand, Knud. *Die Physiologie des Hodens*. Leipzig: Verlag von Curt Kabitzsch, 1933.

Sand, Knud. *Experimentelle Studier over Kønskarakterer hos Pattedyr*. Copenhagen: Steen Hasselbalchs Forlag, 1918.

Sand, Knud. "Experiments on the Endocrinology of the Sexual Glands." *Endocrinology* 7, no. 2 (1923): 273–301. https://doi.org/10.1210/endo-7-2-273.

Sandal, Sigrid. "En særlig trang til å ville forandre sitt kjønn—Kjønnsskiftebehandling i Norge 1952–1982." Master's thesis, University of Bergen, 2017.

Sandal, Sigrid. "Kirurgi og byråkrati." In *Frihet, likhet og mangfold*, edited by Anne Hellum and Anniken Sørlie. Oslo: Gyldendal, 2021, pp. 42–62.

Schiøtz, Aina. *Det offentlige helsevesen i Norge 1603–2003. Bind 2. Folkets helse—landets styrke, 1850–2003*. Oslo: Universitetsforlaget, 2003.

Schiøtz, Aina. *Doktoren. Distriktslegenes historie 1900–1984*. Oslo: Pax forlag, 2003.

Schiøtz, Aina. *Folkets hels—landets styrke, 1850–2003. Det offentlige helsevesen i Norge 1603–2003*. Oslo: Universitetsforlaget, 2003.

Schjelderup, Halfdan. "The Gillies Memorial Lecture 1975." *British Journal of Plastic Surgery* 30 (1977): 59–61.

Schlich, Thomas. *The Origins of Organ Transplantation: Surgery and Laboratory Science, 1880–1930*. Rochester: University of Rochester Press, 2010.

Scull, Andrew. "Contending Professions: Sciences of the Brain and Mind in the United States, 1850–2013." *Science in Context* 28, no. 1 (2015): 131–61. https://doi.org/10.1017/S0269889714000350.

Seip, Anne-Lise. *Veiene til velferdsstaten: norsk sosialpolitikk 1920–75*. Oslo: Gyldendal, 1994.

Sejersted, Francis. *The Age of Social Democracy: Norway and Sweden in the Twentieth Century*. Princeton: Princeton University Press, 2011.

Sekuler, Todd. "Un/Certain Care: From a Diagnostic to a Somatechnic Regime of Care for Medical Transition in Public Hospitals in France." PhD thesis, Humboldt University of Berlin, 2018.

Sengoopta, Chandak. "Glandular Politics: Experimental Biology, Clinical Medicine, and Homosexual Emancipation in Fin-de-Siècle Central Europe." *Isis* 89, no. 3 (1998): 445–73. https://doi.org/10.1086/384073.

Sengoopta, Chandak. *The Most Secret Quintessence of Life: Sex, Glands, and Hormones, 1850–1950*. Chicago: University of Chicago Press, 2006.

shuster, stef m. *Trans Medicine: The Emergence and Practice of Treating Gender*. New York: New York University Press, 2021.

Siim, Birte, and Hege Skjeie. "Tracks, Intersections and Dead Ends: Multicultural Challenges to State Feminism in Denmark and Norway." *Ethnicities* 8, no. 3 (2008): 322–44. https://doi.org/10.1177/1468796808092446.

Silverman, Chloe. *Understanding Autism: Parents, Doctors, and the History of a Disorder*. Princeton: Princeton University Press, 2012.

Silverman, Victor, and Susan Stryker. *Screaming Queens: The Riot at Compton's Cafeteria*. 57 min. USA: Kanopy Streaming, 2005.

Simonsen, Rikke Kildevæld, and Ellids Kristensen. "Transseksuelle i Sexologisk Klinik." In *25 år med Sexologisk Klinik*, edited by Ellids Kristensen and Annamaria Giraldi. Copenhagen: Psykiatrisk Center København, 2011, pp. 69–74.

Skålevåg, Svein Atle. "A Culture of Consensus: Organising Expertise in Norwegian Forensic Psychiatry, Late Nineteenth to Early Twentieth Century." In *Forensic Cultures in Modern Europe*, edited by Willemijn Ruberg, Lara Bergers, Pauline Dirven, and Sara Serrano Martínez. Manchester: Manchester University Press, 2023, pp. 240–60.

Skålevåg, Svein Atle. *Utilregnelighet. En historie om rett og medisin*. Oslo: Pax forlag, 2016.

Skaset, Maren. "Reformtid og markedsgløtt: det offentlige helsevesen etter 1985." In *Folkets helse—landets styrke, 1850–2003*, edited by Aina Schiøtz. Oslo: Universitetsforlaget, 2003, pp. 499–548.

Skinner, Quentin. *Die drei Körper des Staates*. Translated by Karin Wördemann. Göttingen: Wallstein Verlag, 2012.

Skolleborg, Knut. "Plastikkirurger og etikk." *Tidsskrift for Den norske lægeforening* 123, no. 21 (2003): 3088.

Slagstad, Ketil. "Bureaucratizing Medicine: Creating a Gender Identity Clinic in the Welfare State." *Isis* 113, no. 3 (2022): 469–90. https://doi.org/10.1086/721140.

Slagstad, Ketil. "How the Idea of Social Contagion Shaped Trans Medicine." *New England Journal of Medicine* 391, no. 16: 1546–51. https://doi.org/10.1056/NEJMms2407430.

Slagstad, Ketil. "The Political Nature of Sex—Transgender in the History of Medicine." *New England Journal of Medicine* 384, no. 11 (2021): 1070–74. https://doi.org/10.1056/NEJMms2029814.

Slagstad, Ketil. "Psychiatric Practices Beyond Psychiatry: The Sexological Administration of Transgender Life Around 1980." In *Doing Psychiatry in Postwar Europe: Practices, Routines and Experiences*, edited by Gundula Gahlen, Volker Hess, Marianna Scarfone, and Henriette Voelker. Manchester: Manchester University Press, 2024, pp. 312–34.

Slagstad, Ketil. "Society as Cause and Cure: The Norms of Transgender Social Medicine." *Culture, Medicine, and Psychiatry* 45, no. 3 (2021): 456–78. https://doi.org/10.1007/s11013-021-09727-4.

Slagstad, Rune. *De nasjonale strateger*. Oslo: Pax forlag, 1998.

Smith, Matthew. *An Alternative History of Hyperactivity: Food Additives and the Feingold Diet*. New Brunswick: Rutgers University Press, 2011.

Socialstyrelsen. "Beslut om nationell högspecialiserad vård—viss vård vi könsdysfori." 1. December 2020. https://www.socialstyrelsen.se/globalassets/sharepoint-dokument/dokument-webb/ovrigt/nationell-hogspecialiserad-vard-konsdysfori-beslut.pdf.

Sørensen, Øystein, and Bo Stråth, eds. *The Cultural Construction of Norden*. Oslo: Scandinavian University Press, 1997.

Sørensen, Thorkil. "A Follow-Up Study of Operated Transsexual Females." *Acta Psychiatrica Scandinavica* 64 (1981): 50–64.

Sørensen, Thorkil. "Tilblivelsen af Sexologisk Klinik—Samlivsgruppen." In *25 år med Sexologisk Klinik*, edited by Ellids Kristensen and Annamaria Giraldi. Copenhagen: Psykiatrisk Center København, 2011, pp. 10–17.

Sørensen, Thorkil, and Preben Hertoft. "Sexmodifying Operations on Transsexuals in Denmark in the Period 1950–1977." *Acta Psychiatrica Scandinavica* 61, no. 1 (1980): 56–66. https://doi.org/10.1111/j.1600-0447.1980.tb00565.x.

Spector, Scott. *Violent Sensations: Sex, Crime, and Utopia in Vienna and Berlin, 1860–1914*. Chicago: University of Chicago Press, 2016.

"St. meld. nr. 9 (1974–75). Sykehusutbygging m.v. i et regionalisert helsevesen." Oslo: Sosialdepartementet.

"St. meld. nr. 41 (1987–88). Helsepolitikken mot år 2000, Nasjonal Helseplan." Oslo: Sosialdepartementet.

Statens helsetilsyn. *Styring av høyspesialiserte funksjoner innen psykisk helsevern for voksne*. Oslo: 1995.

Statistisk sentralbyrå. "Historisk utvikling av jentenavn." https://www.ssb.no/a/navn/historisk-utvikling-av-jentenavn/, accessed June 23, 2022.

Støa, Karl Fredrik. "Om hormon og hormongransking." *Syn og segn* 66 (1960): 185–93.

Stoff, Heiko. *Ewige Jugend: Konzepte der Verjüngung vom späten 19. Jahrhundert bis ins Dritte Reich*. Cologne: Böhlau Verlag, 2004.

Stoff, Heiko. "Identität und Differenz: Zur Diskursgeschichte der Sexualität zu Beginn des 21. Jahrhunderts." In *Sexualität als Experiment: Identität, Lust und Reproduktion zwischen Science und Fiction*, edited by Nicolas Pethes and Silke Schicktanz. Frankfurt: Campus Verlag, 2008, pp. 27–44.

Stolberg, Michael. "Formen und Funktionen medizinischer Fallberichte in der Frühen Neuzeit (1500–1800)." In *Fallstudien: Theorie—Geschichte—Methode*, edited by Johannes Süßmann, Susanne Scholz, and Gisela Engel. Berlin: trafo, 2007, pp. 81–95.

Stoler, Ann Laura. *Along the Archival Grain: Epistemic Anxieties and Colonial Common Sense*. Princeton: Princeton University Press, 2009.

Stoller, Robert J. "A Contribution to the Study of Gender Identity." *International Journal of Psychoanalysis* 45, no. 2–3 (1964): 220–26.

Stoller, Robert J. "Male Transsexualism: Uneasiness." *American Journal of Psychiatry* 130, no. 5 (1973): 536–39.

Stoller, Robert J. *Sex and Gender: The Development of Masculinity and Femininity*. London: Karnac, 1968.

Stoller, Robert J., Harold Garfinkel, and Alexander C. Rosen. "Psychiatric Management of Intersexed Patients." *California Medicine* 96, no. 1 (1962): 30–34.

Stone, Sandy. "The Empire Strikes Back: A Posttransexual Manifesto." In *Body Guards: The Cultural Politics of Gender Ambiguity*, edited by Julia Epstein and Kristina Straub. New York: Routledge, 1991, pp. 280–304.

Strøm, Axel, Per Sundby, and Yngvar Løchen. *Lærebok i sosialmedisin*. 4th ed. Oslo: Fabritius Forlag, 1973.

Stryker, Susan. "Foreword." *International Journal of Transgenderism* 8, no. 4 (2005): xv–xvi. https://doi.org/10.1300/J485v08n04_a.

Stryker, Susan. *Transgender History: The Roots of Today's Revolution*. 2nd ed. Berkeley: Seal Press, 2017.

Sturdy, Steve. "Knowing Cases: Biomedicine in Edinburgh, 1887–1920." *Social Studies of Science* 37, no. 5 (2007): 659–89. https://doi.org/10.1177/0306312707076597.

Stürup, Georg K. "Male Transsexuals: A Long-Term Follow-Up After Sex Reassignment Operations." *Acta Psychiatrica Scandinavica* 53, no. 1 (1976): 51–63. https://doi.org/10.1111/j.1600-0447.1976.tb00058.x.

Stürup, Georg K. "Transvestisme i klinisk kriminologi." *Nordisk Medicin* 56, no. 35 (1956): 1226–31.

Sundberg, Nomran D. "The Practice of Psychological Testing in Clinical Services in the United States." *American Psychologist* 16, no. 2 (1961): 79–83. https://doi.org/10.1037/h0040647.

Sutton, Katie. *Sex Between Body and Mind: Psychoanalysis and Sexology in the*

German-Speaking World, 1890s–1930s. Ann Arbor: University of Michigan Press, 2019.

Szasz, Thomas S. *The Myth of Mental Illness: Foundations of a Theory of Personal Conduct*. New York: Dell, 1961.

Szasz, Thomas S. *Sex by Prescription*. New York: Anchor Press/Doubleday, 1980.

te Heesen, Anke. "The Notebook: A Paper Technology." In *Making Things Public: Atmospheres of Democracy*, edited by Bruno Latour and Peter Weibel. Cambridge, MA: MIT Press, 2005, pp. 582–89.

Teigen, Karl Halvor. *En psykologihistorie*. 2nd ed. Bergen: Fagbokforlaget, 2015.

Thompson, Patricia Hadley. "Apparatus to Maintain Adequate Vaginal Size Postoperatively in the Male Transsexual Patient." In *Transsexualism and Sex Reassignment*, edited by Richard Green and John Money. Baltimore: Johns Hopkins University Press, 1969, pp. 323–29.

Thompson, Paul. *The Voice of the Past: Oral History*. 3rd ed. New York: Oxford University Press, 2000.

Timm, Annette F. "'I am so grateful to all you men of medicine': Trans Circles of Knowledge and Intimacy." In *Others of My Kind: Transatlantic Transgender Histories*, edited by Alex Bakker, Rainer Herrn, Michael Thomas Taylor, and Annette F. Timm. Calgary, AB: University of Calgary Press, 2020, pp. 71–131.

Timmermans, Stefan, and Marc Berg. *The Gold Standard: The Challenge of Evidence-Based Medicine and Standardization in Health Care*. Philadelphia: Temple University Press, 2003.

Timmermans, Stefan, and Marc Berg. "Standardization in Action: Achieving Local Universality Through Medical Protocols." *Social Studies of Science* 27, no. 2 (1997): 273–305. https://doi.org/10.1177/030631297027002003.

Tomes, Nancy. "Oral History in the History of Medicine." *Journal of American History* 78, no. 2 (1991): 607–17. https://doi.org/10.2307/2079538.

Tomes, Nancy. *Remaking the American Patient: How Madison Avenue and Modern Medicine Turned Patients into Consumers*. Chapel Hill: University of North Carolina Press, 2015.

Töpfer, Frank, ed. *Verstümmelung oder Selbsverwirklichung? Die Boss-Mitscherlich-Kontroverse*. Stuttgart–Bad Cannstatt: frommann-holzboog, 2012.

Tydén, Mattias. *Från politik till praktik. De svenska steriliseringslagarna 1935–1975*. 2nd ed. Stockholm: Almqvist & Wiksell International, 2002.

Ullerstam, Lars. *De seksuelle minoriteter*. Oslo: Pax forlag, 1964.

Ulrichs, Karl Heinrich. *Forschungen über das Räthsel der mannmännlichen Liebe*. Leipzig: Selbstverlag des Verfassers, 1864.

Vargha, Dóra. *Polio Across the Iron Curtain: Hungary's Cold War with an Epidemic*. Cambridge: Cambridge University Press, 2018.

Velocci, Beans. "Standards of Care: Uncertainty and Risk in Harry Benjamin's Transsexual Classifications." *TSQ: Transgender Studies Quarterly* 8, no. 4 (2021): 462–80. https://doi.org/10.1215/23289252-9311060.

Vike, Halvard, Anette Fagertun, and Heidi Haukelien. "Introduction: Welfare State Capitalism, Universalism, and Social Reproduction in Scandinavia." In *The Political Economy of Care: Welfare State Capitalism, Universalism, and Social Reproduction*, edited by Halvard Vike, Anette Fagertun, and Heidi Haukelien. Oslo: Scandinavian University Press, 2024, pp. 9–40.

Vismann, Cornelia. *Akten: Medientechnik und Recht*. Frankfurt am Main: Fischer Taschenbuch Verlag, 2000.

Vogt, Jørgen Herman. "Five Cases of Transsexualism in Females." *Acta Psychiatrica Scandinavica* 44, no. 1 (1968): 62–88. https://doi.org/10.1111/j.1600-0447.1968.tb07636.x.

Waitzkin, Howard, Celia Iriart, Alfredo Estrada, and Silvia Lamadrid. "Social Medicine Then and Now: Lessons from Latin America." *American Journal of Public Health* 91, no. 10 (2001): 1592–601. https://doi.org/10.2105/ajph.91.10.1592.

Wålinder, Jan. *Transsexualism: A Study of Forty-Three Cases*. Gothenburg: Akademiförlaget, 1967.

Wålinder, Jan, Bengt Lundström, and Inga Thuwe. "Prognostic Factors in the Assessment of Male Transsexuals for Sex Reassignment." *British Journal of Psychiatry* 132 (1978): 16–20.

Wålinder, Jan, and Inga Thuwe. "A Law Concerning Sex Reassignment of Transsexuals in Sweden." *Archives of Sexual Behavior* 5, no. 3 (1976): 255–58.

Wålinder, Jan, and Inga Thuwe. *A Social-Psychiatric Follow-Up Study of 24 Sex-Reassigned Transsexuals*. Gothenburg: Akademiförlaget, 1975.

Wallace, Antony F. "The Influence of the Battle of Jutland on Plastic Surgery." In *Clinics in Plastic Surgery: An International Quarterly*, edited by Sharon Romm. Philadelphia: W. B. Saunders, 1983, pp. 657–63.

Wallin, Sanna. *När jag letar efter Max*. Stockholm: Normal förlag, 2007.

Westphal, Carl. "Die conträre Sexualempfindung, Symptom eines neuropathischen (psychopathischen) Zustandes." *Archiv für Psychiatrie und Nervenkrankheiten* 2, no. 1 (1870): 73–108. https://doi.org/10.1007/BF01796143.

WHO. *Education and Treatment in Human Sexuality: The Training of Health Professionals*. Geneva: WHO, 1975.

Winance, Myriam. "Care and Disability: Practices of Experimenting, Tinkering with, and Arranging People and Technical Aids." In *Care in Practice: On Tinkering in Clinics, Homes and Farms*, edited by Annemarie Mol, Ingunn Moser, and Jeannette Pols. Bielefeld: transcript Verlag, 2010, pp. 93–117.

Wintherbauer, Janni Christin, and Ane Rostad Stokholm. *Janni—slik ble mitt liv*. Oslo: Pegasus forlag, 2010.

Young, Allan. *The Harmony of Illusions: Inventing Post-Traumatic Stress Disorder*. Princeton: Princeton University Press, 1996.

Zacher, Judith B. "Plastic Surgery in the Late 1920: Three Points of View." In *Clinics in Plastic Surgery: An International Quarterly*, edited by Sharon Romm. Philadelphia: W. B. Saunders, 1983, pp. 665–67.

Zerubavel, Eviatar. "Timetables and Scheduling: On the Social Organization of Time." *Sociological Inquiry* 46, no. 2 (1976): 87–94. https://doi.org/10.1111/j.1475-682X.1976.tb00753.x.

Zola, Irving Kenneth. "Medicine as an Institution of Social Control." *Sociological Review* 20, no. 4 (1972): 487–504. https://doi.org/10.1111/j.1467-954X.1972.tb00220.x.

Index

Page numbers in italics refer to figures.

www.ingramcontent.com/pod-product-compliance
Lightning Source LLC
Chambersburg PA
CBHW022136020426
42334CB00015B/919